The Law on Health and Safety at Work

M. DEWIS, LL.B.(Lond.)
One of the law examiners of the Institution of Industrial Safety Officers

In conjunction with the Institution of Industrial Safety Officers

MACDONALD AND EVANS

Macdonald and Evans Ltd.
Estover, Plymouth PL6 7PZ

First published 1978

ISBN 0 7121 1247 2

Printed in Great Britain by
Butler and Tanner Ltd.
Frome and London

Foreword

As Chairman of the Education and Examination Committee of the Institution of Industrial Safety Officers I welcome any book which the author has written with the Institution's examination syllabus in mind. This is such a book. Mr. Dewis has slanted his treatment of the subject of health and safety legislation towards the safety professional.

Anyone who is a practising safety professional, or who is studying to become one, should find this a valuable addition to his library. It covers a subject where there is a paucity of available literature, and what there is has often been written for practising lawyers. The book covers a range of legal topics which have seldom before been grouped together in this manner, but all are relevant and of particular interest to the safety professional.

The book should also be of interest to safety representatives who wish to further their knowledge of health and safety legislation within the context of their appointment under regulations. The well-informed safety representative should be an asset to the accident prevention scene.

To those readers who are studying to sit the Institution's examinations I wish every success. A sound knowledge of the topics covered by this book will go a long way to ensuring that success.

J. M. Totterdell
Chairman, Education and Examination Committee
Institution of Industrial Safety Officers

Preface

This book was written with two principal and one subsidiary objective in mind. Firstly, to give safety advisers the legal information necessary to enable them to pass the Institution of Industrial Safety Officers Associate and Full Membership examinations in law. Second, and allied to the first objective, to provide a reference manual on health and safety legislation for the practising safety professional, and to give the company safety adviser and union safety representative detailed guidance on the legal aspects most likely to arise during the course of his work. Health and Safety Executive inspectors may also find useful information in the text. A subsidiary reason behind the writing of the text was to provide those insurers, insurance brokers and liability assessors concerned with industrial safety with a knowledge of health and safety law and insurance, both private and state.

It has been the author's concern throughout to present the text in as straightforward a way as possible, although the principal aim of setting down a body of legal doctrine on health and safety at work makes a certain amount of technical detail inevitable. Chapter 5 in particular, dealing with product liability, may strike readers as being technical, but this is because the subject has grown out of the often complex rules of contract and tort.

The law is stated as at 30th June 1978. Any errors in the text are my own.

June 1978 Malcolm Dewis

Acknowledgments

In the preparation of this text I am grateful to the following publishers, organisations and individuals: to H.M.S.O., London, for permission to reproduce Acts of Parliament wholly or in part, together with Regulations; also for permission to cite Command Papers and Circulars; to Butterworths & Co. Ltd., Kingsway, London, for permission to reproduce judgments from the All England Law Reports, and also to cite extracts from *Employers' Liability at Common Law* by John Munkman and from *Foundations of the Law of Tort* by Glanville Williams and B. A. Hepple; to Sweet and Maxwell Ltd., London, for permission to cite extracts from *Statute Law* by W. Craies; to Weidenfeld & Nicolson Ltd., London, for permission to cite extracts from *Accidents Compensation and the Law* by P. S. Atiyah; to the Law Commission for permission to reproduce extracts from their Working Paper on *Liability for Defective Products*; to Aldine Publishing Co. Ltd., Chicago, U.S.A., for permission to reproduce extracts from *Settled out of Court* by Laurence Ross; to the T.U.C. Press and Information Department, London, for permission to reproduce their 10-day syllabus for the training of safety representatives; to the Central Office of Industrial Tribunals, London, for permission to reproduce extracts from tribunal decisions; to Industrial Relations Services, London, for helpful advice; to the Institution of Industrial Safety Officers, Leicester, for the opportunity to write the book; to Professor Bryan Harvey, Visiting Professor to the University of Aston in Birmingham and formerly Deputy of the Health and Safety Executive and Chief Inspector of Factories, for opinions publicly expressed about the need for professionalism in safety; to John Horner, Principal, Chartered Institute of Insurance College, for helpful advice on insurance aspects; and to Graham Chatfield M.I.I.S.O., formerly Safety/Hygiene Manager, Chrysler U.K. Ltd., for helpful advice in connection with many parts of the script.

Last but certainly not least, I am grateful to the First Diploma Group at Coventry Technical College who were the final catalyst behind the writing of this book; and to my wife Pauline I owe the greatest debt of all for never-ending patience.

List of Contents

To my wife Pauline

That everyone might consult the law of which he stood in need in the least possible time. (Bentham)

Table of Cases

Table of Statutes

Table of Statutory Instruments

Chapter 1

The Sources of Law, Courts and Tribunals and the Interpretation of Statutes

THE SOURCES OF LAW

1. Introduction

A source of law is where the law on a given subject is to be found. There are traditionally two types of source, formal and informal, the distinction being that the former may be cited as authoritative in a court, whereas the latter refers to practices and procedures which are not authoritative in this way. Nevertheless the importance of industrial and commercial practices and procedures must never be overlooked.

2. Industrial standards and codes of practice

Importance

The Report of the Committee on Safety and Health at Work (the *Robens Report*), which precipitated the Health and Safety at Work Act 1974, urged that the problems of industry and industrial safety could best be solved by those who created them: "Our present system encourages rather too much reliance on state regulation, and rather too little on personal responsibility and voluntary, self-generating effort." This serves to underline the crucial future importance of non-statutory codes of practice in the field of workplace safety; a typical set of regulations from the Health and Safety Commission will contain an explicit reference to voluntary codes of practice and standards prepared either by the joint standing safety committees of the industry concerned, or by the British Standards Institutions or even by government departments or employers' organisations. A good example of this sort of "self-government" or "good housekeeping" is the Agriculture (Tractor Cabs) Regulations 1967 which state that tractor safety cabs will meet the requirements of the regulations only if they are able to satisfy the relevant British Standards specification established by the British Standards Institution. As the Robens Report explained, this linking of statutory regulation with detailed guidance has two distinct advantages, firstly avoiding building obsolescence into the regulations, and second providing precise guidance for those on whom duties are laid either by the regulations or by the parent Act.

Another and considerably more important example of this approach is the recently issued Safety Representatives and Safety Committees Regulations 1977.

Advantages and effect

The Health and Safety Commission (*see* below) is empowered under s. 16(1) of the Health and Safety at Work Act 1974 (H.A.S.A.W.A.) to approve and issue codes of practice independently of regulations on the same matter. Much use will no doubt be made of this power, since codes of practice are not subject to the tedium of parliamentary scrutiny, they do not lay down legal duties, and, if necessary, they can be rapidly varied and modified, whereas regulations are subject to parliamentary procedures.

The Robens Committee recommended that such codes of practice should be admissible in proceedings before industrial tribunals. This has now come to pass. The precise effect of a breach of such a code, however, must remain a largely unknown quantity until further case law has accrued. Nevertheless, subs. 6(11) of the Employment Protection Act 1975 states that breach of a code of itself shall not render a person liable to proceedings. Conversely, compliance with regulations which embody a code of practice is not of itself tantamount to freedom from liability at common law. In *Bux* v. *Slough Metals Ltd.*, a 1973 decision of the Court of Appeal, a workman was injured although he had been supplied by his employer with goggles pursuant to regulations. He lost the sight of an eye through failure to wear the eye protection provided. The judge had no hesitation in finding the employer liable for damages at common law in respect of the injury, the fault being unquestionably one of bad supervision, giving rise to liability for common law negligence.

The majority of regulations relating to health and safety at work under the general aegis of H.A.S.A.W.A. will issue from the new law-making body, the *Health and Safety Commission* (H.S.C.). This will include codes of practice. Any proposals made by H.S.C. will be laid before the Secretary of State to present to Parliament for ratification. A current example of this likely to be of immense practical importance is the Safety Representatives and Safety Committees Regulations 1977, which are accompanied by a non-statutory code of practice.

3. Formal sources of law

Common law

Common law has been defined by Osborn as "that part of the law of England formulated, developed and administered by the old common law courts, based originally on the common customs of the country and unwritten.... It is the commonsense of the community, crystallised and formulated by our forefathers and not either local law or the result of legislation." What emerges from this definition is that "common law" refers to a system of law which is essentially judge-made, as distinct from that which is legislative or laid down by Parliament. For many centuries until fairly recent times the bulk of U.K. civil and criminal law consisted of principles laid down in a succession of cases by judges. Parliament did not emerge as the law-maker in the modern sense until well into the nineteenth century. The judge-made nature of much of English law has important consequences today. In particular it means that in certain areas of the law Parliament, when it legislates, does not as such enact any new law but rather legislates to clarify and codify and rid the existing judge-made law of obsolescent survivals. This explains the codificatory nature of much legislation passed in recent years. Examples of legislation passed to clarify judge-made law and/or remove obsolescent survivals, whilst at the same time recognising the common law basis of the law of tort, are the Law Reform (Personal Injuries) Act 1948, the Occupiers' Liability Act 1957 and the Employers' Liability (Defective Equipment) Act 1969, not to mention the general duties section of H.A.S.A.W.A. itself.

(*a*) *Importance of judicial interpretation.* The effect of the judicial tradition in England is that legislation means nothing until interpreted in a court, especially a superior court such as the Court of Appeal or the House of Lords. This fact accounts for many anomalies and apparent absurdities in the history of protective industrial legislation. Nowhere is this better illustrated than by s. 14(1), Factories Act 1961, which provides "Every dangerous part of any machinery, other than prime movers and transmission machinery, shall be securely fenced unless it is in such a position or of such construction as to be as safe to every person employed or working on the premises as it would be if securely fenced". One might be excused for believing that this section was laying down a strict duty and that such machinery had to be securely fenced against the risk of all conceivable kinds of danger. The section has not, however, been interpreted in this way. In *Walker* v. *Bletchley-Flettons Ltd.* du Parcq, J., in a classic judgment said "A part of machinery is dangerous if it is a possible cause of injury to anybody acting in a way in which a human being may be reasonably expected to act in circumstances which may reasonably be expected to occur". This judgment seems to identify the statutory duty with the duty to guard against reasonably foreseeable dangers, as at common law. Similarly in *Carroll* v. *Barclay & Sons Ltd.* (a case occurring under Factories Act 1937) an operative was injured when a driving belt broke in the course of operation. When sued by the employee in respect of his injuries, the factory occupier was held not to be liable as the purpose of the section was to "keep the worker out, not the machine or its product in".

In *Close* v. *Steel Company of Wales* basically the same problem occurred, but with respect to s. 14(1), Factories Act 1961. C, an employee of S.C.W., was injured when the electric drill he was operating shattered and a piece flew into his eye. In an action against his employer for damages it was held that the employer was not liable because the danger was not reasonably foreseeable, such an accident never having happened before. The court moreover reiterated the view expressed in *Carroll* v. *Barclay & Sons Ltd.* that the purpose of the fencing requirements was to prevent the body of the operator coming into contact with the machine and not to protect him from things flying out of the machine.

Another interesting problem was spotlighted by the case of *British Railways Board* v. *Liptrot*, again concerning s. 14(1), Factories Act 1961. This case posed the question whether mobile equipment used in a factory was "machinery" for the purposes of s. 14(1). L was a slinger employed by B.R.B., who occupied a scrap metal yard, itself a factory within the meaning of s. 175(2)(*b*), Factories Act 1961. A mobile crane with a hook or magnet was used to lift scrap metal after sorting and to place it in a railway truck. The crane was mounted on a four-wheeled chassis, the wheels of which were covered with rubber, and the body of the crane was mounted on a vertical shaft fitted in the centre of the chassis. It could move under its own power. The driver, from his cab at the offside of the body, controlled the jib and the lifting and lowering of metal. After loading he had to slew the jib which projected from the body, in order to unload. The body revolved with the jib. There was a gap between the bottom of the body and the chassis where wire sometimes got caught. When this happened the slinger normally told the driver to stop the crane, and the slinger then removed the wire. L saw a piece of wire which was caught but failed to follow the normal practice. While he was in a position where it was impossible for the driver to see him and was trying to free the wire, the driver slewed the jib and L was caught by the revolving body of the crane and was squeezed against one of the chassis wheels. In a claim for damages for negligence in respect of personal injury and for breach of statutory duty L claimed that the gap between the body of the crane and the wheel was a dangerous part of machinery and as such was not securely fenced. It was held by the House of Lords that the crane was part of factory equipment and its mobility did not exclude it from the scope of s. 14(1) and so there was a breach of statutory duty on the part of the factory occupier. During the course of giving judgment Viscount Dilhorne observed "I regret that I am unable to agree with the views expressed in and the *ratio decidendi* of *Cherry* v. *International Alloys Ltd.* In my opinion, the object of the Factories Act 1961 is to make safety provisions for those employed in, or who work in, factories. To that end every dangerous part of any machinery in a factory requires to be securely fenced.... This obligation does not depend on whether the machinery is a fixed part of the equipment of the factory or is on wheels and capable of moving under its own power from one part of the factory to another or outside the factory." Distinguishing the present case from *Cherry* v. *International Alloys Ltd.* (Lister truck not "machinery" for the purposes of s. 14(1), Factories Act 1937), Viscount Dilhorne went on to say "The question was not, as I see it, whether the Lister truck was machinery to which the Act of 1961 applied but whether, as it was part of the equipment of the factory, it contained machinery which required to be fenced. Lorries and cars which came to visit a factory would not, in my opinion, be part of its equipment and the Act of 1961 would not apply to them. I can see no valid reason for excluding any part of the equipment of a factory which contains machinery from the operation of the Act of 1961 on the ground that it is mobile or can be described as a vehicle."

(*b*) *Binding precedent.* This is the legal principle of a case which is binding upon courts confronted in the future with the same problem. Take the facts of a leading case on liability for the consequences of an original act of negligence, the *Wagon Mound* case. Here the crew of a British tankship negligently allowed oil to foul the slipways of Sydney Harbour. On the wharf there was a factory where welding was being done. Sparks from the welding set alight a piece of cotton waste floating on the water. The cotton set alight crude oil from the tanker which in turn burnt down the factory where the welding was being done. In the subsequent action for negligence against the tankship it was held that a person was liable for the consequences of an act of negligence (fouling the slipways) which are reasonably foreseeable but not necessarily liable for all the direct consequences (burning down the factory). Apart from the fact that this decision served to keep insurance premiums within manageable proportions, it forms a binding precedent in remoteness of damage cases: a person is liable only for the foreseeable consequences of his negligent conduct.

(*c*) *Obiter dicta*. These are words or observations said or made *by the way*. As such, they have no direct bearing on the court's decision. Often *obiter dicta* are indistinguishable from the *ratio decidendi*, though sometimes a judge will indicate *obiter* remarks by saying "If, however, the circumstances of the case were different", or "If it were necessary to decide the point, I should be inclined to say ..." Because *obiter dicta* do not form part of the court's decision, they are not legally binding. *Dicta* of the higher courts, however, particularly the House of Lords, are of *persuasive authority*. They may be resorted to by a court in the absence of binding precedent on the point. Otherwise, binding precedent will prevail.

(*d*) *Residual nature of common law in industrial safety*. The law relating to industrial safety sub-divides approximately into two halves—common law and legislation. The consequence of this is that injured workmen may well find that, although they may have no action for breach of a statutory duty, yet there will be a good case at common law. This point is well illustrated by the decision of the Court of Appeal in *Kilgollan* v. *Cooke & Co. Ltd*. K lost the sight of an eye when a small particle of wire flew out of a stranding machine which was not securely fenced, as required by s. 14(1), Factories Act 1937. It was held that, although there was no breach of statutory duty on the part of the factory occupier (since the section was designed to prevent the worker coming into contact with the machine—*see* the rule in *Gorris* v. *Scott*, §**15**, p. 35 nevertheless there was, in the light of such occurrences having happened before, a breach of the employer's duty at common law to take reasonable care not to injure his employees. It is this residual quality of common law that has led to the emergence of the double-barrelled action for injury at work brought by employees against employers.

Legislation

(*a*) *Functions of legislation*. Acts of Parliament lay down the periphery or guidelines of a system, leaving the structure to be cemented in by delegated legislation. This is precisely what has happened in the case of H.A.S.A.W.A. Parliament passed the Act in 1974, the Act itself laying down the general duties upon employers and employees and establishing H.S.C., whose function is to propose regulations relating to health and safety at work; in other words the power to make regulations has been delegated to it. Here it is worth noting that subs. 15(2)(*b*), H.A.S.A.W.A., provides that H.S.C. may exclude or modify any of the provisions of ss. 2–9, i.e. the general duties, and/or any other relevant statutory provisions still in force.

(*b*) *Date of coming into effect*. Legislation takes effect as from the time when it goes on the statute book or as from such date as is specified in the Act itself.

(*c*) *Types of legislation*

(i) *Consolidating statutes*. These are statutes which compress into one general statute several previous special enactments. There are numerous examples of this process in the history of industrial safety legislation, a good instance being the Factory and Workshop Act 1878, which eliminated the distinction between factory and workshop perpetuated by previous special enactments. A more recent example is the Factories Act 1961, which consolidated several amendments to the 1937 Factories Act, e.g. the amendment to s. 25(1), Factories Act 1937 (now s. 28(1), Factories Act 1961), by the Factories Act 1959 in the light of the 1953 decision in *Latimer* v. *A.E.C. Ltd.* to the effect that the need for floors, steps, stairs, etc., to be free from obstruction and substances likely to cause persons to slip did not create a strict duty but rather had to be carried out as far as was reasonably practicable.

(ii) *Declaratory statutes*. These are statutes which put into the form of a civil or criminal code the principles established by case law. They reduce judge-made law to statutory form. The general duties of H.A.S.A.W.A. are an example of this, the rules of negligence being compressed into statutory form and backed up by criminal sanctions.

4. Short history of protective industrial legislation

As is made clear in the Robens Report, the history of occupational safety legislation is to a great extent the history of legislative control over the manufacturing industry. Take, for instance, the definition of "factory" in the series of Factories Acts as referring to a place where articles or parts of articles are either made, repaired, altered, ornamented or broken up—Factories Act 1961, s. 175(1). The advent of the industrial revolution in the late eighteenth century created factory conditions on a scale quite unprecedented before. The powered machinery that characterised the new employment conditions brought with it a tidal wave of human destruction and suffering. Such conditions were particularly noticeable in the textile industries and the mines where young poverty-stricken children became recruits to operate the machines and push the coal carts. It was in an effort to combat these appalling conditions that the first factory legislation was passed in 1802. This sought to regulate the working hours of apprentices in the cotton mills and to posit general standards relating to lighting, heating, ventilation, etc. The enforcement of the Act was left to the magistrates and, partly as a result, became ineffective. Indeed it was not until 1833 that a prototype of the present factory legislation appeared, in an Act that provided for the first time for the appointment of four Inspectors to enforce the law inside factory premises and giving them the right to enter factories and mills. In 1864 an Act was passed extending the power of the inspectorate beyond the textile industry to pottery and matchmaking factories and three years later in 1867 the Workshop Regulation Act introduced more limited health and safety measures into smaller working establishments where fewer workers were employed. In 1878, with the passing of the Factory and Workshop Act, the administration and substantive law relating to factories and workshops was consolidated. A further consolidating measure appeared in 1901, the Factory and Workshop Act, and this heralded the start of a new pattern of legislation, statutes empowering the Home Secretary to draw up *ad hoc* regulations to deal with dangers from industries and processes as they arose. This pattern has continued largely uninterrupted up to the present day and is being perpetuated by H.A.S.A.W.A.

It was the Factories Act 1937 that revolutionised the definition of "factory" by comprehensively redefining the term to include both factories and workshops. As a result of amendments to the 1937 Act in both 1948 and 1959, the 1937 Act was consolidated for the last time by the Factories Act 1961 which retained the wide definition accorded to "factory" by the 1937 Act. By virtue of s. 127, Factories Act 1961, some of the provisions of the Factories Act are extended to building operations and works of engineering construction, since s. 127(4) provides that "any place where such operations or works are carried on" is to be treated as if it were a factory. As a consequence, the Minister had power to make regulations for the safety of workmen employed in building and civil engineering, the end product being the triad of statutory Construction Regulations—the Construction (General Provisions) Regulations 1961, the Construction (Lifting Operations) Regulations 1961 and the Construction (Working Places) Regulations 1966.

As well as factories and building sites, the first inspectorate of mines was established in 1843 and since then comprehensive health and safety measures have followed in respect of mining working conditions, culminating in the Mines and Quarries Act 1954. Agriculture got its first protective statute in 1956 in the form of the Agriculture (Safety, Health and Welfare Provisions) Act. And employees working in offices and shops received their first protective statute in 1963 in the form of the Offices, Shops and Railway Premises Act. Safety on the railways has been regulated, in addition to the Factories Act and the Offices, Shops and Railway Premises Act, by the Prevention of Accidents Rules 1902–1931; shipping by the provisions of the Merchant Shipping Act 1964; and aviation by the Civil Aviation Acts 1949–1971. A more recent arrival on the legislative scene has been the Mineral Workings (Offshore Installations) Act 1971, which is designed to make provision for the health and safety of workers on offshore installations.

Finally comes H.A.S.A.W.A. itself. The concept behind this protective legislation is different from that previously enacted. This Act extends protection to all working situations, that is to all employer–employee relationships with the exception of domestic employment, owing to the difficulty of enforcement. Thus nearly everyone at work is now covered.

5. The legislative role of the Health and Safety Commission
By s. 15, H.A.S.A.W.A., power is given to H.S.C. to issue regulations relating to health and safety at work, and by s. 16 to issue and approve codes of practice.

Section 17 explains the effect in legal proceedings of codes of practice. It states that a failure on the part of a person to observe a provision of a code of practice shall not *of itself* make him liable to civil or criminal proceedings. But a contravention of the code may be regarded as proof in *criminal* proceedings that a requirement or prohibition was breached.

H.A.S.A.WA. is the source from which regulations and codes of practice emanating from H.S.C. will emerge to replace the already existing protective industrial legislation. A time will arrive when all the law on health and safety at work will consist of regulations and codes of practice which are easily workable and amendable so as to be kept up to date.

Since H.S.C. was set up in October 1974 the most important regulations that have appeared, are the Safety Representatives and Safety Committees Regulations 1977. Other important regulations that have been published relate to the packaging and labelling of dangerous substances, and draft regulations on noise are expected shortly.

The Packaging and Labelling of Dangerous Substances Regulations 1978 were laid before Parliament on 1st March 1978; these regulations require suppliers of about 800 dangerous chemicals to provide warning labels on their containers from 1st September 1978, where the capacity is 200 litres or more, and from 1st March 1978, in the case of smaller containers.

Criticism of judicial interpretation in industrial safety
The Robens Report noted that the "results of the application, interpretation and development of legal rules through the process of judicial precedent have not invariably served well the interests of safety". There are two ways of looking at this matter. On the one hand, one can say that the relaxation of strict statutory duties, through the importation of notions of reasonableness and reasonable care on the part of the judges has resulted in fewer successful actions for compensation by injured workmen. A case in point is *Close* v. *Steel Company*

of Wales in 1962, where the House of Lords interpreted s. 14(1), Factories Act 1961 (all other dangerous machinery shall be securely fenced), as a duty to guard against only those risks of injury that were reasonably foreseeable, as is the position at common law. On the other hand, strict draconian standards can lead to patent absurdities. A good example occurred in 1955 in the case of *Summers & Sons Ltd.* v. *Frost*. F was employed by S as a maintenance fitter in a steel works. While he was grinding a piece of metal on a grinding machine, his thumb came into contact with a revolving grindstone and he was injured. The machine in question consisted of horizontal shafts at each end driven by electric motors. The grindstones revolved downwards when the operator faced the machine. An efficient hood had been fitted to the machine, and the only part of the revolving grindstone that was exposed was an arc about 7 in. long, there being a gap of $\frac{3}{8}$ in. between the hood and the revolving arc. It was held by the House of Lords that the grindstone was a dangerous part of machinery for the purposes of s. 14(1), Factories Act 1937 (now s. 14(1), Factories Act 1961), and that there was an absolute duty to prevent such injury as was reasonably foreseeable. The result was that the machine was not securely fenced even though the consequence of securely fencing such a machine was to make it commercially unusable. Needless to say it was left to Parliament to rectify this matter and reverse the effect of this decision by passing the Abrasive Wheels Regulations 1970 which came into force on 2nd April 1971. Regulation 10(2) stated that a guard had to be provided and kept in position at every wheel in motion unless the use of a guard was impracticable owing to the nature of the work or the wheel.

Regulations

Section 76, Factories Act 1961, gave the Minister power to make special regulations until it was repealed by the Factories Act 1961 (Repeals and Modifications) Regulations 1974, which came into effect on 1st January 1975. These regulations were additional to those duties laid down in the Act itself in respect of dangerous trades. Under the powers in the Act the Construction Regulations 1961 and 1966, the Power Presses Regulations 1965, the Abrasive Wheels Regulations 1970, the Electricity Regulations 1944, the Asbestos Regulations 1969 and many more were passed, and the operation of the Woodworking Machinery Regulations 1922 continued.

In the field of industrial safety legislation the relationship of regulations to the parent statute has three main functions. Firstly to make provision for dangers which come to light and are identified only after the passing of the enabling Act: a case in point is the Ionising Radiations (Unsealed Radioactive Substances) Regulations 1968 and the

Ionising Radiations (Sealed Sources) Regulations 1969. Second to give extra teeth to the requirements of the enabling Act: a case in point is the Power Presses Regulations 1965, reg. 3(3) of which states that the regulations are in addition to and not in substitution for other requirements imposed by the Factories Act. Third, and more rarely, to substitute a lesser standard of care, e.g. the Abrasive Wheels Regulations 1970, reg. 10(2) (to the effect that guards need not be provided and kept in position when this is impracticable).

Draft regulations proposed by H.S.C. for submission to the Secretary of State for Employment under s. 50(1) may well be equally if not more effective. More specifically, s. 15(3) provides that health and safety regulations:

(*a*) may repeal or modify any of the existing statutory provisions;

(*b*) may exclude or modify in relation to any specified class of case any of the provisions of ss. 2–9 or any of the existing statutory provisions, e.g. the extant sections of the Factories Act 1961.

Role of the Secretary of State for Employment

These new regulations are likely to be ushered on to the statute book with a minimum of formality. Thus s. 82(3) provides that any power conferred by Part I or Part II or Part IV to make regulations:

(*a*) includes power to make different provision by the regulations for different circumstances or cases and to include in the regulations such incidental, supplemental and transitional provisions as the authority making the regulations considers appropriate in connection with the regulations; and

(*b*) shall be exercisable by statutory instrument which shall be subject to annulment in pursuance of a resolution of either House of Parliament.

This means that the Secretary of State for Employment can introduce expeditiously measures which are the subject of, and are identified in, the Third Schedule of the Act. And although such measures could be rejected by either House of Parliament, the likelihood of this happening must be regarded as remote. Indeed in answer to the criticism that too much delegated legislation makes for arbitrary use of power, we must assume that H.S.C. will act as the custodian of our liberty. This, at any rate, would appear to be the role accorded to it by s. 11(2)(*c*), which enjoins that it shall be the duty of the Commission "to make such arrangements as it considers appropriate for securing that government departments, employers, employees, organisations representing employers and employees respectively, and other persons concerned with matters relevant to any of those purposes are provided with an information and advisory service and are kept informed of, and adequately advised on, such matters".

Likely content of future regulations

The Third Schedule of H.A.S.A.W.A. identifies areas to be covered by new regulations, the scope of which, in terms of legal involvement, is potentially enormous. To take a random sample of such regulations and their content, the Secretary of State for Employment can make regulations regulating or prohibiting the use of any plant or its manufacture or supply; or the carrying on of any process or the carrying out of any operation. Sub-paragraph 1 (2) empowers him to impose requirements with respect to the design, construction, guarding, siting, installation, commissioning, examination, repair, maintenance, alteration, adjustment, dismantling, testing or inspection of any plant. Similarly under sub-para. 1 (4) he can impose requirements with respect to the testing, labelling or examination of any substance. Hence the appearance in late 1976 of the draft regulations relating to the packaging and labelling of dangerous substances. By virtue of para. 3 he may prohibit or regulate the transport of articles or substances of any specified description; also lay down requirements with respect to the manner and means of transporting articles or substances. Paragraph 4 permits him to prohibit the carrying on of an activity except under licence and para. 5 may require a person or premises or thing to be registered as a condition precedent to carrying out certain activities. This seems especially relevant to the application of s. 5, H.A.S.A.W.A. (emission from prescribed premises of noxious substances into the atmosphere).

Of special interest to the safety profession is para. 6(1). This enables the Secretary of State for Employment to require the appointment of persons (or persons with specified qualifications or experience or both) to perform specified functions, and to impose duties or confer powers on persons appointed to perform specified functions. In furtherance of this aim, sub-para. 6(2) empowers the Secretary of State to restrict the performance of specified functions to persons possessing specified qualifications or experience.

6. E.E.C. directives and international labour standards

Three areas of European law are particularly relevant to the future development of industrial relations and industrial safety, firstly the European Social Charter, second Conventions and Recommendations of the International Labour Organisation and third Articles and Directives of the European Economic Community.

The European Social Charter

This charter comprises conventions from the Council of Europe dealing with the conditions of workers; it was signed by a group of member states including the U.K. in 1965, and contains a list of objectives which members of the Council agree to accept, including the right of workers to safe and healthy conditions. Of particular significance is Art. 10, which states that contracting parties undertake:

(*a*) to provide or promote the technical and vocational training of all persons (including the handicapped) in consultation with employers' and workers' organisations, and to grant facilities for access to higher technical and university education, based solely on individual aptitude;

(*b*) to provide or promote a system of apprenticeship and other systematic arrangements for training young boys and girls in their various employments;

(*c*) to provide or promote,

(i) adequate and readily available training facilities for adult workers;

(ii) special facilities for the re-training of adult workers needed as a result of technological development or new trends in employment;

(*d*) to encourage the full utilisation of the facilities provided by appropriate measures, such as

(i) reducing or abolishing any fees or charges;

(ii) granting financial assistance in appropriate cases;

(iii) including in the normal working hours time spent on supplementary training taken by the worker, at the request of his employer, during employment;

(iv) ensuring, through adequate supervision, in consultation with employers' and workers' organisations, the efficiency of apprenticeship and other training arrangements for young workers, and the adequate protection of young workers generally.

A state which is a member can choose to accept ten out of the nineteen Articles and not all the Articles in Part 2 (from which Art. 10 is taken) have to be accepted. The Charter is enforced by an examination by independent experts of reports submitted on a two-yearly basis by the governments concerned. Employers' organisations and trade union organisations provide comment on such reports. Compliance with the Charter is achieved through the adverse publicity which follows when a state is in breach of the Articles. Much British legislation today derives from Articles of the European Social Charter and the Conventions and Recommendations of the International Labour Organisation, two examples being the H.A.S.A.W.A. provisions relating to worker safety representatives, and the emphasis in the Employment Protection Act 1975 on the right to organise through trade unions, to receive maternity pay, and to the rights of workers on company liquidation and on suspension from work on medical grounds. In addition, the law relating to unfair dismissal in the Trade

Union and Labour Relations Act 1974 and the accompanying Schedule, together with the legislation on sex discrimination passed in 1975, owes its existence to International Labour Standards or Recommendations of the International Labour Organisation.

The International Labour Organisation

The difference between the Recommendations and Conventions which the I.L.O. produces lies in the fact that if a member state adopts a Convention, it must effect changes in its own internal law, if necessary, to make the internal law comply with the Convention. A Recommendation, on the other hand, merely lays down a desirable standard to be achieved: it is not enforceable in the same way as a Convention which has been accepted, and as such is enforceable through examination by a body of independent experts, as with Articles of the European Social Charter. From this august source emerged the Convention Concerning Protection and Facilities to be Afforded to Workers' Representatives in the Undertaking of 1972, foreshadowing the regulations proposed to the Secretary of State for Employment by H.S.C. relating to safety representatives and safety committees.

E.E.C. directives and the Rome Treaty

This is a fertile area for new legislation by the U.K. Parliament concerning labour relations and employment conditions generally. The Articles of the Rome Treaty laying down Regulations are automatically binding on member states, whereas Directives are probably not directly enforceable, in the sense that they will not form part of the law of a national state until ratified by its local Parliament. The decision of the European Court of Justice in *Van Duyn* v. *Home Office*, however, would appear to have made serious inroads into this position by establishing that Directives are directly applicable so as to give rights enforceable in the national courts and tribunals of member states. International and continental labour practices will now therefore find their own inexorable way into the fabric of English labour law. The European Court of Justice should be regarded as a formal source of English law; certainly it can be predicted that the future content of health and safety law is likely to be progressively affected by I.L.O.s, Directives and a corpus of judicial law springing from the European Court of Justice. Regulations emerging from H.S.C. will in future almost certainly have a continental flavour.

THE COURTS AND TRIBUNALS

7. Civil and criminal law

Before describing the structure of the courts and the relationship of tribunals to them it will be advisable to make some initial remarks about the division of law into civil and criminal branches, since certain courts exercise a civil and others a criminal jurisdiction.

Criminal law is an amalgam of common law and statute. The court's order may take the form of a fine or imprisonment and both sanctions can be adjusted to meet the gravity or otherwise of the crime in question. This is not the position with damages or compensation, the *civil law* remedy. The court which handles the bulk of criminal matters is the Magistrates' Court, and offences occurring under health and safety legislation are normally dealt with here or more rarely in the higher criminal court, the Crown Court.

Unlike torts or civil offences, crimes are, at least in theory, committed against society itself. Hence most prosecutions are conducted by the Crown through its enforcement officers such as the police, or, in the case of health and safety at work, H.S.E. inspectors. In other areas, such as consumer protection, the prosecutor will be a local authority official. Moreover since a crime is committed against society, defences associated with consent, which might well be valid in tort, are to a large extent non-existent in criminal law. Society cannot permit and condone the commission against it of socially deviant acts, but in tort, where the offence is committed against the person, whether corporate or not, the person is deemed sufficiently rational to exercise freedom of choice as to whether harmful conduct is committed in respect of him. Courts do not easily draw the inference that employees have freely consented to run the risk of injury arising from their employment, but if the inference is inescapable, the defence of consent symbolising freedom of choice will affect the plaintiff's claim for compensation. Persons cannot normally, however, consent to the commission of crimes against them.

The courts dealing with criminal cases are often different from those dealing with civil cases. Whereas most criminal matters are handled by the Magistrates' Court, most civil matters are heard in the county court. Industrial matters or employment issues are normally handled by industrial tribunals, the future role of which is likely to be extensive in this field.

8. Relationship between crime and tort

Insurance cover against torts

Torts are civil wrongs, and are normally handled in the county court. Redress in respect of a tort normally takes the form of damages. Certain conduct can amount to both a crime and a tort, and this fact can have important practical consequences regarding insurance and insurability. Crimes may be negligent or intentional torts—causing death by reckless driving or driving without due care and attention are crimes which are negligent torts, assault is a crime which is an intentional tort. From the point of view of insurance this distinction is crucial. It is a common feature of an insurance policy to agree to cover the insured for court costs and legal expenses arising from the commission of a crime by him: a comprehensive motor vehicle policy will agree to indemnify him against the costs and expenses arising from manslaughter, dangerous driving and causing death by dangerous driving. This is because the essence of the conduct is negligence and this is insurable. Risks amounting to intentional torts, on the other hand, are not insurable. The matter was well put by Bailache, J., in the leading case of *Tinline* v. *Whitecross Insurance Association Ltd.*, where the insured, who was covered by a motor vehicle policy in respect of third-party risks, killed a pedestrian whilst driving at an excessive speed and was found guilty of manslaughter. Here the judge said "The defendants say that . . . where the negligence is so gross and excessive, and the result is that a person is killed and the crime of manslaughter is committed, the policy does not indemnify him against the civil consequences of that crime, because it is against public policy to indemnify a man against the civil consequences of his criminal act. Speaking generally, I think it is true to say that it is against public policy to indemnify a man against the consequences of a crime which he knowingly commits, and in the word 'crime' I include the breach of any statutory duty which renders a man liable to fine or imprisonment." It was held that the insurer was liable to pay.

An example of a comprehensive accident policy which could not be relied upon to indemnify the policyholder occurred in the interesting case of *Gray* v. *Barr*. In this case B had killed G's husband with a shotgun and was later acquitted of manslaughter. G. now sued B for damages and B brought his insurance company, the Prudential Assurance Co. Ltd., in as co-defendant. B's wife had taken out a "hearth and home" policy which agreed to indemnify her and/or any member of the household (this included B) "against all sums which should become legally liable to be paid as damages in respect of bodily injury to any person caused by an accident". It was held by the Court of Appeal that although G's action against B succeeded on the grounds of negligence, yet B could not expect to be indemnified by his insurer because his initial act of threatening G's husband with a gun was intentional and constituted an assault, an intentional tort.

The consequence of this approach is, or seems to be, that the more evil the motive behind the conduct and the more heinous the conduct in question, the more unlikely it is that the harm or loss will be underwritten. Hence if a motor car driver is injured through the negligent driving of another motorist, or even killed in circumstances amounting to manslaughter, he or his dependants can be compensated by insurance. If, however, the facts show that the driver deliberately "aimed" his car at the other (e.g. after a lover's tiff), compensation should not be forthcoming. This may not, however, be the true position, as is made clear by the decision of the Court of Appeal in 1964 in *Hardy* v. *Motor Insurers' Bureau*, where a security officer at a metal company tried to stop a van which had been stolen from the company's car park. He managed to put one hand on the door, but the driver, who was uninsured, drove off at high speed, dragging the security officer along with him. The security officer obtained judgment against the driver in a civil action which was unsatisfied. He then sued the Motor Insurers' Bureau, who refused to pay on the ground that the driver's liability was a liability for the consequences of his own wilful act and as such not required to be covered by a policy of insurance under the Road Traffic Act 1960, s. 203. This section states that a driver must take out an insurance policy "in respect of any liability which may be incurred by him or ... in respect of the death of or bodily injury to any person caused by or arising out of, the use of the vehicle on the road". It was held by the court that the Motor Insurers' Bureau was liable to pay. During the course of giving judgment Lord Denning, M.R., observed:

"The policy of insurance, which a motorist is required by statute to take out, must cover any liability which may be incurred by him arising out of the use of the vehicle by him. It must, I think, be wide enough to cover, in general terms, any use by him of the vehicle, be it an innocent use or criminal use, or be it a murderous use or a playful use.... Of course, if the motorist intended from the beginning to make a criminal use of the vehicle—intended to run down people with it or to drive recklessly and dangerously—and the insurers knew that was his intention, the policy would be bad in its inception. No one can stipulate for iniquity. But that is never the intention with which such a policy is taken out.... The question only arises when the motorist afterwards makes a criminal use of the vehicle. The consequences are then these: if the motorist is guilty of a crime involving a wicked and deliberate intent, and he is made to pay damages to an injured person, he is not himself entitled to recover on the policy. But if he does not pay the damages, then the injured third party can recover against the insurers under s. 203 of the Road Traffic Act 1960 ..."

The result of *Hardy's* case would appear to be that there is a difference of substance between manslaughter and motor-manslaughter, and that in the *Gray* v. *Barr* type of situation, where the preliminary assault was intended, a contract of insurance made on this basis would be void from its inception.

Insurance cover against crimes
This is really cover against the court costs involved and against the expenses of legal representation on both sides. It is impossible, and contrary to public policy, to give cover against the penalty for the crime, e.g. the fine. This last fact is of immense importance and should not be overlooked when taking out insurance in respect of breaches and prosecutions under H.A.S.A.W.A. (*see* §**31,** p. 205).

9. Types of criminal offence

Criminal offences can be categorised into two separate groups—those which require for the purpose of conviction a *wrongful intent to accompany the act* and crimes of *strict liability*. By and large, crimes requiring wrongful intent are of common law origin, whilst crimes of strict liability are of statutory origin. In both cases there must be a guilty act or at least an attempt at one. In the case of crimes of statutory origin, the terminology of the statute or the regulation is the key to whether the offence is one requiring wrongful intent or one of strict liability: where words such as "wilfully", "knowingly", etc., are used in the Act, presumptively the offence requires wrongful intent. Significantly, there is also a halfway house between the two: *negligence*; and this halfway house is germane to an understanding of criminal liability under H.A.S.A.W.A. The leading case is the 1970 House of Lords' decision in *Sweet* v. *Parsley*, which is generally regarded as having established the following crucial rule in the interpretation of statutory crimes—that an *element of fault* is an essential ingredient of every offence unless it can be shown that Parliament elected to enact otherwise. In other words, there is a presumption against strict liability.

Offences under H.A.S.A.W.A.

(*a*) *Terminology of the Act.* That this is the position with respect to offences occurring under H.A.S.A.W.A. is self-evident from the terminology of the Act. The Act prescribes that the general duties be carried out "as far as is reasonably practicable", clearly importing the common law notion of reasonable care as distinct from the imposition of an absolute duty. Moreover, s. 37, H.A.S.A.W.A., laying duties upon directors and managers and other corporate officers, is couched in parallel language. Hence offences under H.A.S.A.W.A. are neither of wrongful intent nor of strict liability, but rather offences where liability is based on negligence. Here it is worth noting that whilst vicarious liability is an example of strict liability, the crime for which a person is alleged to be vicariously liable may not itself be one of strict liability. Where, for example, a company official fails to carry out a particular duty as far as was reasonably practicable, the company may also be guilty of an offence. In other words, once negligence has been established on the part of the official, automatically that guilt can be attributed to the company itself.

(*b*) *Vicarious liability*. This is in no sense comparable with its counterpart in civil law. In civil law the doctrine of vicarious liability has been entrenched for many years now, for reasons of pressing practicality designed to ensure that injured parties, either within the course of their employment or not, will receive compensation at the hands of the negligent party or his insurer. In criminal law, on the other hand, the notion of one person paying a penalty for the offence of another is inconsistent with public policy, and issues of compensation arise much less frequently in criminal cases. Nevertheless there is a limited doctrine of vicarious liability here, applicable mainly to criminal offences of statutory origin, which occurs in two principal types of situation. Firstly, where a statute prescribes that an offence may be committed by a person himself or by his *servant* or *agent*. Second, where the terminology of the statutory provision is such as to be capable of covering acts or omissions of both principal and agent, as where it says that a person is guilty of an offence if, under certain circumstances, he *sells* or *uses* articles or substances. An employer can be said to sell or use in a contractual sense, even if his employee does the actual selling or using, though if the statute said "knowingly" sell or use, normally the guilt of the employee could not be attributed to the employer, except in the cluster of anomalous cases relating to the licensing laws.

The likelihood of vicarious liability arising in connection with the commission of offences under H.A.S.A.W.A. cannot be regarded as impossible. Regulations passed by the Secretary of State for Employment could specifically refer to offences being committed by "selling" or "using" certain articles and substances. If an employee in contravention sold or used a particular article or substance his employer or the company could therefore conceivably be vicariously liable, if the regulation created an offence of strict liability, as where the words "knowingly" or some similar terminology did not appear. It has been pointed out, however, that duties under H.A.S.A.W.A., unless the contrary is expressed in regulations, probably create offences liability for which depends on negligence or something akin thereto, as the words "as far as is reasonably practicable" seem to import notions of reasonable care. This is consistent with the position in *Sweet* v. *Parsley*. Is there, therefore, a likelihood that an employer or company could become vicariously liable for the commission of an offence liability for which lies in the negligent act or omission of the employee, e.g. an offence committed by an employee contrary to s. 7, H.A.S.A.W.A.? The answer would appear to be negative, at least if cases decided on the 1961 Factories Act can be regarded as having laid down criteria. The decisions in *Potts* v. *Reid* in 1943 and *Gallagher* v. *Dorman Long Ltd.* in 1947, occurring under the 1961 legislation or its predecessors, have established that there is for these purposes no connection between civil and criminal liability. In *Potts* v. *Reid* the court said, "Criminal and civil liability are two separate things. . . . The legislation [1937 Factories Act] might well be unwilling to convict an owner who failed to carry out an absolute statutory duty of a crime with which he was not himself directly concerned, but still be ready to leave the civil liability untouched." Following on from these remarks, the court in both cases held that, although a workman had committed a criminal offence and was guilty, s. 161 came to the aid of the factory occupier if he could prove that he used *all due care* to enforce the Act, though significantly this did not prevent the workman getting compensation in respect of any injury arising from the commission of the criminal act or from the breach. Similarly in *Wright* v. *Ford Motor Co. Ltd.* in 1966 an occupier of a factory was charged with failure to fence a dangerous part of machinery when an employee had been injured in machinery which was normally fenced but where the lock on the gate of the fence had been forced by another employee. Here it was held (*a*) that the employer was not vicariously liable under s. 155(1) Factories Act 1961, for the offence of the unknown employee under s. 143, and (*b*) he was not liable under s. 155(2) which makes an employer liable in addition to the employee committing the offence, since it had not been proved that the employer had failed "to take reasonable steps to prevent the contravention".

The other situation possibly involving vicarious liability occurs in relation to s. 36, H.A.S.A.W.A. This concerns offences which are committed by persons other than the body corporate, or in addition to the body corporate. In order for there to be a conviction under s. 36, it is assumed, following the case of *Tesco Ltd.* v. *Nattrass* under similarly worded provisions of the Trades Description Act 1968, that the offence must be committed by a person other than anyone who could be construed as the alter ego of the company, i.e. other than a member of the board of directors or other than a person to whom the board of directors had delegated the performance of a statutory duty: Section 36 therefore applies to offences committed by hands of the company rather than the brain. Despite this, and despite the fact that the section does not rule out the prosecution of a director personally or the body corporate itself in addition to the other person, nevertheless, if the offence committed by the other person were one of selling or using, etc., it is difficult to see how the company could not be vicariously liable as distinct from being liable as a principal under s. 37.

10. Torts

Torts are civil wrongs redressable in a civil court by unliquidated damages. Certain torts may correspond to criminal offences as well, e.g. assault. The action in tort is always at the suit of a person, whether corporate or not, rather than an action brought by the police on behalf of the Crown. The principal area of tortious liability is negligence, where liability presupposes want of reasonable care in commission or omission. There are also other areas where the liability is strict, though not normally absolute, where a person will be liable despite the fact that he took reasonable care or was not at fault. Vicarious liability, an example of quasi-joint liability occurring in the case of employer–employee relationships, is an instance of such strict liability.

Purpose of tortious liability

Tort actions normally take place in the County Court or the Queen's Bench Division of the High Court, depending upon the amount of damages or compensation claimed. In practice, actions for damages in excess of £2,000 are heard in the High Court. The law of tort is underpinned by two criteria, on the one hand the desire to provide compensation for damage done or loss caused irrespective of subjective factors, and on the other the desire to punish. For the purposes of health and safety at work legislation and actions for injury at work generally, it is the former approach, the compensatory one, which is relevant. The consequence of this desire to compensate is that the injured plaintiff must be compensated irrespective of the absence of moral guilt on the part of the defendant or despite the plaintiff's personal shortcomings, an example of the latter being afforded by the case of *Nettleship* v. *Weston* in 1971 (*see* §**19**, p. 44). Here, the real basis of liability on the part of the defendant is that his conduct *caused* the plaintiff's injury. As damages here are compensatory, insurance is normally bought by employers in respect of industrial injuries to their employees and by road vehicle users in respect of accidents which may occur on the highway. In recent times this sort of insurance has become compulsory and, as a result, has had a considerable effect upon the development of tort liability (*see* §**70**, p. 282).

Distribution of loss

In the field of industrial injuries and road accidents, tort serves as a vehicle for social insurance. It has grown into a loss distribution mechanism, particularly in respect of the Road Traffic Acts 1930–1972, the Employers' Liability (Compulsory Insurance) Act 1969 and the Employers' Liability (Defective Equipment) Act 1969. Moreover vicarious liability has accelerated the transition from a system of liability based on personal fault to one of risk allocation. This process has culminated in the establishment of the Royal Commission on Civil Liability and Compensation for Personal Injuries under the chairmanship of Lord Pearson (*see* §**78**, p. 288).

11. The courts

Courts exercising civil jurisdiction

There are courts of *first instance* where actions are initiated and courts of *appeal* which hear appeals from courts of first instance on points of law, generally. In civil actions where damages or injunctions are claimed, the two courts of first instance are the County Court and the High Court. As far as protective industrial safety legislation is concerned, and actions against the employer at common law, the great majority of these cases is concerned with liability in tort and the resulting compensation. All first instance cases in tort begin either in the County Court or the Queen's Bench Division of the High Court. If damages of £2,000 or more are claimed the action should begin in the High Court unless the parties agree otherwise. The action is initiated by a plaintiff against a defendant. It should be remembered, however (in the light of remarks made earlier about tort being in some respects a synonym for loss distribution), that the defendant is not necessarily the person who will end up by paying the damages. The person who pays is rather an unseen defendant, namely, the insurer.

If a defendant or unseen defendant is dissatisfied with the decision of either the County Court or the High Court, he may lodge an appeal with the Court of Appeal (Civil Division) which may allow or dismiss it. A further appeal may be lodged with the House of Lords, provided that either the Court of Appeal or the House of Lords gives leave. There is no appeal machinery beyond the House of Lords. Unsuccessful litigants must hope for, or endeavour to promote, changes in the law through Parliament.

Courts exercising criminal jurisdiction

The principal courts exercising criminal jurisdiction are the Magistrates' Courts and the Crown Courts, with appeals lying to the Court of Appeal (Criminal Division) and finally to the House of Lords. Appeal lies to the House of Lords on a point of criminal law when the Court of Appeal or the House of Lords gives leave and from the High Court when there is a point of law of general public importance involved in the issues spotlighted by the case.

(*a*) *Breaches of health and safety legislation.* Most prosecutions for breaches of health and safety legislation, i.e. breaches of the general duties in ss. 2–9, H.A.S.A.W.A., or of any other relevant statutory provision, will take place before the magistrates. Where there is a conviction under H.A.S.A.W.A. or any other relevant statutory provision, the maximum penalty that can be ordered on summary conviction will depend upon the offence committed. If there is an offence committed in relation to s. 33(1)(*d*), (*f*), (*h*) or (*n*), or of s. 33(1)(*e*), in so far as this consists of a contravention of a requirement imposed by an inspector under s. 20, the offence now carries a maximum penalty of £1,000 by virtue of s. 31(1), Criminal Law Act 1977 (these new penalties which can be ordered by the magistrates became operational as from 8th September 1977). This sub-section has the effect of updating s. 33(2), H.A.S.A.W.A.

As for offences triable on indictment in the Crown Court, it is only those offences caught by s. 33(3)(*b*)(*i*), H.A.S.A.W.A., which can lead to imprisonment for not more than two years and/or an unlimited fine. The offences referred to by s. 33(3)(*b*)(*i*) are those set out in s. 33(4). Not all indictable offences under H.A.S.A.W.A. therefore carry the penalty of imprisonment, and it should also be remembered that penalties, because they rapidly become out of date, are continually being updated, and this will apply to health and safety as well as to other criminal offences (*see* § **29,** p. 194). Moreover, although certain offences are triable on indictment, the case will normally begin in the Magistrates' Court for the purpose of deciding whether the case should be sent to the Crown Court, the so-called committal proceedings.

Unlimited fines can be high. In a recent Scottish case, for instance, there had been contravention of H.A.S.A.W.A. by a brewery company (the occupier of the building where work was going on) and by a contractor who, whilst working on the building, was allowing his employees to work with white asbestos without providing them with any protective equipment or clothing and in blatant disregard of the Act's provision: the brewery was fined £10,000 and the independent contractor £6,500. This severity is likely to continue, especially where dangerous chemical processes, dangerous machinery, or asbestos are concerned. The breach of the Asbestos Regulations 1969 in this case proved very expensive.

(*b*) *Unauthoritative nature of past criminal cases.* In the past criminal cases occurring under the Factories Acts or other similar protective legislation or under regulations did not yield much authoritative interpretation. Most cases were dealt with before the magistrates with little if no scope for legal argument, and appeals to the High Court and from thence to the Appeal Courts proper were rare.

This was not the position with civil actions, where appeals were much more frequent, but how far interpretations in civil cases could be invoked to influence the outcome of criminal cases is doubtful. Issues before the magistrates are decided more often on the facts than on application of legal principle and doctrine. If the observations made in *Potts* v. *Reid* are relevant, the criminal action is in any case entirely separate from the civil one. This trend in favour of the unauthoritative nature of criminal cases is not likely to continue under H.A.S.A.W.A., because the general duties of the Act have to be carried out "as far as reasonably practicable", and this necessarily implies the importation of previous case law to elucidate the meaning of the expression. Liability in criminal cases may no longer be the cut and dried issue that it was before. Crown Courts acknowledge the doctrine of binding precedent, at least as far as decisions of superior courts, such as the Court of Appeal and House of Lords, are concerned, and all this seems to point to a body of authoritative criminal law growing up under H.A.S.A.W.A.

(*c*) *Subordinate role of criminal law in health and safety matters.* The Robens Report recommended that dangers at work be put right or eliminated or their incidence minimised *on the spot.* This implied (as has now come to pass) the granting of greater executive power to inspectors whose job it was to visit factories and places of work and, if necessary, bring subsequent prosecutions. Also implicit in the recommendation was a strong criticism of the previous system under which guilty persons had been brought before the courts for punishment. H.A.S.A.W.A. recognises that this process is time-consuming and does not in itself go far enough in reducing the level of accidents at work. It therefore adopts the premiss that the role of criminal sanction shall be ancillary to that of the improvement and prohibition notice which inspectors may now serve. Executive action now comes before the judicial process. Give inspectors power to get dangerous machinery and dangerous processes rectified or to eliminate dangers on the shop floor by halting machinery or even destroying it, and the number of industrial injuries and accidents will fall; only if executive action is disobeyed, e.g. a notice is not complied with, should the criminal law be invoked and accused persons brought before the courts for punishment (*see* §**29**, p. 186).

(*d*) *Can criminal courts award compensation?* Prior to 1973, if a person wished to obtain compensation as a result of having suffered damage or injury arising out of a criminal offence, he would have to initiate proceedings himself in a civil court. When, however, the relevant sections of the Criminal Justice Act 1972 came into operation on 1st January 1973 this duplication of court proceedings was minimised. Section 1 of that Act gave criminal courts power to order a person convicted of a criminal offence to pay compensation in respect of any personal injury, loss or damage caused by the offence. This compensation is in addition to any fine or other penalty imposed by the court and amounts are limited to £400 in the case of compensation awarded by magistrates. If later civil proceedings are commenced, any compensation paid as a result of the criminal action and by order of the criminal court must be brought into account in assessing damages in the civil action. This power to award compensation is now contained in the Powers of Criminal Courts Act 1973. It should also be remembered that a fair proportion of criminal injury cases are compensated on an *ex gratia* basis through the machinery of the Criminal Injuries Compensation Board (*see* § **25**, p. 151).

12. The role of industrial tribunals in industrial safety
Industrial tribunals were first created under the Industrial Training Act 1964, which established training boards to provide industrial training within various industries and gave power to Industrial Training Boards (I.T.B.s) to impose levies on employers in order to raise finance for such training. A person or company assessed to a levy may appeal against it on the ground that he falls outside the scope of the I.T.B. applying the levy, or that the levy has been incorrectly calculated. Appeal was to an industrial tribunal and such tribunals first became appeal bodies as from 31st May 1965.

The role of industrial tribunals since that date has been a continually expanding one, reaching further and further into the field of labour law. Under the Redundancy Payments Act 1965 tribunals were given a further jurisdiction to hear claims relating to redundancy payments, and the view was expressed by the then Minister of Labour that he "would not rule out the possibility that as the tribunals became more established and gained experience, further functions relating to industrial relations might be given to them...."

During the 1970 Conservative government, power was given to industrial tribunals under the Contracts of Employment Act 1972 to inquire into matters relating to the written particulars of employment required by that Act; when the Labour government came to power in 1974, tribunals were empowered to hear questions of unfair dismissal (Trade Union and Labour Relations Act 1974), and complaints of unlawful discrimination (Sex Discrimination Act 1975). A vast new range of powers to hear cases was conferred by the Employment Protection Act 1975, ranging from those concerning guarantee payments, medical suspension payments and maternity pay through to rights to organise in a union, time off with pay for union activities and duties, as well as time off for training and to look for work. When the Industrial Tribunals (Improvement and Prohibition Notices Appeals) Regulations 1974 were passed, tribunals also gained the right to hear appeals against improvement and prohibition notices under

H.A.S.A.W.A. The future further development of what are, in essence, labour courts is evidenced by s. 109, Employment Protection Act 1975, which empowers the Lord Chancellor to accord jurisdiction to industrial tribunals in matters relating to damages for breach of contracts of employment and other contracts connected with employment. The Industrial Tribunals Regulations 1965, as amended subsequently, provide for the appointment of a president, who must be a barrister or solicitor of at least seven years' standing; the Lord Chancellor draws up a panel of barristers and solicitors (of seven years' standing) from whom a chairman is selected for the sitting of a tribunal; panels of persons have also been drawn up by the Secretary of State for Employment after consultations with employers' organisations and associations and with workers' organisations. From each of these panels the president nominates a member to make up a tribunal of three. This predominantly lay representation on industrial tribunals reflects the traditional scepticism of organised labour and employers' associations towards the judiciary; trade unions in particular have always regarded the law and law courts as an obstacle to their objectives.

Proceedings in industrial tribunals

These vary, depending upon the statutory jurisdiction which the tribunal is exercising. Generally such proceedings are governed by the Industrial Tribunals (Labour Relations) Regulations 1974, as amended by the Industrial Tribunals (Labour Relations) (Amendment) Regulations 1976. It should be noted, however, that although these regulations govern proceedings relating to matters occurring under the Contracts of Employment Act 1972, the Equal Pay Act 1970, the Trade Union and Labour Relations Act 1974, the Sex Discrimination Act 1975 and the Employment Protection Act 1975, they do not apply to proceedings relating to appeals against improvement or prohibition notices under H.A.S.A.W.A. (*see* § **29,** p. 27). Appeals against notices issued pursuant to H.A.S.A.W.A. are governed by the Industrial Tribunals (Improvement and Prohibition Notices Appeals) Regulations 1974. Proceedings in such appeals are substantially similar to those where the tribunal is exercising another statutory jurisdiction; apart from the time limit involved for the lodging of an appeal against an improvement or prohibition notice, health and safety tribunals do not differ from other industrial tribunals. Decisions are reached on a tripartite basis equally, although the tribunal's written decision is signed only by the chairman. The lay members of the tribunal are picked by the C.B.I. and the T.U.C., and although in theory they could outvote or vote against the chairman, this does not normally take place. A clerk, who is not legally qualified, is attached to the tribunal. Parties to the hearing may be represented by counsel or by a solicitor or by a representative of an employers' association or of a trade union, or by any other person whom the party appealing desires should represent him.

Parties to the hearing can make opening statements, call upon witnesses, give evidence, cross-examine and generally address the tribunal. Ordinary rules of evidence are inapplicable, and where one party is legally represented but the other is not the chairman will see to it that the unrepresented party is not harassed by counsel. Legal aid is not available for representation before industrial tribunals, and the decision of the tribunal may be that of a majority.

Power of review

The power of industrial tribunals to review their decisions is a limited one. The 1974 Regulations make it clear that review is possible only in the following circumstances:

(*a*) the decision was the result of an error of the tribunal;

(*b*) the decision was made in the absence of a party or other person entitled to be heard;

(*c*) a party did not receive notice of the proceedings;

(*d*) new evidence has come to light since the hearing;

(*e*) the interests of justice demand a review.

Appeals against tribunals

Normally appeals from industrial tribunals go to the Employment Appeal Tribunal where questions of law only are involved. In the case of appeals against improvement and prohibition notices under H.A.S.A.W.A., however, the appeal goes to the Queen's Bench Division of the High Court on a point of law only. The sort of issue which might conceivably precipitate an appeal from a local industrial tribunal to the High Court would be whether the tribunal had correctly interpreted a legal expression such as "reasonably practicable", or if it had failed to take into account a contrary ruling of the High Court or even of the Court of Appeal or the House of Lords. As yet there is little case law relating to appeals beyond health and safety tribunals, and so the likelihood of failure to take cognisance of a previous inconsistent precedent of a higher court is as yet remote.

STATUTORY INTERPRETATION

13. The constituent parts of a statute

A statute normally consists of eight constituent parts, of which six—the long and short titles; the preamble (if any) but only if the enacting words and sections, etc., are obscure; interpretation clauses (with some reservations); enacting sections and sub-sections; and Schedules (with some reservations)—are treated as being an integral part, and construed as part, of the Act, and two—marginal notes; headings and sub-headings—are deemed not to be part of the statute. H.A.S.A.W.A. is used as the example.

The title

The *long title* of H.A.S.A.W.A. is "An Act to make further provision for securing the health, safety and welfare of persons at work, for protecting others against risks to health or safety in connection with the activities of persons at work, for controlling the keeping and use and preventing the unlawful acquisition, possession and use of dangerous substances, and for controlling certain emissions into the atmosphere; to make further provision with respect to the employment medical advisory service; to amend the law relating to building regulations, and the Building (Scotland) Act 1959; and for connected purposes". The *short title* of H.A.S.A.W.A. is "the Health and Safety at Work Act 1974".

The preamble

Rarely seen today; H.A.S.A.W.A. has no preamble.

Enacting words

For example, "Be it enacted by the Queen's most Excellent Majesty, etc."

Interpretation clauses and adaptation to Scotland clauses

Interpretation clauses are of two kinds:

(*a*) where the word defined is declared to *mean* something. Here the definition is explanatory and restrictive;

(*b*) where the word defined is declared to *include* something. Here the definition is extensive.

It is important to note that where an interpretation clause extends the meaning of a word, this does not take away its ordinary meaning.

Enacting clauses and sub-clauses or sections and sub-sections

For example, Section 2(1): "It shall be the duty of every employer to ensure, so far as is reasonably practicable, the health, safety and welfare at work of all his employees."

Schedules

H.A.S.A.W.A. has ten, including the schedule of repeal.

In the leading case of *A.G.* v. *Lamplough* the court said "The Schedule is as much a part of the statute as any other part, but if the enactment in a Schedule contradicts an earlier clause, the clause prevails against the Schedule".

Marginal notes
"General duties of employers to their employees (s. 2)", for example. In the leading case of *Chandler* v. *D.P.P.* Lord Reid said "In my view sidenotes cannot be used as an aid to construction. They are mere catchwords and I have never heard that an amendment to alter a sidenote could be proposed in either house. So sidenotes cannot be said to be enacted in the same sense as the long title or any part of the body of the Act".

Headings and sub-headings
These occur especially in long Acts like H.A.S.A.W.A. An example of a heading in H.A.S.A.W.A. is "Part 1 Health, Safety and Welfare in connection with Work, and Control of Dangerous Substances and certain emissions into the Atmosphere". An example of a sub-heading from Part 1 of the Act would be "General Duties". In *R.* v. *Hare* the court said "Headings of sections and marginal notes form no part of a statute. They are not voted on or passed by Parliament but are inserted after the Bill has become law. Headings cannot control the plain meaning of the enactment, though they may in some cases in the light of preambles, if there is any ambiguity in the meaning of the sections on which they throw light".

The constituent parts of a statute which are construed by the courts as part of the statute are known as *intrinsic aids*, so as to distinguish them from *extrinsic aids*, such as reports of debates in Parliament, White Papers, views of government departments, etc. (*see* § **14**, p. 31). Extrinsic aids are traditionally resorted to only when the meaning of the statute is not clear.

14. Construction of words in statutes

The basic rules are as follows:

Explicit terminology
Where the terminology is clear and explicit, effect must be given to it, irrespective of the consequences, because the words speak the intention of the legislature.

Context
Words are to be construed in their context and are to be given their primary meaning. This means their popular or normal grammatical meaning. Only technical words should be given a technical meaning.

The intention of Parliament

The intention of Parliament should not be speculated upon. Rather, its intention should be gathered from the words it has enacted.

Evasion

Evasion in itself is not unlawful. But it should be noted that here "evasion" has two meanings:

(*a*) an illegal or underhand dealing (which is illegal);

(*b*) a deliberate attempt to avoid something unpleasant (which is lawful in itself). A person may go to his lawyer to ask him how to keep out of the way of an Act of Parliament, if he wants advice, for example, as to how to arrange his financial affairs in such a way that a tax does not apply to him; or he may call in a health and safety consultant for advice upon how to avoid the consequences of s. 6, H.A.S.A.W.A. This is all perfectly legal. What is illegal is when a man goes to his lawyer and says that he wishes to escape from the consequences of an Act, even though he has already done something which brings him within its scope.

Sense at the passing of the Act

Words must be construed in the sense they originally bore at the time the Act was passed.

The Golden Rule

It was noted earlier that words must be construed in their plain and literal sense so as to give them their primary meaning. If, however, this leads to patent absurdity or inconsistency, the courts will apply the so-called Golden Rule. This was probably best expounded by Lord Wensleydale in *Grey* v. *Pearson*, when he observed "The grammatical and ordinary sense of words is to be adhered to, unless that would lead to some absurdity or inconsistency with the rest of the intrument, in which case the grammatical and ordinary meaning may be departed from so as to avoid the inconsistency or absurdity".

Elementary presumptions in statutory interpretation

Parliament is deemed, in the absence of express words to the contrary:

(*a*) not to legislate inconsistently with the content of common law;

(*b*) not to legislate contrary to international law;

(*c*) not to wish to take away individual rights;

(*d*) not to impose a tax or lay down a penalty.

Words of limitation

Words of limitation should not be read into a statute if this can be avoided, although often it cannot, especially where the *ejusdem generis* rule applies. In *R.* v. *Edmundson* Lord Campbell put the matter as follows: "Where there are general words following specific words, the general words must be confined to things of the same kind as those specified." A good example of the *ejusdem generis* rule occurred in *Brownsea Haven Properties Ltd.* v. *Poole Corporation*, where the words "in any case" following "in all times of public processions, rejoicings or illuminations and in any case where the streets are thronged or liable to be obstructed" were confined to public processions and not extended to cover circumstances of ordinary traffic conditions.

Interpretation clauses

In addition, most modern statutes contain an interpretation clause, i.e. their own little dictionary (*see* s. 53, H.A.S.A.W.A.).

Extrinsic aids

These are aids to statutory interpretation which exist outside the four corners of the statute itself (hence they are referred to as extrinsic). They should be invoked only when the meaning of the statute is not clear or discloses a patent ambiguity.

(*a*) *The rule in Heydon's case.* The position was well put in the original case on the subject, *Heydon's* case in 1584, where the court laid down the following rules for construing statutes whose meaning is patently ambiguous:

(i) What was the common law before the making of the Act?

(ii) What was the mischief and defect for which the common law did not provide?

(iii) What remedy has Parliament resolved to cure the defect?

(iv) What is the true reason of the remedy?

Then the judges shall make such construction as will suppress the mischief and advance the remedy. For obvious reasons this last rule is sometimes referred to as the mischief rule.

(*b*) *Use of the rule in Heydon's case.* The rule in *Heydon's* case requires that we endeavour to discover the cause of, and the necessity for, the Act. In order to do this, it is permissible to examine:

(i) the state of the law when the Act was passed;

(ii) the history of the Act, both political and social, and the reasons which led to its being passed;

(iii) the meaning of the statute by reference to other statutes *in pari materia* (dealing with the same or related subject-matter). Thus in ex parte *Copeland*, the court said "Upon a question of construction arising from a subsequent statute on the same branch of law, it is perfectly legitimate to use the former Act, though repealed";

(iv) when construing consolidating Acts, prior statutes repealed but reproduced in substance, since these are regarded as being *in pari materia* and so decisions on a repealed statute are treated as applicable to substantially identical provisions which it has consolidated. Hence in *I.R.C.* v. *Hinchy*, Lord Reid observed, "In construing a consolidating Act it is proper to look at the earlier provisions which it consolidated".

(*c*) *Interpretation Act 1889*. Worthy of note here is s. 38(1), Interpretation Act 1889, which states "Unless a contrary intention appears, when an Act repeals and reproduces, with or without modification, any provisions of a former Act, references in any other Act to the provisions thus repealed are to be read as references to the provisions thus re-enacted".

(*d*) *Extrinsic aids which cannot be used*. Generally speaking, the courts prevent themselves from inquiring into the Parliamentary history of an Act. Thus, it is not permissible:

(i) to examine debates in Parliament, hence statements of a Secretary of State in proposing to Parliament a measure which eventually becomes law are not to be looked at in construing a statute; similarly a report of commissioners cannot be studied because, as Lord Wright observed in *Assam Railways & Trading Co. Ltd.* v. *I.R.C.*, it does not follow that their recommendations were accepted;

(ii) to use memoranda prefixed to Bills to gather the meaning of subsequent Acts;

(iii) to use Practice Notes to construe a statute;

(iv) to refer to a White Paper in interpreting a statute;

(v) to regard explanatory notes issued by a government department as authoritative of what an Act means;

(vi) to regard the views expressed by government departments as authoritative of the Act's meaning, though as the court pointed out in *I.T.C.* v. *Pemsel*, where there is a continuous flow of similar legislation, e.g. the Finance Acts using similar, not to say identical, terms, it may be assumed that Parliament meant that those words be construed in their received sense.

Textbooks

Generally, these may not be referred to as authoritative statements of the law. This serves to emphasise the essentially judge-made quality of common law, as distinct from the essentially jurist-made law of Continental legal systems. In a very limited number of situations, however, textbooks, particularly old textbooks, may be referred to as authoritative statements of what the law was at a given time in the past.

15. Duty-creating statutes

Performance

In modern times many statutes have been passed which lay duties upon persons. Generally the performance of a statutory duty may either be carried out by the person individually, or delegated. The fact that performance has been delegated does not as such mean that the person upon whom the duty was placed is relieved of liability, criminal or civil. In many cases he will be liable for any offence of the delegate, as is made clear in *Allen* v. *Whitehead* (*see* § **30,** p. 200).

When faced with a duty-creating statute, one immediately asks what is the sanction for the breach stipulated to compel performance of the duty. This varies according as the Act lays down powers or duties, the difference being underlined by the use of the words "may" and "shall", the former referring to powers which are conferred upon persons and bodies which the courts cannot enforce, the latter referring to things people must or must not do under the threat of legal sanction. H.A.S.A.W.A. falls into the latter category—it lays down duties disobedience to which entails criminal consequences.

Sanctions

These are the unpleasant consequences which will ensue if the Act is not abided by. They are most frequently of a criminal nature but can also be civil, i.e. imposed in a civil action. Most mandatory statutes make express provision for the imposition of criminal sanctions if the duties of the Act are not carried out, either absolutely or as far as is reasonably practicable, though this is not invariably true. For instance, the Wills Act 1837 lays duties upon persons who wish to make a valid will but does not provide any criminal sanction in the form of imprisonment or a fine for failure to do so. The sanction here is really non-legal, that the will may turn out to be invalid. H.A.S.A.W.A., however, and its predecessor the Factories Act 1961, does make specific provision for criminal sanction in the event of disobedience (*see* § **29,** p. 193).

Breach of statutory duty

(*a*) *Do actions lie for breach?* If a person disobeys, or is in breach of, a statutory duty, does this entitle another person suffering injury or damage as a result to recover compensation in respect of his injury or damage? This is known commonly as an action for breach of statutory duty, and although the law reports are replete with examples of such cases going back effectively to the old Factory and Workshop Act of 1878, it must be emphasised that today, and ever since 1972, when the Employers' Liability (Compulsory Insurance) Act 1969 came into operation, industrial injuries which disclose a breach of statutory duty on the part of an employer, or factory occupier or both, are normally governed by liability insurance. The effect of this is to subordinate the importance of legal action in a civil court. On the assumption that there is no question of the employer's policy being invalid for contractual reasons (*see* § **55**, p. 261), this is normally a straightforward matter handled by the insurance company who pay the compensation to the injured worker. Nevertheless if the insurance company elect not to pay or the injured party considers that the compensation offered is insufficient, he may be able to sue the employer or factory occupier in respect of a breach of statutory duty.

(*b*) *Nature of action for breach.* Three types of situation are possible, depending on the type of statute. The position was well summarised in *Wolverhampton New Waterworks Co.* v. *Hawkesford*, where it was said:

(i) there is that class of statute where there is liability existing at common law which is only remedied by the statute with a special form of remedy; thus, unless the statute contains words expressly excluding the common law remedy, the plaintiff can proceed either under statute or under common law;

(ii) there is that class where the statute has created a liability but has given no special remedy for it; here the party may proceed at common law;

(iii) there is that class where the statute creates a liability not existing at common law and provides also a particular remedy for enforcing it. Here the party must adopt the form of remedy given by the statute.

The Factories Act, and, probably, H.A.S.A.W.A., falls into class (ii) as regards civil liability or an action for compensation for breach of statutory duty, and into class (iii) as regards criminal liability, specifying new criminal penalties in respect of offences which did not exist at common law. The position as regards the action for damages for breach of statutory duty corresponds with the action for negligence at common law, subject to the proviso that, in the former case, the duty, or rather the scope of the duty, is defined by the statute whereas at common law the duty imposed upon a person is one of taking reasonable care. Otherwise the action for breach of statutory duty and the common law action for negligence are identical, in so far as once the existence of a duty has been established it is incumbent upon the plaintiff to show as well that:

(i) the defendant failed to carry out the duty; and

(ii) the defendant's breach in failing to carry out the duty was the *cause* of the plaintiff's injury (*see* § **20**, p. 55).

Does an employer or factory occupier, once he is convicted in criminal proceedings of a breach of s. 14(1), Factories Act 1961, or s. 2, H.A.S.A.W.A., become automatically liable for payment of damages in a civil action brought by a workman injured as a result of the breach? In other words, does breach *simpliciter* give rise to civil action? This is a question best answered by an appeal court, but if *Potts* v. *Reid* and *Gallagher* v. *Dorman Long* (*see* § **9**, p. 53) are to be regarded as good law the answer must be in the negative.

Construction of penal statutes

Whatever the final answer, it is invariably true that penal statutes, like the Factories Act 1961 and H.A.S.A.W.A., are construed by the courts strictly against the draftsman and in favour of the person upon whom it is sought to place an obligation. This has three important consequences when deciding whether there is an action for statutory duty or a breach of the specific duty created by the statute. The rule is that a breach occurs when:

(*a*) the injury suffered by the workman was of the kind contemplated by the statute; and

(*b*) the injured person was within the ambit or expressed scope of the statute; and

(*c*) the damage was within the contemplated injury (by the statute).

This is known as the rule in *Gorris* v. *Scott*, a leading case on the interpretation of penal statutes. In 1874 an order made under the Contagious Diseases (Animals) Act 1869 prescribed that sheep being carried to British ports had to be put into pens and the floor of such pens had to be furnished with footholds. The order was passed to prevent disease spreading because the sheep were in overcrowded conditions. The defendant's ship failed to meet the order, and the plaintiffs' sheep were washed overboard and lost. The plaintiffs sued in respect of their loss for breach of this statutory duty. It was held that their case disclosed no cause of action. In the course of giving judgment Kelly, C.B., observed:

"... that although when penalties are imposed for the violation of a statutory duty, a person aggrieved by its violation may sometimes maintain an action for the damage so caused, that must be in cases where the object of the statute is to confer a benefit on individuals, and to protect them against the evil consequences which the statute was designed to prevent, and which have in fact ensued; but that if the object is not to protect individuals against the consequences which have in fact ensued, it is otherwise; that, if, therefore, by reason of the precautions in question not having been taken, the plaintiffs had sustained that damage against which it was intended to secure them, an action would lie, but that when the damage is of such a nature as was not contemplated at all by the statute, and as to which it was not intended to confer any benefit on the plaintiffs, they cannot maintain an action founded on the neglect."

Later on in his judgment the judge concluded "But, looking at the Act, it is perfectly clear that its provisions were all enacted with a totally different view; there was no purpose, direct or indirect, to protect against such damage; but, as is recited in the preamble, the Act is directed against the possibility of sheep or cattle being exposed to disease on their way to this country. . . ." Another instructive case was *Knapp* v. *Railway Executive* in 1949. A motorist hit a level-crossing gate which was not properly and securely fastened. As a result the gate swung back over the line and struck a passing train and the driver of the train was injured. The train driver sued the motorist for damages and this action was later settled, but the motorist then sued the Railway Executive for breach of statutory duty (under the Brighton and Chichester Railway Act 1844) for failing to secure the gate properly. The court ruled that his action must fail. They said that the purpose of the 1844 Act was to protect road users from dangers associated with the railway, with the result that the driver of a train employed by the railway could not bring an action. This, in turn, meant that the motorist had no right of action either.

One could say that these cases argue that no new statutory torts should be created in the absence of unequivocal authority to the contrary. This approach, however, mirrors only one side of the picture. In the field of industrial safety legislation the courts since the passing of the Factory and Workshop Act 1878 have argued strenuously in favour of the presumption of civil liability arising out of statutory breach. In the leading case of *Groves* v. *Lord Wimborne* in 1898 the plaintiff sued the defendant for breach of a duty to fence certain machinery, the duty being placed upon the occupier by the Factory and Workshop Act 1878. As a result of a breach of this duty the plaintiff suffered injury. During the course of giving judgment and reversing the decision of the court of first instance, the Court of Appeal made the following observations which, as far as actions for breach of statutory duty are concerned in the field of industrial safety legislation, have stood the test of time.

"The Act . . . in question, which followed numerous other Acts *in pari materia*, is not in the nature of a private legislative bargain between employers and workmen . . . but is a public Act passed in favour of the workers in factories and workshops to compel their employers to do certain things for their protection and benefit. . . . Could it be doubted that, if s. 5 stood alone, and no fine were provided by the Act for the contravention of its provisions, a person injured by a breach of the absolute and unqualified duty imposed by that section would have a cause of action in respect of that breach? Clearly it could not be doubted. That being so, unless it appears from the whole 'purview' of the Act, to use the language of Lord Cairns in the case of *Atkinson* v. *Newcastle and Gateshead Waterworks Co.*, that it was the intention of the Legislature that the only remedy for breach of the statutory duty should be by proceeding for the fine imposed by s. 82, it follows that, upon proof of a breach of that duty by the employer and injury thereby occasioned to the workman, a cause of action is established. The question, therefore, is whether the cause of action which prima facie is given by s. 5 is taken away by any provisions to be found in the remainder of the Act. It is said that the provisions of ss. 81, 82 and 86 have that effect, and that it appears thereby that the purview of the Act is that the only remedy, where a workman has been injured by a breach of the duty imposed by s. 5, shall be by proceeding before a court of summary jurisdiction for a fine under s. 82, which fine is not to exceed 100*l*. In dealing with the question whether this was the intention of the Legislature, it is material, as Kelly C.B. pointed out in giving judgment in the case of *Gorris* v. *Scott* to consider for whose benefit the Act was passed, whether it was passed in the interests of the public at large or in those of a particular class of persons. The Act now in question, as I have said, was clearly passed in favour of workers employed in factories and workshops, and to compel their employers to perform certain statutory duties for their protection and benefit. It is to be observed in the first place that under the provisions of s. 82 not a penny of the fine necessarily goes to the person injured or his family."

Later in the course of judgment, the judge observed:

"It seems to me that the fine is inflicted by way of punishment of the employer for neglect of the duty imposed by the Act, and must be proportionate to the character of the offence. This consideration and the fact that whatever penalty the magistrates inflict does not necessarily go to the injured workman or his family lead me to the conclusion that it cannot have been the intention of the Legislature that the provision which imposes upon the employer a fine as a punishment for neglect of his statutory duty should take away the prima facie right of the workman to be fully compensated for injury occasioned to him by that neglect. . . ."

It seems that H.A.S.A.W.A. has opted for the presumption in favour of new statutory torts, at least as far as regulations passed pursuant to it are concerned. Sub-section 47(2) states that breach of a duty imposed by health and safety regulations ". . . shall, so far as it causes damage, be actionable except in so far as the regulations provide otherwise".

Delegation of statutory duty

The complexity of modern industry requires that an employer delegate to a competent person the performance of certain duties imposed on him by statute. In the field of industrial safety this will probably mean delegation to a competent safety officer. If a workman is injured because of the negligence of that competent person, can an employer raise the defence that he delegated the performance of the statutory duty to him? At one time, following the decision in *Harrison* v. *N.C.B.* in 1951, it was felt that this would be a defence, but since the abolition of the doctrine of common employment by the Law Reform (Personal Injuries) Act 1948 employers have become automatically vicariously liable for the negligence of their employees towards other employees (for the position in criminal law of the employer who has delegated a duty, *see* §§ **30,** p. 200).

Burden of proof in actions for breach of statutory duty

The burden of proving a case in a civil action is on a balance of probabilities, i.e. that it is more likely than not that the injury occurred in the way alleged. To this extent it is a lesser burden of proof than in criminal cases. The law was stated in *Bonnington Castings Ltd.* v. *Wardlaw* in 1956. A workman contracted pneumoconiosis as a result of being exposed to two sources of dust during the course of his employment. One source of dust was under the control of his employer whilst the other was outside the employer's control. Evidence showed that while at work as a steel-dresser he was exposed to silica dust from the pneumatic hammer with which he worked and at that time there was no known means of preventing the escape of dust from pneumatic hammers. He had also been exposed to dust from the swing grinders used in the factory and it was established that the dust from the swing grinders could have been rendered harmless. The workman, however, proved that the swing grinders were not kept free from obstruction, a fact which constituted a breach of the Grinding of Metals (Miscellaneous Industries) Regulations 1925. The question arose as to whether this breach constituted the cause of the plaintiff's illness. It was held that the onus was on the plaintiff to show that the breach materially contributed to his condition, i.e. that it was more likely than not that the breach caused the condition. The plaintiff's claim succeeded because exposure to dust from the swing grinders made a material contribution to his physical condition. Another instructive case is *Gardiner* v. *Motherwell Machinery & Scrap Co. Ltd.* In 1961 G contracted dermatitis as a result of working in conditions likely to give rise to it. In the course of giving judgment Lord Reid said "When a man who has not previously suffered from a disease contracts that disease after being subjected to conditions likely to cause it, and when he shows that it starts in a way typical of disease caused by such conditions, he establishes a prima facie presumption that his disease was caused by those conditions". Hence it seems reasonable to assert that if there is a breach of, say, s. 63, Factories Act 1961 (measures to prevent the accumulation of dust and fumes), and a workman then contracts lung dust, it should not be difficult to show a causal connection between the breach of s. 63 and the disease contracted by the workman.

16. Synopsis of case law on the Factories Act

The case law attracted by the Factories Act 1961 and its predecessors has tended to establish three main points. Firstly, that civil actions generally lie in respect of a breach of the safety provisions but that the position with respect to the health and welfare provisions is nowhere so clear. Second, that owing to the existence of the rule in *Gorris* v. *Scott* statutory obligations of a penal nature are construed strictly, with the practical consequence that if the action for breach of statutory duty fails, the injured plaintiff may well still have an action at common law; common law liability here is of a residual nature, mopping up areas not covered by statute. Third that when interpreting an ostensibly strict duty, common law concepts, particularly foreseeability, have crept into the court's decision. This is especially true of s. 14.

Distinction between safety and health and welfare provisions

The decision in *Groves* v. *Lord Wimborne* established that an action would lie for damages for breach of the safety provisons. It is more difficult to generalise in respect of health and welfare, but in *Nicholson* v. *Atlas Steel Ltd.* a widow and children of a deceased workman who had died of pneumoconiosis were held entitled to damages for breach of s. 4, since the company, as a result of failure to provide adequate ventilation in the workshop where silica dust was given off, had exposed the workman to the risk of this condition.

It seems best to follow the view that the welfare provisions are essentially of two types, those which are of a purely welfare nature and those which concern both health and safety. This observation was made by the court in *Reid* v. *Westfield Paper Co. Ltd.*, where it was held that damages could be got for a breach of s. 58, dealing with the provision of adequate washing facilities, if it could be shown that the plaintiff's dermatitis was attributable to the lack of such facilities.

Residual nature of common law

Kilgollan v. *Cooke & Co. Ltd.* is a good example of this. The plaintiff was injured when a particle of wire flew out of a stranding machine which was not securely fenced. It was held (interpreting s. 14(1), Factories Act 1937, strictly) that the section was designed to prevent the worker coming into contact with the machine rather than to prevent his being hit by things flying out of the machine. Irrespective of this, the plaintiff was entitled to damages at common law for negligence. This case pioneered the double-barrelled action against the employer; indeed, as long as the employer's obligation is commensurate with the duty to take reasonable care in respect of his employees, as expressed in the leading case of *Wilsons & Clyde Coal Co. Ltd.* v. *English*, one may feel justified in saying that most matters which are the subject of statutory duty, e.g. heating, lighting, etc., will or may give rise to a common law duty. This would also appear consistent with the court's remarks in *Wolverhampton New Waterworks Co.* v. *Hawkesford*, especially as regards statutory duties of category (i) (*see* § **15**, p. 34).

The concept of foreseeability

The first two sections of the general safety provisions of the Factories Act 1961, ss. 12 and 13, state that prime movers and transmission machinery must be securely fenced because they are presumed to be dangerous. Section 14, however, is different. Sub-section 1 states that every dangerous part of machinery other than prime movers and transmission machinery shall be securely fenced unless it is safe by construction or position. In other words machinery other than prime movers and transmission machinery is not presumed to be dangerous. If, however, by reason of the operations carried out by the machine, a dangerous part cannot be secured by a fixed guard, it is enough to provide an automatic device which will prevent the operator coming into contact with that part of the machine (s. 14(2)). This obligation appears absolute. Case law, however, has interpreted the duty as one of guarding absolutely against those dangers which are reasonably foreseeable. This trend is noticeable in cases like *Walker* v. *Bletchley-Flettons Ltd.*, *Eaves* v. *Morris Motors Ltd.*, *Burns* v. *Terry & Sons*, *Smith* v. *Chesterfield & District Co-operative Society*, and *Close* v. *Steel Company of Wales* (*see* § **3**). In *Close's* case the appellant, an employee of the respondent, was operating an electric drilling machine, when the bit of the drill shattered and a particle flew into his eye, causing him injury. He claimed damages under s. 14(1), Factories Act 1937 (now s. 14(1), Factories Act 1961). Evidence showed that such an accident had not happened before but that bits had sometimes shattered. It was held by the House of Lords that the employer was not liable because there was no breach of s. 14 (1) since the danger arising from the use of the bit was not reasonably foreseeable.

This case seems to indicate that if the accident or a similar one has happened before, the fencing requirement is absolute unless the part of the machinery is safe by position or construction. In fact, this is only partly true because, by virtue of s. 76, Factories Act 1961, the Secretary of State for Employment used to be able to make regulations which had the effect of modifying or extending the provisions of the Act. An example of the exercise of this power occurred in the case of the Woodworking Machinery Regulations 1922 which stipulate the type of guard to be used on a circular saw. Thus in *Miller* v. *William Boothman* in 1944 a workman was injured whilst using a circular saw. This saw complied with the Woodworking Machinery Regulations 1922. Nevertheless the plaintiff argued that he was entitled to damages, since the obligation laid upon the employer by s. 14(1) was absolute. It was held that if the Secretary of State for Employment had by regulation modified the scope of the Act, the Regulations prevailed, with the result that the plaintiff was not entitled to recover. Similar modification has also taken place with respect to abrasive wheels following the passing of the Abrasive Wheels Regulations 1970 consequent upon the decision in *Summers & Sons* v. *Frost* in 1955. It should not be forgotten, however, that regulations may be construed as having the effect of strengthening the provisions of the Act, e.g. the Power Presses Regulations 1965 (*see* Appendix IV).

Repeal

This is where a statute is revoked or its scope is curtailed, a situation often associated with a change of government. The incoming government may feel that the legislation passed by its predecessor is repugnant to the objectives which brought it to power, that it does not reflect the wishes of the electorate, and may decide to remove all or part of it. This is known as repeal, and there are two types, express and implied.

(*a*) *Express repeal.* This is more common than implied repeal and normally consists of annexing a schedule of repeal to the statute. In the case of H.A.S.A.W.A. 1974, it will be found that the schedule of repeal (Sched. 10) repeals several statutes and parts of statutes.

(*b*) *Implied repeal.* This, though less common (since it is standard practice for statutes to contain a repeal schedule), emphasises Parliamentary legal sovereignty, a point well illustrated by the leading case of *Vauxhall Estates Ltd.* v. *Liverpool Corporation.* The Minister of Health had made a Street Improvement Scheme Order in respect of a certain area of Liverpool. The Order consisted of the compulsory purchase of property, as a result of which the question of how much compensation should be paid to affected landowners arose. By virtue of the Acquisition of Land (Assessment of Compensation) Act 1919, the landowners would get just over £2,300, but if the Housing Act 1925 applied the landowners would receive about half this amount. The

1919 Act stated that other statutes inconsistent with it were not to have effect, but it was held by the court that, despite the terminology of the 1919 Act, the later Act repealed the earlier Act to the extent of the inconsistency, with the result that compensation was to be assessed at the lower rate under the Housing Act 1925.

The implication is that any later statute will impliedly invalidate any previous statute or part of it which is inconsistent with it. Were the position otherwise, the courts would not be applying the doctrine of the legal sovereignty of Parliament. Parliament cannot bind its successors, no matter how much it might like to do so.

Chapter 2

The Criteria of Compensation

COMMON LAW

17. Accidents and compensation

If a person is injured as a result of the fault or negligent conduct of another, he can generally get compensation in respect of such an injury. If he is unable to establish fault or negligence, however, he will generally be entitled to compensation only if the accident happened at work, and the criteria applicable to compensation are then closely associated with social security benefits, compensation being commensurate with loss of earnings and loss of faculty. These are the two main channels of compensation—negligence dates back effectively to the advent of the industrial revolution, whilst social security is of much more recent origin, in effect initiated by Lloyd George's National Insurance Act 1911 and the post-war national insurance scheme introduced by Sir William Beveridge. There is a third source of compensation of later origin still, dating back only to 1964—the Criminal Injuries Compensation Scheme. Compensation consists of *ex gratia* payments, and is confined to situations where injury arises through the commission of a criminal offence upon a person. Finally there is compensation payable in respect of an injury insured under a personal accident policy. This sort of insurance, unlike national insurance, is voluntary and is mostly used by self-employed persons and this may be the only way in which a person can gain compensation for loss of earnings and disability caused by a disabling disease which is attributable neither to the fault of another nor to the nature of the person's employment. If such personal accident insurance has not been taken out, the person contracting the disease will normally go uncompensated, unless he qualifies for supplementary benefit.

18. Damages and social security

Damages in tort

Tort damages, though in theory and sometimes in practice paid by defendants in civil actions, are normally handled by private liability insurance. The key to compensation of this nature is indemnity and the damages are in the form of a lump sum. Such insurance in respect of tort liability may be compulsory; employers are under a legal obligation to take out this sort of insurance cover for industrial injuries. Before any damages or insurance moneys can be paid out in respect of liability fault must be proved, hence such insurance is frequently referred to as liability insurance.

State compensation

State compensation for industrial injuries is part of the national insurance scheme and contributions are compulsory. Benefit does not depend on proof of fault or negligence, though not all causal connection between employment and injury has been severed from the state industrial injuries' scheme, in that the injury must arise *out of* the employment. Payment is in the form of periodical handouts or a pension rather than a lump sum, a fact which has advantages over the tort system in times of inflation. The state also does not exercise subrogation rights through the claimant against private insurance, and both private and state systems exist alongside one another; this permits of overlapping which can result in double indemnity for industrial injuries.

19. Insurance and legal liability

Legal liability has tended to assume the availability of liability insurance, and once this sort of insurance was made compulsory with the passing of the Employers' Liability (Compulsory Insurance) Act 1969, the process was accelerated. Illustrative of this are the remarks of Lord Denning in the 1971 case of *Nettleship* v. *Weston* where, referring to a situation where a learner driver had injured another motorist, he explained "Thus we are moving away from the concept 'No liability without fault'. We are beginning to apply the test—'On whom should the risk fall?' Morally the learner driver is not at fault but legally she is liable to be because she is insured and the risk should fall on her."

Foreseeability

A large number of other cases similarly show that the concept of reasonable foreseeability, which is the basis of negligence, is being stretched to such limits that it is barely distinguishable from strict liability; the reason appears to be the universal availability of a variety of different types of insurance policies. In 1963, in *Videan* v. *British Transport Commission*, a trolley driver who was an employee of B.T.C. mowed down and fatally injured the station-master, who had jumped from the platform on to the line in a heroic and successful attempt to rescue his son who had trespassed on to the permanent way in order to play. It was held by the Court of Appeal that the trolley-driver should have foreseen that his negligent driving might bring about an emergency and that someone might be endangered if the trolley approached too fast and without a proper look-out being kept. The presence of the station-master on the track was foreseeable by the driver and, as he was not a trespasser, a duty was owed to him.

Can it seriously be suggested that the presence of the station-master was reasonably foreseeable? Or can it seriously be suggested that, in the absence of an employers' liability policy, the presence of the station-master would have been reasonably foreseeable, as required for the act to be negligent? It seems doubtful. In practice at that time most large employers, included among whom would be the respondent, would have an employers' liability policy providing indemnity in respect of injuries suffered by an employee and arising out of the employment. This sort of insurance cover was precipitated by the Law Reform (Personal Injuries) Act 1948, which abolished the doctrine of common employment (whereby one employee had to sue another employee, and not his employer, in respect of injury suffered at work and such insurance for employees is now compulsory by virtue of the Employers' Liability (Compulsory Insurance) Act 1969).

Another case worth noting is *Chadwick* v. *British Railways Board*. In 1967 C went to help at the scene of the Lewisham train disaster and suffered nervous shock as a result. He sued B.R.B. for negligence in that they had caused his injury, and recovered substantial damages. During the course of giving judgment the court said "In the present case, the defendants were negligent towards their passengers. As a result, passengers were injured and put in peril. All of that could reasonably have been foreseen. It could also have been foreseen that somebody might try to rescue passengers and suffer injury in the process, and in my opinion the defendants owed a duty to Mr. Chadwick, who was within the area of contemplation." The concept of reasonable foreseeability has again been stretched to become strict liability, which is likely to happen as a result of an employer having taken out a public liability policy offering indemnity to third parties other than employees.

The eggshell skull rule

The next set of circumstances illustrating the effect of insurance upon negligence occurred in 1962, in the case of *Smith* v. *Leech Braine, Ltd.*, and concerns the so-called eggshell skull rule. S was a galvaniser at L.B.'s factory, and he received a burn on the lip from a splash of molten metal; this injury was the result of the employers' carelessness in not providing adequate protection. The burn led to a cancer (to which S was susceptible as he had previously worked in a gas works) and three years later he died. It was held by the court that the employers were liable. During the course of giving judgment Lord Parker observed "It has always been the law of this country that a tortfeasor takes his victim as he finds him. . . . The test is not whether these employers could reasonably have foreseen that a burn would have caused cancer and that he would die. The question is whether these employers could reasonably foresee the type of injury he suffered, namely, the burn."

Smith's case has recently been followed in *Robinson* v. *Post Office*. In the course of his employment R slipped when coming down a ladder because oil had been negligently left on the rungs. He sustained a minor wound, 3 in. long, on his left shin. His employer admitted liability for this injury. Some eight hours later, however, after finishing work, the employee visited his doctor to receive treatment for the wound and was given an injection of anti-tetanus serum as a result of which, nine days later, he developed encephalitis, causing brain damage and permanent disability. Medical evidence showed that the use of this particular serum carried with it some risk to the patient, especially to those who had had a previous dose of the serum. R had had a previous injection of the serum some years earlier, but the doctor, although he knew this, judged the circumstances surrounding the accident to be fit for a further injection. It was held that the doctor had acted reasonably in deciding to use the serum again; standard practice was to give patients a test dose then wait for half an hour to see whether any reaction resulted. In this case, however, the doctor did a test of his own which was judged to be unsuitable and so negligent. The patient showed no reaction till nine days later. It was held that although the doctor had been negligent his negligence had not caused the encephalitis, because even if the proper test had been done the full injection would still quite properly have been given and the end result would have been the same. The question before the Court of Appeal was therefore whether the Post Office was liable for damages resulting not merely from the initial wound but also from the encephalitis, and it was held that the Post Office was liable for all the injuries because it was foreseeable that if oil was negligently allowed to be on the ladder, an employee was likely to sustain the original wound and that such a wound would require medical attention. Applying the rule in *Smith* v. *Leech Braine Ltd.*, the court ruled that the employer must take the employee as he found him, i.e. with an allergy to a second dose of anti-tetanus serum, and so was liable for all the injuries flowing from it.

Employers' liability

These two important cases underline the significance of the Law Reform (Personal Injuries) Act 1948: an employer may be vicariously liable for negligent injury caused by one employee to another (following the abolition of common employment) and also (since the decision of the House of Lords in *Wilsons & Clyde Coal Co. Ltd.* v. *English*) an employer is directly liable under a non-delegable personal duty for injury caused to his employees through unsafe system of work, place of work, etc. To meet this type of contingency, most employers, even in 1962 when *Smith*'s case was decided, would have taken out an employers' liability policy; by 1974, when *Robinson*'s case was decided, most employers (excluding nationalised industries) were legally obliged to do so.

When *Westwood* v. *Post Office* was decided in 1974, it appeared that the fault principle had been well and truly jettisoned, at least as far as fatal injuries were concerned. W was employed by the Post Office as a technician at a telephone exchange. One day he went on to the flat roof of the building for his mid-morning break; the door to the roof was locked so he entered the lift motor room and went out through a window on to the roof. The door to the lift motor room was normally locked but on this occasion it was slightly ajar. On the door was a large clear notice: "Only the authorised attendant is permitted to enter." W was not an authorised attendant. The lift motor room contained machinery which would be in motion when the lift was in use; in the floor was a trap door which was not of sound construction. While returning from the roof W jumped down from the window on to the trap door, which gave way, and he fell through to a concrete floor below, suffering severe injuries from which he died. The Court of Appeal held that W was a trespasser whilst he was in the lift motor room and proceeded to apply the reasoning of Lord Reid in *Herrington* v. *British Railways Board* when he said "... the question whether an occupier is liable in respect of an accident to a trespasser on his land would depend on whether a conscientious, humane man with his knowledge, skill and resources could reasonably have been expected to have done or refrained from doing before the accident something which would have avoided it. If he knew before the accident that there was a substantial probability that trespassers would come I think that most people would regard as culpable failure to give any thought to their safety." The court accepted the view that the Post Office did not know before the accident that there was a substantial, or any, probability that W would use the room for an unauthorised purpose and in disregard of the notice: W went into the room at his own risk. On appeal to the House of Lords, however, the judgment of the Court of Appeal was reversed, the House of Lords taking the view that an employer owed a statutory duty to provide safe premises to an employee who was trespassing. The court found that the lift motor room into which W had trespassed was a place in office premises, as defined by the Offices, Shops and Railway Premises Act 1963, or, alternatively, premises which must be treated as forming a part of office premises, since the activity for the purposes of which the lift motor room existed was a telephone exchange and it was of unsound construction.

On the question of trespass their Lordships were unable to distinguish between the case before them and *Uddin* v. *Associated Portland Cement Manufacturers Ltd.* (in which an employee had been injured whilst doing an unauthorised act, and his employer was still found guilty of negligence, damages being partly reduced on the grounds of the appellant's contributory negligence). The inescapable fact was, so the Lords said, that the lift motor room had not been constructed as the Act prescribed and this was the cause of the employee's death, the Act containing no provisions restricting protection to those injured in the course of their employment. As to whether W had been contributorily negligent, Lord Kilbrandon said 'the sole act of negligence giving rise to this accident was the respondent's breach of statutory duty. Any fault on the part of the deceased was a fault of disobedience, not a fault of negligence, because he had no reason to foresee that disregard of the order to keep out of the lift motor room would expose him to danger. It would indeed not have done so, had it not been that, unknown to him, the respondents were in breach of their duty to take care for the safety of those employed in the premises." The real reason behind the decision, it is suspected, is the existence of an employers' liability policy, designed to make provision for injured employees (§ **61,** p. 272). Indeed, as will be seen later, the provisions of the Employers' Liability (Compulsory Insurance) General Regulations 1971 do not appear to encourage employers who are policy-holders to take reasonable care for the protection of their employees. The courts, instead of talking frankly about insurance, tend to speak in terms of the accident being the "exclusive fault" of the employer, or more rarely the employee.

20. Negligence

This is probably the most important tort and it is used here in the sense of an independent tort rather than as an element in torts other than negligence. Negligence is about careless or unintentional conduct either in commission or omission; consequential rather than direct injury is its pedigree. Most industrial injuries result from negligent misconduct. Another way of putting the matter is to say that where the consequences of an act, or more rarely an omission to act, are positively desired or intended or foreseen as certain, the conduct is intentional and the injury is direct, whereas if the injurious consequences are merely foreseen as a possibility, the injury is indirect or consequential and the conduct is unintentional. Suppose A hits B in the face, this is a direct injury inflicted upon B assuming A either desired or intended to hit B, and is known as assault. On the other hand, suppose that A supplied B with a tool for his job which later proved defective and injured him. Here the injury is unintentional or consequential, i.e. not intended or desired. This sort of conduct constitutes negligence. Negligent conduct is insurable and so it might be said that negligence is insurable carelessness; the legal definition, however, was given as follows by Lord Wright in *Lochgelly Iron & Coal Co. Ltd.* v. *McMullan*: "In strict legal analysis negligence means more than heedless or careless conduct, whether in omission or commission; it properly connotes the complex concept of duty, breach of that duty and damage, loss or harm thereby resulting."

Three-tiered concept

It is apparent from this case that negligence consists of three interrelated elements:

(*a*) proof of a duty owed by the defendant to the plaintiff;
(*b*) breach of that duty; and
(*c*) damage, loss or harm resulting to the plaintiff from the breach.

All these ingredients must be established, otherwise generally the action will fail, though where the maxim of *res ipsa loquitur* applies, some facts will be inferred (*see* § **20,** p. 62).

Duty of care

One of the most difficult things is to decide whether the defendant owed the plaintiff a duty to take reasonable care to see that the plaintiff was not injured by the conduct of the defendant or by those for whom the defendant was responsible, e.g. his employees. Here there are certain well-defined areas where case law has established that one category of person owes a duty to take reasonable care to another—road users vis-à-vis road users, employers vis-à-vis employees, manufacturers vis-à-vis users of their products and occupiers vis-à-vis lawful visitors to their premises. Not infrequently, however, situations arise where there is no well-established category of liability, and so one is forced to inquire whether or not there exists a universal principle of liability sufficient to cover all situations of consequential loss or harm. In this respect the judgment of Lord Atkin in *Donoghue* v. *Stevenson* is generally regarded as authoritative. It was in this case that his Lordship enunciated the famous *neighbour rule* when he observed "You must take reasonable care to avoid acts or omissions which you can reasonably foresee would be likely to injure your neighbour. Who then in law is my neighbour? The answer seems to be persons who are so closely and directly affected by my act that I ought reasonably to have them in contemplation as being so affected when I am directing my mind to the acts or omissions which are called in question."

What emerges from *Donoghue* v. *Stevenson* is that the essence of negligence is foreseeability or, more particularly, the foreseeability of the average man or workman. Subjective foresight on the part of the defendant is generally of no importance. What matters is average foreseeability. Would Mr. Average have done or not done the act in question during the course of his job or trade? If the judge finds that the defendant deviated or fell below the norm, despite personal shortcomings and the fact that he could not be expected to take the same sort of care in avoiding accidents as Mr. Average or the Reasonable Man, the plaintiff will be compensated, either by the defendant personally or by his employer or by his employer's insurer.

The "reasonable man" test in itself, however, is too vague to be useful in identifying duty of care situations. Other criteria, therefore, will have to be considered. It is safe to say that in deciding whether a duty of care exists the following four tests are applicable. Firstly, is there a degree of probability that damage or harm will be done? Second, what will be the extent of the harm if the risk eventuates? Third, did the object to be achieved by the conduct have a social utility value? Fourth, what is the burden in terms of cost, time and inconvenience of taking effective precaution against the risk?

(*a*) *Degree of probability*. An event becomes reasonably foreseeable, or there is a degree of probability of harm or damage occurring, when a similar event has happened in the past causing injury. Thus in *Smith* v. *Crossley Bros.*, a case on employers' liability, two apprentices, whilst playing a nasty practical joke upon another apprentice, injected compressed air into his anus, causing him serious injuries. His subsequent action against the employer failed because the conduct of the two apprentices was not reasonably foreseeable, in that such an occurrence had never transpired before to the knowledge of the employer. In *Hudson* v. *Ridge Manufacturing Co. Ltd* six years later in 1957, on the other hand, H was employed by R.M.C. and during the course of his job was injured as a result of a prank being played upon him by C, a co-employee. Evidence showed that C had been in the habit for the previous four years of indulging in dangerous horseplay and skylarking and his employers had warned him about this on several occasions. It was held that the employer was liable to H for his injuries. During the course of giving judgment the court said "If a fellow workman ... by his habitual conduct is likely to prove a source of danger to his fellow-employees, a duty lies fairly and squarely on the employers to remove the source of danger." One of the reasons why the employers were liable was that they knew, or ought reasonably to have known, of such incidents happening, since they had happened before.

(*b*) *Extent of harm. Hayley* v. *London Electricity Board* in 1965 well illustrates this aspect of negligence. L.E.B. had excavated a trench in the pavement and erected a 2-ft.-high fence to protect members of the public from the excavation. H, a blind man, fell into the trench and was injured. At first instance the court ruled that the defendants were not liable and this was affirmed on appeal to the Court of Appeal, who took the view that there was adequate protection for an ordinary passer-by, and, in addition, the place where the trench was was not a place that many blind people were likely to use. On a further appeal to the House of Lords, this court reversed the decision of the lower courts and held that L.E.B. was liable. Especially instructive is the judgment of Lord Reid who spoke as follows:

"The respondents argue that they were only bound to have in mind or to safeguard ordinary able-bodied people and were under no obligation to give particular consideration to the blind or infirm. If that is right, it means that a blind or infirm person who goes out alone goes at his peril. He may meet obstacles which are a danger to him but not to those with good sight because no-one is under any obligation to remove or protect them. And if such an obstacle causes him injury he must suffer the damage in silence. . . . I could understand the respondents' contention if it was based on an argument that it was not reasonably foreseeable that a blind person might pass along that pavement on that day; or that, although foreseeable, the chance of a blind man coming there was so small and the difficulty of affording protection to him so great that it would have been in the circumstances unreasonable to afford that protection. Those are well recognised grounds of defence. But in my judgment neither is open to the respondents in this case. . . . In deciding what is reasonably foreseeable one must have regard to common knowledge. We are all accustomed to meeting blind people walking alone with their white sticks on city pavements. No doubt there are many places open to the public where for one reason or another one would be surprised to see a blind person walking alone, but a city pavement is not one of them. And a residential street cannot be different from any other. . . . There is evidence in this case about the number of blind people in London and it appears from Government publications that the proportion in the whole country is near one in five hundred. By no means all are sufficiently skilled or confident to venture out alone but the number who habitually do so must be very large. I find it quite impossible to say that it is not reasonably foreseeable that a blind person may pass along a particular pavement on a particular day."

This case demonstrates that the extent of injury or damage which might befall a blind person is considerably greater than that which could happen to a sighted person. Statistical evidence of the numbers of blind persons in London was also taken account of in determining whether the likelihood of injury was reasonably foreseeable. It is hoped that this tendency to consult statistical information is one that will continue, as in the past judges rarely referred to such sources, regarding them as untouchable extrinsic aids. Lord Reid felt no such compunction.

Another case where the use of statistical data helped to make legal history was *Berry* v. *Stone Manganese Marine Ltd.*, the first case where liability was imposed for deafness. B became deaf as a result of working with pneumatic hammers for a number of years. These pneumatic hammers gave off a noise level of between 115 and 120 decibels, compared with the tolerable limit of 90 decibels. Although there were other factors involved, culminating in the employer's being held liable, including poor supervision, the statistical information relating to decibel rating clearly served to show that the deafness was reasonably foreseeable. *Paris* v. *Stepney B.C.*, which concerned the employer's obligation at common law towards a one-eyed workman (*see* § **35**, p. 223), is equally instructive.

(*c*) *Social utility value.* Here the courts compare the harm which may eventuate from the conduct with the inconvenience or harm that would result if the conduct did not take place. Suppose that a factory manufactured products which were made on machines which had to be guarded. Suppose further that it was established that the machines could not be securely guarded against the operator being injured from all dangers without being rendered commercially useless. The consequence of this might be that the factory would cease production and a large number of jobs be lost. In such circumstances the court might well think that the social utility value of the plant outweighed the likelihood of injury involved in the use of machines which had been guarded against most risks of injury.

(*d*) *Time, cost and inconvenience.* This matter was put authoritatively by Lord Reid in *Marshall* v. *Gotham Co.* in 1954. The Metalliferous Mines Regulation Act 1872 required the roof of a mine to be made secure so far as was reasonably practicable. The roof in question in the gypsum mine was normally tested with a hammer and dangerous or unsafe parts brought down, a procedure differing from that in coal mines where the roof is supported by props. The roof of the mine had been tested in the usual way but came down owing to the existence of a rare geological condition. Evidence showed that systematic support might have minimised the damage done but would not have prevented it. Finding for the mine occupiers, Lord Reid in the course of his judgment said "The danger was a very rare one. The trouble and expense involved in the use of the precautions, while not prohibitive, would have been considerable. The precautions would not have afforded anything like complete protection against the danger, and their adoption would have had the disadvantage of giving a false sense of security."

If the time, cost and inconvenience involved in installing a safety device outweigh the likelihood of risk of injury occurring from its not being installed, it is therefore not negligent to fail to install the device.

Breach of a duty of care

This means that the defendant failed by his act or omission to take care not to injure another whom it was reasonably foreseeable he would injure. Here the effect of criminal liability arising from a breach of a statute and used as a basis of action for breach of statutory duty (i.e. an action for statutory negligence) should be considered, and the remarks of Lord Diplock in *Boyle* v. *Kodak Ltd.* seem particularly relevant. "When considering the civil liability engrafted by judicial decision upon the criminal liability which has been imposed by statute, it is no good looking to the statute and seeing from it where the criminal liability would lie, for we are concerned only with civil liability. We must look to the cases." As in the cases of *Potts* v. *Reid* and *Gallagher* v. *Dorman Long Ltd.* (*see* § **9**, p. 35), the point being made here is that the breach of statutory duty *simpliciter* giving rise to criminal liability does not entitle an injured employee to compensation. He must show that the breach caused the injury, though the statutory provision (breach of which gives rise to criminal liability) establishes the necessary duty of care for the purposes of civil action for breach of the statute.

Effect of breach

It has been the law now for many years that physical injury is redressible by compensation. What about mental injury or nervous shock? The courts have sometimes said that mental injury carelessly caused is not necessarily reasonably foreseeable. In the leading case of *Bourhill* v. *Young* Y, a motor cyclist, was travelling at a high speed when he crashed into a car and was killed. B, who was pregnant at the time, was getting down from a tramcar when she heard but did not see the crash. Later she saw some blood left on the road after Y's body had been removed from the scene of the accident. B suffered nervous shock which, she alleged, caused her child to be stillborn. Her action against Y's executors for negligence failed. During the course of giving judgment in the House of Lords, Lord Russell said:

"Can it be said that [the motor cyclist] could reasonably have anticipated that a person, situated as was [the plaintiff], would be affected by his proceeding towards Colinton at the speed at which he was travelling? I think not. His road was clear of pedestrians. The [plaintiff] was not within his vision, but was standing behind the solid barrier of the tramcar. His speed in no way endangered her. In these circumstances I am unable to see how he could reasonably anticipate that, if he came into collision with a vehicle coming across the tramcar into Glenlockhart Road, the resultant noise would cause physical injury by shock to a person standing behind the tramcar. In my opinion, he owed no duty to the [plaintiff] and was, therefore, not guilty of any negligence in relation to her."

Nevertheless if the plaintiff is within the "foreseeable impact" area, i.e. physically near to the scene of the accident, so as to be able to see or hear it, the outcome may be different. In the 1970 case of *Hinz* v. *Berry*, H had left her husband and children in a car at a layby and had crossed the road to pick bluebells. Suddenly B ran into the car whilst driving his own car. H heard the crash and turned round to see her husband and children lying injured, the former fatally. As a result she became ill from nervous shock and sued B for damages. The Court of Appeal decided that she was entitled to recover.

Economic injury or loss

Pure economic injury or loss is not normally redressible in negligence. In *Spartan Steel & Alloys Ltd.* v. *Martin (Contractors) & Co. Ltd.* M's employees were digging up a road when they carelessly damaged an electric cable which supplied power to S.S.'s factory. S.S., who relied upon electricity for their furnaces, were deprived of power for 14 hours and in the event suffered loss of three kinds: (*a*) damage to metal in the furnace when the power failed; (*b*) loss of profit on the "ruined" metal; (*c*) loss of profit on four other melts which would have been carried out had the power not failed. In the 1973 Court of Appeal decision it was held that the plaintiffs were entitled to recover profits only under the first two heads.

A similar result had been reached earlier in the case of *Weller & Co.* v. *Foot and Mouth Disease Research Institute.* F.M.D.R.I. carried out experiments on their land concerning foot and mouth disease. They had imported an African virus which escaped and infected cattle in the vicinity. In the event two cattle markets had to be shut down and W, who were auctioneers, sued for damages in negligence for loss of business. It was held in the High Court that F.M.D.R.I. owed no duty of care to W, who were not cattle owners. These cases are consistent with the view that it is not a tort, even less negligence, to deprive a person of the opportunity of making a profit. In this respect the decisions are to be welcomed, as loss of opportunity of making a profit is better dealt with in claiming damages for breach of contract.

Causation

This third integral element in negligence is in some ways the most important. It means simply that the breach of duty of care, existing at common law or in statute or both, must have *caused*, i.e. materially contributed, to the plaintiff's injury (*see* § **15**, p. 38). That the element of causation is a limiting device designed to keep claims for negligence within manageable proportions is self-evident from the words of Thesiger, J., in *S.C.M.* (*U.K.*) *Ltd.* v. *Whittall & Son Ltd.*: "In order to limit liability ... the courts sometimes say either that the damage claimed was 'too remote' or that it was not 'caused' by the defendant's carelessness or that the defendant did not owe a duty of care to the plaintiff."

(*a*) *The causative link.* The causative element has been responsible for the failure of a good many actions for breach of common law and statutory duty. *Bonnington*'s case (*see* § **15**, p. 38) showed that proof is on a balance of probabilities, i.e. that it is more likely than not that the breach caused the disease or injury, and the following cases should be looked at with care. In *Bailey* v. *Ayr Engineering & Constructional Co. Ltd.* in 1958 B was an apprentice electrician employed by A.E.C. and was working in an open shaft in a block of flats which were under construction. He was injured by a heavy piece of falling masonry forced out of the shaft by some iron railing which had got out of control whilst being handled by two other workmen. B alleged that A.E.C. was in breach of reg. 90, Building (Safety, Health and Welfare) Regulations 1948 which prescribed that such shafts should protect workmen from being hit by any falling "material or article". It was held by the Court of Appeal that the employer was in breach of the regulations but nevertheless B was not entitled to damages. During the course of giving judgment the court said:

"Regulation 90 is designed to protect men working at sites where the men habitually do work from what I may call falling materials and debris of that kind and not from a collapse of the structure, whether it be a chimney, a roof, or (as in this case) part of the shaft. If in fact this plaintiff had been hit by a hammer or a loose brick falling down the shaft the ... defendants would have been liable to him for breach of that regulation. Accordingly this is a case where they cannot say they were not in breach of the regulation at all, because it was their duty to take steps to put up some umbrella, albeit only an umbrella designed to prevent materials and debris and tools from falling down. Be that as it may, and assuming, as I do, that they were in breach, it was not that breach which caused the injuries in this case, but the fact that this block of masonry, something which is completely outside the regulation, fell in the way I have described. ..."

Another case which illustrated the operation of causation in negligence was *McWilliams* v. *Sir William Arrol Ltd.* in 1962. M's husband, who was employed as a steel erector with A, fell 70 ft. from a steel tower which he was helping to build. The fall killed him and his widow subsequently claimed damages from A for negligence and from the occupiers of the shipyard for breach of statutory duty (s. 26(2), Factories Act 1936 (applicable in Scotland)), for failure to provide a safety belt for use by a steel erector. During the course of the proceedings, the following matters came to light and had a substantial bearing on the outcome of the case:

(i) if a safety belt had been worn, it would have prevented the spiderman from falling;

(ii) safety belts had been available on site until about three days before the accident took place;

(iii) M's husband had been an experienced steel erector and it was very unlikely that he would have worn a safety belt even if one had been to hand;

(iv) the employers had not instructed or encouraged M's husband to wear a safety belt;

(v) there was a widely established practice amongst steel erectors of not wearing safety belts, as they slowed down the pace of the work.

It was held by the House of Lords that, even supposing that the employer and the occupier were in breach of their obligations, nevertheless this was not the *immediate cause* of the dead steel erector's injuries. In particular, the evidence contained in (v) to the effect that, even if provided, safety belts were not usually worn and would not have been worn by the deceased, meant that the deceased's own conduct in failing to wear a safety belt, even if provided, was the exclusive cause of his injury. This is known as the doctrine of *hypothetical causation*: the employer would not have been liable for the employee's injury, even if the employer had been in breach of his statutory or common law obligations. The employer's breach, if there was one in this case, would have been too remote a cause of the injury.

A similar line of argument was followed in 1964 in *Doughty* v. *Turner Manufacturing Co. Ltd.* D was employed by T.M.C. in their factory and one day was injured by molten liquid when an explosion took place in a galvanising cauldron. Evidence showed that a co-employee had allowed the lid of the cauldron to fall into the liquid and a few minutes later the eruption took place. D admitted that an explosion was unforeseeable but that injury from splashing, caused by a lid falling into the cauldron, was foreseeable. In the court of first instance judgment was given for D, but on appeal the court found for the employer. During the course of giving judgment Harman, L.J. said:

"So it is said here that a splash causing burns was foreseeable and that this explosion was really only a magnified splash which also caused burns and that, therefore, we ought to follow *Hughes* v. *Lord Advocate* and hold the defendants liable. I cannot accept this. In my opinion, the damage here was of an entirely different kind from the foreseeable splash. Indeed, the evidence showed that any disturbance of the material resulting from the immersion of the hardboard was over an appreciable time before the explosion happened. This latter was caused by the disintegration of the hardboard under the great heat to which it was subjected and the consequent release of the moisture enclosed within it. This had nothing to do with the agitation caused by the dropping of the board into the cyanide. I am of opinion that it would be wrong on these facts to make another inroad on the doctrine of foreseeability which seems to me to be a satisfactory solvent of this type of difficulty. I would allow the appeal."

(For further cases, *see* § **21,** p. 69.)

(*b*) Novus actus interveniens (*new act or cause supervening*). This expression refers to the situation where the original act or omission causing the damage or injury is broken up or interrupted by the act of a third party. Whether this new supervening act will be construed as a proximate cause of injury depends on whether the act or the supervening conduct of the third person is reasonably foreseeable in the light of the original act or omission. If it is reasonably foreseeable, the perpetrator of the original act or omission is still liable for the ultimate loss or harm. Conversely, if the supervening conduct of the third party is not reasonably foreseeable, damages will have to be paid by the third party. Several cases, such as *Donoghue* v. *Stevenson*, are particularly instructive. D had been injured, suffering nervous shock and physical illness as a result of seeing the disintegrated remnants of a snail in the base of an opaque ginger-beer bottle which a café proprietor had sold to her friend. She sued the manufacturer for negligence (she could not sue the retailer in contract, as she had not purchased the ginger beer) and on appeal to the House of Lords judgment was given in her favour. During the course of giving judgment Lord Atkin explained the law as follows: "... a manufacturer of products, which he sells in such a form as to show that he intends them to reach the ultimate consumer in the form in which they left him with no reasonable possibility of intermediate examination, and with the knowledge that the absence of reasonable care in the preparation or putting up of the products will result in an injury to the consumer's life or property, owes a duty to the consumer to take reasonable care."

The implication of Lord Atkin's remarks is that if the defect allegedly causing injury should have been spotted by someone carrying out an intermediate inspection, failure to carry out a careful examination which should have revealed the defect will constitute a *novus actus interveniens*, and the original manufacturer will not be liable to the person injured. As regards the design and manufacture of machinery and other articles for use at work, specific obligations giving rise to criminal liability are now laid on manufacturers by s. 6, H.A.S.A.W.A. (*see* § **51**, p. 251). Nevertheless, the principles relating to *novus actus interveniens* will still continue to apply in civil cases where an employee of X has been injured as a result of defective equipment supplied to X by Y, a producer. We can therefore say almost certainly that if the machinery is new and a defect in it causes injury to an employee of X when using the machine properly, the producer should be liable to the employee, unless it was reasonable and so foreseeable for the immediate employer to inspect the equipment.

In many cases such an inspection will be deemed reasonably foreseeable on the part of employers whose employees use equipment which is often complicated and dangerous, and where injury occurs as a result of product defect in such circumstances, the Employers' Liability (Defective Equipment) Act 1969 makes specific provision for the employer being liable even though the product defect is attributable to the fault of the producer. This is a form of "channelling" not untypical of strict liability with the liability being "channelled" to the employer, subject to the retention of indemnity exercisable against the producer or the person at fault (*see* § **47**, p. 247). It is even more reasonable to expect an inspection to be carried out on behalf of the employer-user of the product where the equipment to be used by his employees is second-hand. In *Pearce* v. *Round Oak Steel Works Ltd.* in 1969 a second-hand machine was bought for use in R.O.S.W.'s factory and later a piece fell off due to metal fatigue in a bolt. It was held that the purchase of this sort of second-hand machinery presupposed an intermediate inspection on the part of R.O.S.W., and since this had not taken place, R.O.S.W. was liable to P, the injured workman. In an effort to avoid injuries of this kind s. 2(2) (*a*), H.A.S.A.W.A., places a duty of periodical inspection and testing upon all employers, whether the equipment is new or not, there being a duty not only to provide safe equipment but also to maintain it. Failure to carry out or arrange for the carrying out of such inspections by safety engineers will, it seems, constitute a *novus actus inter-*

veniens and so make the employer liable to an injured employee. This situation may change if a regime of product liability follows from the recommendations of the Pearson Commission or the Strasbourg Convention or the E.E.C. Draft Directive (*see* § **47**, p. 246).

In the original case on *novus actus*, *Scott* v. *Shepherd* in 1773, which is difficult to reconcile with the principle in *Donoghue* v. *Stevenson* (which equates the distributor's or user's or employer's act of inspection with a *novus actus* if reasonably foreseeable), the intervening act of the third party was a reasonably foreseeable consequence of the original act of negligence and was held not to be a *novus actus*. In *Scott* v. *Shepherd*, SH threw a lighted squib which landed on the stall of Y in the market place. W, in order to protect the property of Y, threw it away and it landed on the stall of R. R threw it to another part of the market place and it landed in SC's face, causing him serious injury and putting out his eye. It was held that the original perpetrator SH was liable to SC because the conduct of all the intervening parties had been reasonably foreseeable consequences of SH's original negligent act. Similarly in *Philco Radio Corporation* v. *Spurling* in 1949 S delivered in error some packing cases containing inflammable film scrap to P.R.C.'s premises. The cases gave no warning as to their contents. Later P.R.C.'s employees opened the cases and a foreman saw that the contents were inflammable. He ordered the scrap to be replaced and forbade smoking in the area. The foreman then telephoned S and arranged to have the cases delivered to their proper destination. Before the cases were taken away a typist employed by P.R.C. set light to the scrap whilst lighting a cigarette and it exploded causing a good deal of damage. When sued, S said that the immediate cause of the injury was the typist's act of negligence which served to break the chain of causation. It was held by the court that S was negligent in not seeing to it that such dangerous material was not properly delivered. Although negligent the typist's act did not break the chain of causation.

Philco Radio Corporation v. *Spurling* seemed to indicate that in the case of goods or material dangerous in themselves, the defendant responsible for sending or delivering or transporting them can never plead the defence of *novus actus interveniens*. *Baker* v. *Willoughby*, however, is an example of a case that went the other way, showing that the *novus actus* doctrine is very much alive. In 1964 B was in an accident on the highway caused by W's negligent driving, though B had been found one-quarter contributorily negligent towards his injuries. B received severe injuries to his left leg; much later, after spending a long time in hospital, B began employment with a scrap metal dealer and in 1967, during the course of his employment, B was the innocent victim of an armed robbery and received gunshot wounds which required the immediate amputation of his left leg which was already defective as a result of the previous accident. In 1968 the question of the amount of damages that should be paid to B for the injury received in 1964 came before the court for assessment. It was held by the Court of Appeal that W was only liable for consequences of the original injury to B's left leg up to 29th November 1967 (when the armed robbery occurred).

This case reflects wise decision-making in law, otherwise insurance premiums would reach absurdly high levels. Despite this decision, however, cases like *Smith* v. *Leech Braine Ltd.* and *Robinson* v. *Post Office* (*see* § **19**, pp. 45, 46) decided on the basis of the eggshell skull rule show that the doctrine of eggshell skull is inconsistent with the *novus actus* rule. Cases like *Smith* and *Robinson* go to show why premiums might rise to such high levels.

Product design

Although s. 6, H.A.S.A.W.A., imposes liability upon designers for failing to take reasonable care in the design of articles and substances for use at work, it is unlikely that actions for negligence involving design are likely to occupy much of the courts' time. Certainly in the past design has not been a subject for actions of negligence. The reason for this is one both of policy and constitutional propriety. The courts feel that if they decided that motor-manufacturers, for instance, were guilty of negligence in the design of their cars when a certain proportion of persons had been killed or injured whilst driving cars round bends (and it is conceivable that design as well as highway conditions plays a considerable part in the cause of such accidents), this would involve motor-manufacturers in the expenditure of vast sums of money to eliminate or minimise the incidence of such defects in design, and this cost would inevitably have to be passed on to the consumer and could well mean a considerable increase in the price of cars. Such arguments could apply equally to the designers of highways and hospitals and school buildings, which leads conveniently into the second argument against judges awarding damages for negligence in design, namely that increases in car prices or in the rates (necessary for safer highways) is a matter best left to the appropriate Secretary of State or Minister, who could deal with individual problems of design by means of regulations. For courts to pronounce motor-manufacturers and highway designers negligent would amount to a usurpation of the powers of the Ministry to regulate the design and construction of motor vehicles or the design and construction of highways. This is a constitutional matter tied up with the separation of powers. Of course, such regulations may provide for articles to be designed or constructed in a particular way so as to be safe, or for labelling or packaging to be done to achieve the same result. The regulations on the packaging and labelling of dangerous substances are an example of this. Nevertheless the courts do from time to time support an action for negligence arising from faulty design or bad, inadequate or non-existent labelling: *Vacwell Engineering Ltd.* v. *B.D.H. Chemicals Ltd.* in 1969 is a case in point. The defendant supplied to the plaintiff a chemical called boron tribromide. Ampoules of the chemical contained the words "harmful vapour". On contact with water, in the course of normal industrial use, the chemical caused an explosion, killing a person and damaging the plaintiff's premises. It was held that the defendant was liable for negligence. The labelling was misleading and inadequate.

As we saw earlier in *Donoghue* v. *Stevenson*, the duty of manufacturer to ultimate user is to take reasonable care in preparing the product to see that it will not foreseeably injure anyone who may use it in the way intended or designed. This element of *proper use* is important. Failure to use the product in the way stipulated may break the chain of causation existing between manufacturer and ultimate user. In cases where new products are being marketed the manufacturer should keep up to date as regards any new information concerning the potentially dangerous nature of his product or drug. In this respect he will have to carry out research or have it done for him at government or commercial research centres, this being incumbent upon him by virtue of s. 6, H.A.S.A.W.A. If the manufacturer believes that the product is safe when first marketed and if any other reasonable manufacturer would believe the same (owing to absence of authoritative information to the contrary), the manufacturer will not become liable to any person who foreseeably uses the product and is injured, if the manufacturer had kept himself informed of any new development regarding the propensities of the product. If, however, before the injury or disease induced by the product had happened, it had come to light that the product or some of its properties might be carcinogenic, the only way in which the manufacturer could probably discharge his duty would be to take the product off the market or drastically modify it or its method of use. If the risk of cancer was comparatively minor, written warnings of the possible risk involved to users of the product should be given. This, which emerged from *Wright* v. *Dunlop*, has now been put into statutory form in s. 6, H.A.S.A.W.A. As far as negligence is concerned whether a product should be discontinued or a mere warning given with it depends on the gravity of the injury which would arise from one or other of these measures not being taken.

Considering the apparent reluctance of the courts previously to give damages for negligence where there has been faulty design, the introduction of s. 6, H.A.S.A.W.A., placing new obligations upon designers, raises the question of how far actions for negligence can and will be brought against designers for bad design. In theory they can and will, because any regulations passed under s. 6 affecting designers give rise to civil liability, unless the regulations state otherwise, this much being clear from s. 47(2), H.A.S.A.W.A. *Caveat* designer (*see also* § **51,** p. 204).

Where proof of negligence is unnecessary

(*a*) *The doctrine of* res ipsa loquitur. There are circumstances which make it unnecessary for the plaintiff to prove negligence. This occurs where the thing or event causing the harm speaks for itself and is known as the doctrine of *res ipsa loquitur*. It purports to place the burden on the defendant of proving that he or his employees had taken reasonable care in carrying out the activity which gave rise to the injury. If the defendant can show that he took such care, he will not be adjudged negligent. Before this rule can apply, however, certain conditions must be satisfied. The law was stated by Erle, C.J., in *Scott* v. *London & St. Katherine's Docks Co. Ltd.* in 1865 when the judge said "There must be reasonable evidence of negligence. But where the thing is shewn to be under the management of the defendant or his servants, and the accident is such as in the ordinary course of things does not happen if those who have the management use proper care, it affords reasonable evidence, in the absence of explanation by the defendants, that the accident arose from want of care." The facts of *Scott*'s case are a classic illustration of this rule. The plaintiff was a customs officer and was carrying out his employment when, as he passed the defendant's warehouse, several bags of sugar fell on to his head, injuring him. The defendant could not account for how the accident happened. On appeal it was held that the facts constituted reasonable evidence of negligence. The usefulness of the operation of this rule to plaintiffs cannot be overstated.

Scott's case shows that four conditions must be satisfied if the rule is to operate:

(i) the event causing the accident must be one which does not happen in the "ordinary course of things";

(ii) the event causing the accident would not have happened if proper care had been used by the defendant or by his employees or by anyone else for whom he might be responsible;

(iii) the stituation leading to the accident was within the defendant's management;

(iv) there is absence of explanation on the part of the defendants as to how the accident happened or could have happened.

If the facts can be matched against all these conditions the plaintiff is well on his way towards a successful action and there have been several recent illustrations of the rule. In *Henderson* v. *Jenkins & Sons* in 1970 J's lorry careered down a hill and killed H's husband. The reason for the accident was that the brakes on the lorry failed, owing to an escape of brake fluid through a hole in a pipe which had corroded away. This corrosion occurred at a spot in the pipe where it was impossible to inspect it visually. When sued the defendants argued that the cause of the accident was a latent defect which had developed without any fault on their part in that the corrosion was not discoverable by the exercise of reasonable care and skill. Evidence showed that the pipe was inspected once a week, that the brake pedal had been depressed whenever the pipe was inspected to monitor possible leaks from the pipe and that none had been found. Moreover it was shown that neither the Ministry of Transport's rules nor the maker's advice required anything more than visual inspection of the pipe. In the High Court and the Court of Appeal it was held that reasonable care had been taken by the defendants in maintaining the vehicle, but the House of Lords by a majority decision reversed the lower courts' judgment and found for the plaintiff. During the course of giving judgment the court uttered the following words of caution to any goods transporter:

"Yet the kind of load this lorry had been carrying in the past was something which had to be known in order to assess the measure of the duty of reasonable care resting on the [defendants]. For the corrosion of the pipe was caused by some chemical agent. Had the lorry, therefore, been carrying chemicals of any kind? Or had it operated under conditions where salt (also a corrosive agent) might come into contact with the pipe? Or had it at some time been adapted for carrying cattle and done so? If any of these things were the case then clearly visual inspection of the pipe in situ would not have been enough. It should have been removed at intervals so that the whole of it, and not merely part of it, could be examined. ... It was, therefore, incumbent on the [defendants], if they were to sustain their plea of latent defect undiscoverable by the exercise of ordinary care, to prove where the vehicle had been and what it had been carrying whilst in their service and in what conditions it had operated. Only then could the standard of reasonable care be ascertained. ..."

Similarly in 1971 in *Bennett* v. *Chemical Construction* (*G.B.*). *Ltd.* some heavy panels fell on to B, injuring him. The contractor did not call any evidence to show why and how the panels came to fall. It was assumed that the accident could only have occurred through the negligence of those working on the panels. The Court of Appeal ruled that this was a classic case of negligence on the part of the employer.

The *res ipsa loquitur* cases are all examples of injuries occurring in circumstances which make the cause of the injury inexplicable. If the defendant can offer an explanation, or the facts of the case suggest the cause of the accident, this rule cannot benefit the plaintiff. In *Barkway* v. *South Wales Transport Co. Ltd.* in 1950 a widow brought an action against a bus company in respect of the death of her husband, who had been a passenger on one of its buses and who had been killed when the bus was involved in an accident owing to a tyre bursting. During the course of proceedings the House of Lords was told about the system of inspection operated by the company which they thought to be excellent. Despite this, however, the company was found to be negligent because it did not require drivers to report incidents which might have contributed to a hidden tyre fracture. With respect to the question of *res ipsa loquitur* the court decided that it was not applicable, since the facts giving rise to the accident were known. In other words condition (iv) above was not present, in that there was an explanation as to how the accident happened.

Actions in negligence, requiring the plaintiff to establish cause, underline the adage of English law that he who alleges must prove. Since *res ipsa loquitur* is an exception to this rule, it has been regarded as a judicial backwater. In recent times, however (and the fact that the whole concept of fault liability is under the microscope cannot be merely coincidental), the doctrine is reasserting itself: take the facts of the 1976 appeal decision in *Tesco Ltd.* v. *Ward*. Whilst shopping in T's store in Liverpool, W slipped on some yoghourt which had spilled on to the floor, as a result of which she suffered injury and sued T in negligence. In the court of first instance W was awarded damages and T appealed. The Court of Appeal ruled that it was the duty of the shopkeeper to use reasonable care to see that the shop floor which was under his management was reasonably safe. The trial judge was justified in ruling that the accident would not have happened had the shopkeeper taken reasonable care, since the accident was one which, in the ordinary course of things, does not happen if floors are kept clean and spillages dealt with as soon as they occur. It was not for the plaintiff to show that the spillage had been on the floor an unduly long time and that there had been opportunities for the management to clean it up which were not taken. Rather it was for the plaintiff to show that there had occurred an event which was unusual and which, in the absence of explanation, was more consistent with fault on the part of the shopkeeper than with the absence of fault. Once the plaintiff had done that, it was then incumbent on the shopkeeper to adduce evidence that he had taken reasonable care. This done, he could then escape liability if he could show that the accident must have happened, or on a balance of probabilities would have been likely to have happened, in spite of the existence of a proper and adequate system to provide for the safety of customers. In the present case, however, the only precautions which were taken were to have the floor brushed five or six times a day and instructions to the staff that, if they saw any spillages on the floor, they were to stay where the spill had taken place and call someone to clean it up. The precautions in question were not enough to enable the appellant to say that there was a proper and adequate system to provide for the safety of customers.

(*b*) *The future of the doctrine*. *Tesco Ltd.* v. *Ward* demonstrates a change of attitude on the part of the judiciary. The previous position whereby one who was injured at work had to prove negligence in the employer seems to be giving way to a desire on the part of the courts to shift the burden of proof in such cases, with the result that industrial negligence claims may begin to approximate to strict liability. The courts seem now to be more interested in placing the risk of injury on the head and shoulders of organisations carrying out potentially hazardous activities and operations with a view of profit, and it is significant that such organisations will normally be covered by both an employers' and public liability insurance policy.

Time limit for bringing actions in negligence

The time limit in which actions for negligence must be brought is normally *six years* as from the time the plaintiff suffered damage or harm, or ought reasonably to have become aware of the damage or harm which he eventually suffered. The six-year rule, however, does not apply to cases of personal injury. In the case of personal injuries the Law Reform (Limitation of Actions) Act 1954, as subsequently amended, limits the period to *three years*. In other cases (i.e. not cases of personal injury) s. 2, Limitation Act 1939, applies to all actions in contract and tort (with the exception of specialty contracts, where the period of time is 12 years).

Actions for damages against the employer in respect of injuries suffered at work are normally brought in tort, the basis of the claim being negligence, though there are now signs that the courts might be prepared to treat the claim as based upon breach of an implied term of a contract of employment (remarks in *Lister* v. *Romford Ice & Cold Storage Co. Ltd.*). Notwithstanding the possible double pedigree of the action, it is thought that as far as time limits are concerned tort offers more than contract. With actions in contract time begins to run as from the date of the breach of the contract, whereas in tort the action does not start until such time as the damage or harm has occurred. This position was established in 1972 in the Court of Appeal in *Dutton* v. *Bognor Regis U.D.C.*, where it was held that builders of houses owed a duty of care to subsequent purchasers who might suffer loss as a result of being forced to put right defects in the house caused by negligently laid foundations. The Limitation Act says that actions for such damage shall not be brought after the expiration of six years from the date when the cause of action arose, but a problem arises in that the time lag between the negligent conduct (here laying foundations) and the date when the damage became visible can be considerable (and in the case of a house it may have been transferred to successive owners). Be that as it may, a 1976 Court of Appeal case has now established that the cause of action starts to run from the date when the damage manifested itself or would have manifested itself to a reasonable purchaser. In *Sparham-Souter* v. *Town & Country Developments* (*Essex*) *Ltd.* in 1976, a case reiterating the principle of *Central Asbestos Co. Ltd.* v. *Dodd* four years earlier (*see* below), some houses were built between December 1964 and September 1965. The council had inspections made as it was required to do by its bye-laws and in December 1965 it issued certificates to the effect that the work had been inspected and that they were of the opinion that the work did not contravene any of the bye-laws. S-S contracted to buy two houses from the builder in October and November 1965, and the houses were conveyed in November 1965 and January 1966. Two or three years after the owners had moved in, cracks appeared in the brickwork as a result of defective foundations, which did not comply with the council's bye-laws. On 22nd October 1971 S-S sued the builder and the council for negligence. It was held by the court that a cause of action arises at the earliest when a tort has been committed, i.e. when the plaintiff suffers damage as a result of the negligent act. In this type of case, the earliest that a person can become aware of his ability to bring action is when he buys the house. Even then the damage may not have manifested itself. The cause of action runs from the date when he became aware of the damage or ought, by exercising reasonable care, to have become aware of the damage. This happened in 1968 and time began to run from that date.

Sparham-Souter's case is relevant when considering *Central Asbestos Co. Ltd.* v. *Dodd.* This case is particularly instructive in spotlighting the reasons behind the passing of the Limitation Act 1963, which amended the Law Reform (Limitation of Actions) Act 1954. The 1972 decision came as manna from heaven to workmen wishing to claim in respect of work-induced conditions that had not manifested themselves within the three-year period required by the 1954 Act. D was employed by C.A.C. in a workshop where various processes connected with asbestos were carried out. D started to work there in 1952 and remained employed there until 1965. Throughout the whole of that period C.A.C. was in breach of the regulations designed to protect workmen against asbestos dust and as a result of those breaches D was exposed to asbestos dust and contracted asbestosis, a condition leading irreversibly to increasing incapacity and eventual death. D was diagnosed as suffering from the condition in January 1964 when he was advised by the Pneumoconiosis Medical Panel not to go on working with asbestos, but he did not feel ill at the time and continued working with C.A.C., though in March 1964 he was awarded a disablement benefit based on 10 per cent disablement. In September 1965 he left C.A.C. on his doctor's advice. By that time he knew (*a*) that he had contracted asbestosis; (*b*) that in 1964 C.A.C. had been convicted for allowing such conditions to prevail; and (*c*) that the nature of the condition he had contracted rendered it necessary for him to stop working in the asbestos industry. But he did not know that he had a good cause of action against C.A.C., having been told by the works' manager (who was also suffering from asbestosis) that if he caught asbestosis there was a panel which would award compensation but that it was impossible to maintain an action against his employer. When, however, he learnt that a co-employee was suing C.A.C. in April 1967 he decided to do likewise and initiated action in October 1967. C.A.C. argued that the action was statute-barred. It was held by the House of Lords in a majority decision that the employer could not rely on the Limitation Act 1939 and that the claim was not statute-barred. During the course of giving judgment Lord Reid dealt with the history of the Limitation Act 1963 (on which C.A.C. sought to rely) in the following way:

"The appellants plead that this action is barred by the Limitation Act 1963. Before coming to examine the terms I must recall the circumstances in which it was passed. Under the previous law time ran from the date when the damage was suffered. In *Cartledge* v. *E. Jopling & Sons Ltd.* the appellant workman had contracted pneumoconiosis by inhaling noxious dust before 1950. In this disease, as in asbestosis, the sufferer's lungs may have been damaged many years before any symptoms develop or even many years before any X-ray or other examination can disclose that there is anything wrong So the men only discovered that they had the disease at various dates from 1955 onwards; and this House was compelled by the terms of the statute to reach the absurd result that a man's claim may be time barred before it is possible for him to know that he has suffered any damage....

"Obviously one of the purposes of the 1963 Act was to remedy the defect in the law brought to light in *Cartledge*'s case. But that cannot have been the only purpose.... This at least is plain. The Act extends the three years' time limit in cases where some fact was for a time after the damage was suffered outside the knowledge of the plaintiff, if that fact was 'material' and 'decisive'. Before a person can reasonably bring an action he (or his advisers) must know or at least believe that he can establish (*a*) that he has suffered certain injuries; (*b*) that the defendant, or those for whom he is responsible, has done or failed to do certain acts; (*c*) that his injuries were caused by those acts or omissions; and (*d*) that those acts or omissions involved negligence or breach of duty.

"In the present case the first three of these were all known to the respondent more than twelve months before this action was brought, but the fourth was not; he only got to know of it some six months before the writ was issued. The question for decision is whether the fact that the appellant's acts involved or amounted to negligence or breach of duty is or can be a 'material' or 'decisive' fact within the meaning of the Act."

After referring with approval to the above test Lord Morris in his judgment said:

"The respondent passed the statutory tests. Although he knew in January 1964 that he was suffering from pneumoconiosis and after that time knew that his disease had been caused by the inhalation of asbestos dust when employed by the appellants (as he was until September 1965), he first had actual or constructive knowledge that his disease was attributable in law to the appellant's negligence and/ or breach of statutory duty when he consulted solicitors about the end of April 1967.... In these circumstances though his cause of action arose more than three years before the issue of his writ on 6th October 1967, the provisions of the 1963 Act deprived the appellants of a defence (which was the only defence that they could put forward) by pleading s. 21, Limitation Act 1939."

The Limitation Act 1975

Critical observations on the part of Lord Reid in *Central Asbestos Co. Ltd.* v. *Dodd* about the abstruseness of the terminology of the 1963 Limitation Act have led to a welcome change in the law, and on 1st September 1975 a new Limitation Act became law. This statute was passed principally to accommodate cases like *Dodd's* where work-related or induced diseases, e.g. pneumoconiosis, manifest themselves only after years of employment. What the 1975 Act says is that any action for negligence, nuisance or breach of contractual or statutory duty in respect of personal injuries must be brought within three years of the date:

(*a*) on which the cause of action occurred (e.g. in the case of injury, when the injury happened); or

(*b*) if later, of the plaintiff's knowledge.

The Act goes on to define the expression "plaintiff's knowledge" as follows.

"References to a person's date of knowledge are references to the date on which he first had knowledge of the following facts—

(*a*) that the injury in question was significant, and

(*b*) that that injury was attributable in whole or in part to the act or omission which is alleged to constitute negligence, nuisance or breach of duty, and

(*c*) the identity of the defendant, and

(*d*) if it is alleged that the act or omission was that of a person other than the defendant, the identity of that person and the additional facts supporting the bringing of an action against the defendant."

The Act goes on to say that the same limitation period applies to civil actions whether brought by the injured person himself, or if he is killed or dies, on behalf of his estate by his dependants. This means that if the injured person dies before his action is time-barred, his estate has a further three years in which to bring the action. Moreover the decision of the High Court in *Buck* v. *English Electric Co. Ltd.* in 1978 establishes that the fact that the injured or deceased workman does not institute proceedings within the statutory period because, being in receipt of wages and a disability pension, he does not wish to "sponge" on his employer, is not a legally valid reason for failing to institute the proceedings within such period. Notwithstanding this obstacle, the action may proceed if it appears to the court that it would be equitable for it to proceed, and this was what happened in *Buck's* case.

21. Employers' liability at common law

In *Lister* v. *Romford Ice & Cold Storage Co. Ltd.* in 1957 Lord Radcliffe observed:

"The existence of the duty arising out of the relationship between employer and employed was recognised by the law without the institution of an analytical enquiry whether the duty was in essence contractual or tortious. What mattered was that the duty was there. . . . Since, in any event, the duty in question is one which exists by imputation or implication of law and not by virtue of any express negotiation between the parties, I should be inclined to say that there is no real distinction between the two possible sources of obligation. But it is certainly, I think, as much contractual as tortious."

Advantages of the contractual approach

The decision in *Matthews* v. *Kuwait Bechtel Corporation* in 1959 shows that for English personnel working abroad in dangerous jobs for foreign companies, it is an advantage for the employee to show that the contract was made in England and, if he is injured, to sue his foreign employer for breach of a contractual term in circumstances where it might be difficult to sue the employer in tort (there might not even be a law of tort or the equivalent in the host country). The advantage to the employer of the contractual approach is that, if the employee is in breach of a contractual term which results in injury to another employee (for which the employer will have to pay, by virtue of the doctrine of vicarious liability), the employer can exercise a reciprocal right of indemnity as against the negligent employee. This was the position in *Lister*'s case in 1957. The plaintiff was employed by the Romford Ice & Cold Storage Co. as a lorry driver. His father was also employed by the same company and one day, whilst reversing the lorry, he negligently injured his father. His father sued the company and was awarded £1,600 damages. These damages were paid by the insurers of the company and, exercising their right of subrogation, i.e. exercising any rights the insured might have had against the defendant, they claimed the sum of £1,600 back from the plaintiff, on the grounds that he owed a duty towards his employer to perform his contract and his job with reasonable care and skill, this duty being an implied term of his contract of employment. It was held that the employee had failed in his duty and had to repay the insurer.

Restraint in exercise of contractual rights

A committee reported to the then Minister of Labour that *Lister*'s case had served to endanger good industrial relations. In consequence the British Insurance Association agreed that, when insuring for employers' liability, they would not sue an employee of an insured employer in respect of injury caused to a co-employee, unless (*a*) there was collusion or (*b*) there was wilful misconduct on the part of the employee causing the injury. This attitude has recently been indorsed by the courts in *Morris* v. *Ford Motor Co. Ltd.* in 1973. F.M.C. had placed a contract for cleaning services at one of their plants with C.I.S. for whom M worked. Whilst engaged in this work at the factory M was injured through the negligence of R who was driving a fork-lift truck. Thereupon M claimed damages from F.M.C. for the negligence of their employee. However, C.I.S. had entered into a contract of indemnity with F.M.C., agreeing to indemnify F.M.C. for all losses or claims for injury arising out of the carrying out of the cleaning work in their factory. C.I.S. accepted that they were bound by the terms of this contract but contended that they should be subrogated against R, on the basis that, following *Lister*, F.M.C. would have been in a position to do so. Giving judgment Lord Denning, M.R., was emphatic that to depart from the position in the report following

Lister's case was to pave the way for unnecessary strikes. He therefore ruled that C.I.S. could not be subrogated to F.M.C.'s rights against their negligent employee.

The employer's liability in negligence

At present it is safe to assert that there are certain duties owed by an employer to his employees at common law and that breach of such duties causing injury to the employee will make the employer liable for negligence. These common law duties exist in addition to any statutory duties placed upon the employer or a particular type of employer, e.g. a factory occupier. This means that employees can bring double-barrelled actions against their employers, and this is a trend which is likely to continue under H.A.S.A.W.A. regarding regulations made pursuant to it, unless the regulations excise the common law action.

(*a*) *Direct and indirect liability*. At common law all employers are the subject of obligations relating to the safety of their employees. This is often not the position with legislation, which has tended to lay duties upon specific categories of employers, e.g. factory occupiers had duties enjoined upon them by the Factories Act 1961. H.A.S.A.W.A. has also seen fit to follow the common law principle of ubiquity of obligation between employer and employee.

This notwithstanding, the nature of the employer's liability is dual. It may be direct, but it can be indirect. Direct liability refers to situations where the employer is himself personally liable to the employee through failure to implement or for having implemented badly a duty personal to him. An example of this might occur where an employee was injured as a result of the place of work or the system of work being unsafe. Indirect liability, on the other hand, is concerned with situations which give rise to vicarious liability in the employer. This happens where one employee negligently injures another employee in the course of his employment, the injury arising out of the employment. Both these types of injury to employees must be insured against under pain of criminal sanction, this being laid down in the Employers' Liability (Compulsory Insurance) Act 1969.

(*b*) *Duties owed*. In the leading case of *Wilsons & Clyde Coal Co. Ltd.* v. *English* in 1938 the House of Lords formulated the four main duties owed by employers to employees. There is a duty to provide and maintain:

(i) a safe place of work and safe means of access to and egress from it;

(ii) proper and safe plant and appliances;

(iii) a safe system of work;

(iv) competent co-employees.

Although in the wake of the Law Reform (Personal Injuries) Act 1948 (which eliminated the doctrine of common employment whereby an employer escaped liability if one employee injured another employee in the course of their employment) requirement (iv) has lost much of its former significance, the remaining three requirements are still part and parcel of the common law. It might indeed be said that it was the intention of the draftsman of s. 2, H.A.S.A.W.A. to give criminal sanction to the rules of common law. In the *Wilsons* case E, who was employed in a coal mine owned by W.C.C.C., was badly injured when he was going along an underground haulage road at the end of the day shift. This happened as a result of being struck by some haulage plant which had been set in motion. E argued that the employer was negligent in allowing haulage plant to be set in motion at a time when the day shift was coming to the surface. The owners of the coal mine, on the other hand, contended that they were entitled to delegate the duty to maintain a safe system of work to a qualified manager who was technically competent to manage, and that they had done so. The House of Lords disagreed and during the course of giving judgment Lord Wright uttered words which have petrified into legend and have meant the difference between a successful action and no compensation for many injured workmen. So important is this judgment to an understanding of the common law and H.A.S.A.W.A. that it is worth recalling Lord Wright's exact words.

"I do not mean that employers warrant the adequacy of plant, or the competence of fellow employees, or the propriety of the system of work. The obligation is fulfilled by the exercise of due care and skill. But it is not fulfilled by entrusting its fulfilment to employees, even though selected with due care and skill. The obligation is three-fold—'the provision of a competent staff of men, adequate material, and a proper system and effective supervision. . . .' I think the whole course of authority consistently recognises a duty which rests on the employer and which is personal to the employer, to take reasonable care for the safety of his workmen, whether the employer be an individual, a firm, or a company, and whether or not the employer takes any share in the conduct of the operations. . . ."

What emerges from an analysis of this part of the judgment is that the duty is not absolute, as a contractual one might be. The employer does not *warrant* the safety of the workplace. It does not have to be, indeed it cannot be, absolutely safe against the risk of all types of hazards. The duty is based on negligence and requires that an employer take precautions against foreseeable danger. In certain areas mentioned by Lord Wright dangers are more likely to be foreseeable to the man at work—he refers to safe place of work, safe plant and above all a safe system. Sections 2(1) and 2(2), H.A.S.A.W.A., apparently require the employer to do no more and no less.

Lord Wright goes on to state that the duty is personal to the employer and cannot be delegated to either a competent employee or to a manufacturer or some other third party. Whether this is as true now as it was in 1937 is open to doubt. Once Parliament had passed the Law Reform (Personal Injuries) Act 1948, with the result that an employer became liable for injuries suffered by one employee owing to the negligence of another, there was little, if any, need for the retention of the separate category of duties (i)–(iii), the so called "personal" duties. The inference here is that these duties were a reserve weapon with which to hit employers whilst the doctrine of common employment was still in force (during which time employers were not liable for negligent acts committed against employees by other employees). The Law Reform Act 1948 recognised the reality that most accidents at work are caused by the negligence of one employee towards another and so vicarious liability attached itself to conduct that was negligent. The employer's duties towards his employees, if breached, gave rise to actions in negligence. And since negligence was the antithesis of strict liability it was reasonable to assume that the employer's duty was delegable. This matter was examined in *Davie* v. *New Merton Board Mills Ltd.* in 1956, a case which concerned one of the commonest situations involving injury at work. D was employed by N.M.B.M. His job consisted of working on a machine and in the course of his employment he had to use a drift (a tapered bar about 12 in. long). This drift had to be struck with a hammer. One day D was using the drift, which appeared to be nearly new, but on the second stroke of the hammer the drift broke and a piece flew off, hitting D and putting out the sight of his eye. The employer showed that he had obtained the supply of drifts from the Baldwin Co., who were reputable suppliers and who had obtained them from reputable toolmakers in Sheffield. It was held by the House of Lords that the employers had exercised reasonable care and skill and were not liable for a fault of manufacture or supply.

This case shows that blame was redressible at source and that employers were not liable for defects of manufacture. This meant that the employer could, if he could show that he had selected his supplier with reasonable care, delegate the duty to take reasonable care for his employees to a manufacturer. And this had the consequence that the so-called "personal" duties ceased to be personal to the employer and could be delegated to others, e.g. a manufacturer or supplier. This position would seem to be in accordance with the common law rule relating to manufacturers, as outlined by Lord Atkin in the leading case of *Donoghue* v. *Stevenson* in 1932, which provided that manufacturers of products had to take reasonable care in the making and marketing of them to see that those who might foreseeably use them, assuming that they were being used properly, would not be injured. In the case of products supplied for use at work, however, whether new or second-hand, there is a duty on the employer-user to provide and maintain safe plant and equipment, which involves periodical inspections of such products by trained personnel that, as far as actions in negligence against the producer by the user's employee are concerned, may well constitute a *novus actus interveniens* and make the employer liable (*see* § **20,** p. 57).

If as a result of failure to use reasonable care a detectable fault in the product is overlooked, with the consequence that an employee is injured when using it, the liability lies with the user or employer of the injured workman (as to how far the above situation may have been altered by s. 6, H.A.S.A.W.A., *see* § **51,** p. 251). The interaction of s. 6 with s. 2, H.A.S.A.W.A., seems to point the way towards joint liability on the part of producer and user to employees who are injured when using products properly at work. Section 6, however, is merely concerned with the imposition of criminal liability upon producers and others involved in production and distribution, and until Parliament implements the proposals of the Pearson Commission in favour of a system of product liability based on the Strasbourg Convention on Product Liability, the position in civil law of producers to injured employees remains uncertain.

Employers' Liability (Defective Equipment) Act 1969

This piece of legislation passed by the 1966 Labour government was conceived in response to *Davie*'s case, it being felt that from the employee's point of view the decision was retrograde since it meant that there would be occasions when employees would be forced to pursue their remedies against manufacturers based in foreign countries and as such difficult to sue, or against manufacturers who had ceased to trade or gone into liquidation. The legislation reversed the effect of the decision in *Davie*'s case. Section 1 (1) provides that:

"(*a*) where an employee suffers personal injury in the course of his employment in consequence of a defect in equipment provided by his employer for the purposes of the employer's business; and

(*b*) the defect is attributable wholly or partly to the fault of a third party (whether identified or not);

the injury shall be deemed to be also attributable to negligence on the part of the employer (whether or not he is liable in respect of the injury apart from this sub-section), but without prejudice to the law relating to contributory negligence and to any remedy by way of contribution or in contract or otherwise which is available to the employer in respect of the injury."

This meant that employers often had to shoulder the liability for injury to employees which was really the fault of the manufacturer; what we have in the U.K. is a system the reverse of that in the U.S.A. where, owing to the emphasis on product liability, producers have had to contend with astronomical insurance premiums and discontinue certain commercial lines. A happy medium somewhere between the two systems, arguing for a form of joint insurance on the part of employer and manufacturer in situations like *Davie*'s case, seems to offer the best solution, whereas the outcome of the Employers' Liability (Defective Equipment) Act 1969 appears to be to consolidate the principle in *Wilsons & Clyde Coal Co. Ltd.* v. *English* that the duties of an employer are personal to him and not delegable to a manufacturer (*see* p. 72).

Specific duties of the employer at common law

There are four specific obligations, namely provision of a safe place in which to work, of safe plant and equipment to work with, of a safe system of work and of competent staff. All these duties are part of the general obligation upon all employers at common law to take reasonable care for the health and safety of their employees. Often the test of what is reasonably foreseeable depends on the general practice in the trade, or whether a similar injury has happened before. These are guide lines for assessing liability, but they are not conclusive.

(*a*) *Safe place of work.* As early as *Brydon* v. *Stewart* in 1855 this duty had been fixed upon employers at common law. In this case, because the ventilation in a mine was bad some miners left the pit before the end of the shift. They were on their way up the pit shaft when a stone fell from the side of the shaft, killing one of them. Evidence showed that the cage in which they were going up to the top was open. It was held that the owner of the mine was liable since the means of access to and from the pit were unsafe.

Another case which illustrates well this aspect of the employer's duty is *Paine* v. *Colne Valley Electricity Supply Co. Ltd. and British Insulated Cables Ltd.* C.V.E.S.C. were suppliers of electricity and generated high-voltage current to kiosks constructed for them by B.I.C.L. There were transformers in the kiosks to reduce the voltage at which it was distributed to consumers. Kiosks consisted of three cubicles separated by a screen of insulating material. At the back of the cubicle there was a part measuring about 6 in. which had been left unscreened. P, while working inside this cubicle, inadvertently put his elbow through this gap, and came into contact with a live wire. As a result he was killed. It was held that:

(i) there was a breach of statutory duty on the part of C.V.E.S.C. for having failed to fence;

(ii) there was a breach of common law duty on the part of C.V.E.S.C. for having failed to provide their employee with a safe place in which to work; and

(iii) the manufacturer was not liable to the employee, since there was opportunity for inspection on the part of the employer, which put the employee outside the contemplation of the manufacturer.

In other words, there had been a *novus actus interveniens* which had destroyed the proximate relationship of manufacturer and user's employee. (Whether the facts of *Paine* would be decided similarly today would depend upon what instructions the manufacturer and supplier had issued to the user-employer and whether the latter had followed them. Assuming that he had done so, the chain of causation would exist as between supplier/manufacturer and the user's employee. It is thought that the employer would not be liable here, if he had done what was required of him by s. 2, H.A.S.A.W.A. This sort of situation would constitute the vital stepping stone on the road to a system of product liability.)

A place of work can be indoors or outdoors and can apply to a vehicle as well as to a building. In *Bradford* v. *Robinson Rentals Ltd.* in 1966 B suffered frostbite after being required to drive 500 miles in freezing conditions in a van that had no heater and a leaking radiator. It was held that the frostbite was a reasonably foreseeable consequence. The effect of this decision is that persons who are required to work out of doors or spend a substantial part of the working day out of doors can require to be supplied with protective clothing against the elements. But it is only the foreseeable risk which has to be guarded against and sometimes this element of reasonableness has operated to dilute the strict or even absolute essence of statutory requirements. The classic case illustrating this situation is *Latimer* v. *A.E.C. Ltd.*, which resulted in a modification of s. 25, Factories Act 1937, by s. 4, Factories Act 1959, to the effect that, as far as reasonably practicable, floors, steps, stairs and passages, etc., should be kept free from obstructions (this contrasts with the strict nature of the first part of what is now s. 28 (1), Factories Act 1961, requiring that such floors, etc., be properly constructed and maintained). L worked as a milling machine operator at A.E.C.'s works which were approximately 15 acres in extent. One day the factory became flooded as a result of an exceptionally heavy thunderstorm. This caused the rain water to get mixed with an oil liquid which was used for cooling the machines. After the rain water had drained away a slippery film of oil was left on the surface of the factory floor. A.E.C. placed sawdust over the floor but, owing to the acreage involved, were unable to cover the whole floor surface. L slipped on a part of the floor which had not been covered with sawdust and sustained injuries. It was held by the House of Lords that A.E.C. was not in breach of s. 25(1), Factories Act 1937 (now s. 28(1), Factories Act 1961), in that the words "all floors, stairs, passages and gangways must be of sound construction, and properly maintained" did not extend to a "transient and exceptional" condition, such as a torrential thunderstorm, though the court did add that it would include a dangerously polished surface (as a result of *Latimer*'s case the following words were added by the Factories Act 1959, and now form part of s. 28(1), Factories Act 1961: "and, so far as is reasonably practicable, be kept free from any obstruction and from any substance which is likely to cause persons to slip").

(i) *Working away on premises of a third party*. An employer must expect that certain work will be done by his employees away from premises within his control or occupation. This will occur where electricians employed by X, a firm of electrical contractors, are sent to wire up a factory belonging to Y. Here the fact that employees are not doing work on X's premises does not mean that X's duty to provide safe working premises and a safe system of work has come to an end. On the contrary, X, as long as he remains the employer, must provide a safe way of doing the job, though he is not expected to guarantee the safety of the structure of the third-party premises on which his men are working. A leading case is *McQuilter* v. *Goulandris Bros. Ltd.* Workmen were sent to repair a ship belonging to a third party. As part of the job they were required to walk along an unlighted deck. M, having tripped over a ring-bolt, fell into an uncovered hatchway and died from his injuries. It was held that his employers were liable. Lord Guthrie uttered the following warning to his employers, which has since been regarded as a classic statement of law. "The fact that the work had to be carried out on the premises of a third party did not absolve an employer from his duty of exercising reasonable care for the safety of his workmen. The duty must still be fulfilled, although its scope is circumscribed by the fact that the work was being done on premises not within the possession and control of the employer. As the structure of the premises is outwith his control, and any defects therein beyond his power to rectify, his care for his men could only be exercised within the limits imposed by those circumstances. But he was still under the duty of exercising reasonable care to safeguard them against dangers which he should anticipate and which he had power to avert." In this case, therefore, the employer should have provided lighting.

The statement of principle in this case reflects the correct legal position. It must be accepted that, because the premises are in the control of a third-party occupier, there are practical limits upon what the employer can do. If the job is an isolated one, he cannot be expected to inspect the premises, as was made clear in the judgment of the Court of Appeal in *Wilson* v. *Tyneside Window Cleaning Co.* in 1958. Alternatively, if the job is a long operation, e.g. sub-contract scaffolding on a large civil engineering or R.I.B.A. project, it is advisable that a safety officer employed by the firm of sub-contractors should inspect the site both before work begins and regularly while it is in progress. In the case of isolated jobs, the employer's liability extends to safe plant and equipment and a safe way of doing the job. He will not be liable for structural defects in the premises unless they are obvious and associated with the job to be done on the premises. For example, if a painter employed by a firm of painting contractors is sent to paint a house or a shop which has rotten window frames on the third floor, it is incumbent on the employer to warn and instruct his employee to be on the look-out for this sort of hazard, as was made clear in 1952, in *General Cleaning Contractors Ltd.* v. *Christmas.* In the above hypothetical case, if the painter decides to go ahead despite the danger and paint the window frames and suffers injury, without having reported back either to his employer for further instructions or having drawn the occupier's attention to the defect, he will be without redress. In *Wilson* v. *Tyneside Window Cleaning Co.* W, who was a very experienced window cleaner, was required by his employer to clean windows at the Newcastle Brewery. W had been doing this work at the brewery on and off for about 10 years. On one quarterly cleaning W with another employee started to clean a window which was 12 ft. from the ground and which faced a street with a very steep incline. He took hold of a handle with his right hand under the sash, and put the fingers of his left hand on the bottom bar of the sash, so as to be able to pull the sash down. As he gave a pull on the handle it came away in his hand. He lost his balance, fell off the ladder and was seriously injured. Evidence showed that the woodwork surrounding the sash window was rotten and that W knew this. His claim for damages for negligence against his employer failed, it being held by the Court of Appeal that his employers had taken reasonable care for his safety, because they had instructed him not to clean windows which might not be safe and further that the risk was an obvious one to a man like the plaintiff who was very experienced at his job.

If, on the other hand, the defect in the premises is not apparent to the employee but should have been known about by the occupier, e.g. a plumber being electrocuted by a faulty electrical circuit, the occupier will be liable to the employee, assuming that the defect is apparent or one that he should have known about. Whether in the above case the occupier should have known about a faulty electrical circuit will depend on whether he actually knew or was told or, alternatively, whether he was an electrician and so should have known. In such a case the employer would not be liable. In *Cilia* v. *James* in 1954 C's husband was killed as a result of being electrocuted in an empty house which had just been bought and to which the deceased employee, a plumber, had been sent to repair a sink. His death was the result of a faulty electrical circuit in the house. It was held that the widow was without redress, as the hazard was one which was so exceptional that neither the employer was expected to know about it as part of the safe system of work nor the occupier as a structural defect in the house.

The upshot of this case law is that if defects in the premises are apparent to the employee, he should report back to his employer, who should advise the occupier that work will not proceed till the latter has taken steps to remove the defect. If, however, these precautions are not taken and an employee is injured, assuming that the defect causing the injury is part and parcel of the way of doing the job on the third-party premises, it is the liability of the employer; whereas if the defect causing the injury is associated with the structural condition of the premises, the occupier is liable. This was made perfectly clear in 1952, in the leading case of *General Cleaning Contractors Ltd.* v. *Christmas*. An employee of G.C.C. was a very experienced window cleaner. One day whilst cleaning windows at the Caledonian Club in London he fell, injuring himself badly, when a defective sash dropped on to his hand and made him lose his hold. He was obtaining hand and foot holds by means of the window frames and sills, his employer had made safety belts available, but the employee was not using one because there were no fittings on the club building to which he could attach them. It was held by the House of Lords that his employer was liable because the employer should have instructed employees to test for defective sashes and should have provided wedges to stop sashes dropping. There was thus a breach of safe system of work. The fact that the employer supplied safety belts was no defence, since he should have anticipated that certain work would have to be done where a safety belt could not be used. On the other hand, the Caledonian Club was not liable, since they were entitled to expect that an experienced window cleaner would look out for defective sashes and take the necessary precautions, or make sure that his employer took such precautions.

It is difficult to distinguish between cases like *Wilson* v. *Tyneside Window Cleaning Co.* and *General Cleaning Contractors Ltd.* v. *Christmas*, in so far as in the first case the employer was absolved of liability while in the second, on similar facts, he was found liable. The fact that the occupier was absolved from liability in both cases seems right, in that if the Occupiers' Liability Act 1957 assumes the common law, as it is generally supposed to do, the occupier would be liable only for "unusual" dangers. The matter was authoritatively put by Willes, J., as long ago as 1866 in the leading case of *Indermaur* v. *Dames*. An employee gasfitter was sent by his employers to test some gas-burners at a sugar refining factory. Whilst there he fell into an unfenced shaft and was injured. Finding the factory occupiers liable Willes, J. (a good judge to follow), observed "And with respect to such a visitor at least, we consider it settled law, that he, using reasonable care on his part for his own safety, is entitled to expect that the occupier shall on his part use reasonable care to prevent damage from unusual danger, which he knows or ought to know ..." As to the meaning of the expression "unusual" danger, this was considered by the House of Lords in *London Graving Dock Ltd.* v. *Horton*, where they said that a danger is unusual if a particular kind or class of invitee would not normally encounter such hazard on such premises, e.g. a plumber being injured by a faulty electrical circuit or a painter by a chimney that proved unsafe. Conversely a rotten window sash is not *per se* an unusual danger to an experienced window cleaner nor a tall chimney to a steeplejack.

Whether it is possible, however, to reconcile the two cases from the point of view of employers' liability is more doubtful. Certainly from the point of view of workforce safety the decision in *Christmas*' case is more welcome, and would probably be followed today in the light of the employer's duty to instruct and train under s. 2(2), H.A.S.A.W.A., and of the employee's duty to take the necessary precautions, e.g. wearing safety belts, for his own safety under s. 7, H.A.S.A.W.A. However, there is the argument that an employer cannot be expected to stand over his skilled employees like a wet nurse. Indeed most employers would not wish to do so. If, therefore, the employer has carried out his statutory obligations to provide safety equipment and carried out his common law obligation of initiating a safe system of work, if necessary with proper and effective supervision, assuming that the trained employee fails to carry out the instructions and is injured, it is difficult to believe that the fault is not exclusively his own. A typical instance of this occurred in *Ginty* v. *Belmont Building Supplies Ltd.* in 1959. An employee of B.B.S. was sent to do work on an asbestos roof. He was aware that this roof was in a defective state and that he should not work on the roof without boards. In fact, there was a statutory duty imposed on him to that effect by (what is now) reg. 36 (1), Construction (Working Places) Regulations 1966. His employer had provided a ladder and boards and these were available to him. The employee went on to the roof but without using the boards, as a result of which he fell through and was badly injured. It was held by the Court of Appeal that he could not recover any damages, since the cause of the injury was exclusively his fault, there being no fault on the part of his employer who had made the boards available for use.

Another well-known case in this context is *Johnson* v. *Croggan & Co. Ltd.* in 1954. J was employed by C as a steel erector. He was put in charge of the erection of a steel roof on some farm premises belonging to the occupier, the second defendant in this case. J had been told to get all the necessary equipment, which included ladders, from the head contractor. He obtained some equipment from the contractor but not the ladders. Instead he merely obtained a light fruit picking ladder which was on the farm premises. J subsequently stepped on to the top rung of this ladder which broke into three parts. He fell to the ground and was injured. He sued C for common law negligence and/or breach of reg. 29(1), Building (Safety, Health and Welfare) Regulations 1948, on the basis that the ladders were not of adequate strength. It was held by the court that C was entitled to delegate the statutory duty to J, a skilled steel erector, which they had done and the negligence of the plaintiff and his breach of statutory duty was the exclusive cause of the injury which he suffered. Nor was the farm occupier liable for the loan of the ladder (as for the liability of hirers of equipment to users, *see* § **49,** p. 249).

(ii) *Joint occupation as between employer (building owner) and main contractor.* The courts treat occupation as a multiple thing. Joint occupation is relevant to safe place of work, and this problem is especially relevant to the construction industry where a head contractor is in occupation, physically speaking, of a site owned by a building owner or an employer. If an injury were to befall an employee of the head contractor or an employee of a sub-contractor, could the argument in favour of joint occupation be invoked, so as to make both building owner and head contractor liable? And even to make all three including the sub-contractor liable? Although ultimately, by virtue of the provisions of the particular kind of building contract, the matter will be dealt with on an indemnity basis, with the sub-contractor indemnifying the main contractor and the latter indemnifying the building owner, as in the case of the R.I.B.A. Form of Main and Sub-contracts, there would seem little to prevent an employee of a sub-contractor suing his immediate employer, the head contractor and the building owner. In *A.M.F. International Ltd.* v. *Magnet Bowling Ltd. and Trentham* a sub-contractor was allowed to proceed against the head contractor and the building owner for their having negligently allowed his equipment to be damaged, the court taking the view that both were in occupation.

One must not confuse the employer's duty on the one hand to set up a safe system of work and a safe place of work, with the duty of an occupier or joint occupiers on the other hand, to take reasonable care to all lawful visitors. On the first count a head contractor would owe a duty only to his immediate employees, a point confirmed in 1972 in *Smith* v. *George Wimpey Ltd.*, where the court decided that reg. 3(1)(*a*), Construction (Working Places) Regulations 1966 (which imposes on a contractor, who is undertaking any of the operations or works to which the 1966 Regulations apply, the duty of complying with such of the requirements of specified regulations "as affect him or any workman employed by him"), renders a contractor liable in damages to any workman employed by him who is injured as a result of a breach of the regulations, but to no-one else.

An action, on the other hand, brought by an employee of a sub-contractor under the Occupiers' Liability Act 1957 could be brought against both the head contractor and the building owners. The Occupiers' Liability Act in other words is designed to deal with problems outside the immediate employer–employee relationship. In the hypothetical case mentioned above the sub-contractor's employee could proceed against his immediate employer, the sub-contractor, at common law and under 3(1)(*a*), Construction (Working Places) Regulations 1966, and against the head contractor under the Occupiers' Liability Act 1957, together with the building owner. The fact that the building owner is not in physical occupation would seem irrelevant. It has been made clear on many occasions that occupation is a synonym for control. If, therefore, the building owner – head contractor relationship is one of licensor – licensee and the building owner can extract indemnities from the head contractor, there is little doubt that the building owner, being in control, is also in occupation. The injured sub-contractor's employee, therefore, can proceed along two avenues, one based on employment, the other on occupation, for his redress. Indeed, although the decision in *Smith* v. *George Wimpey Ltd.* establishes that, for the purposes of the allocation of duties under the Construction (Working Places) Regulations 1966, a main contractor is not under a duty to the employee of a sub-contractor, this in no way operates to alter the position at common law with respect to vicarious liability. This point was well illustrated by *Smith* v. *Vange Scaffolding & Engineering Ltd.* in 1970, where it was held that, because the head contractor's employee had been negligent (along with the sub-contractor's employee), the head contractor was vicariously liable for the injury caused to another employee of the sub-contractor, along with the sub-contractor himself who was also vicariously liable for the negligence of his employee (*see* p. 226).

(iii) *Working away with two employers*. Employees may have to work away for another employer. When this occurs the services of the employee are transferred from his permanent employer to a temporary employer. In these circumstances, which employer is responsible for the employee's safety and which employer will be vicariously liable for any injuries negligently caused by the transferred employee? The answer depends on whom the relationship of employer–employee is with. At one time this relationship used to be based simply on control—could the employer tell his workman when and how to do his job? As industrial life has become more complicated it is increasingly clear that this test is not proving satisfactory—could a member of the governing body of British Airways tell an experienced pilot how to operate his aircraft? Clearly not. It was for this reason that Denning, L.J., in *Stevenson, Jordan & Harrison* v. *Macdonald & Evans Ltd.* in 1952 uttered the following formula to enable a distinction to be drawn between employees and independent contractors, employers normally not being liable for torts and injuries committed by independent contractors: "One feature which seems to run through the instances is that, under a contract of service, a man is employed as part of the business, and his work is done as an integral part of the business; whereas, under a contract for services, his work, although done for the business, is not integrated into it but is only accessory to it."

Since then the approach has become more general and based on the reality of the situation. The first move away from the unreality of the control test was signposted by the decision of MacKenna, J., in *Ready Mixed Concrete Ltd.* v. *Minister of Pensions and National Insurance*, where the judge laid down three conditions for a contract to be a contract of service rather than for services (giving rise to the relationship of employer and independent contractor):

1. the employee agrees that, in return for wages or a salary, he will provide his skill and labour;
2. the employee agrees that, to a certain extent, he will be subject to the employer's control in deciding what job is to be done, the means to be employed in doing it, the time it is to be done and the place where it is to be done;
3. the other provisions of the contract must be consistent with its being a contract of service, e.g. holidays, sick pay, etc.

In this case the court had to consider whether, for the purposes of national insurance, owner drivers employed by the manufacturers of concrete to deliver concrete were employees or independent contractors. The contract of employment described the drivers as independent contractors and further stated that:

1. the lorries were to be painted in the company's colours;
2. the lorries were to be marked with a number provided by the company;

3. the lorries had to carry one of the company's mixing machines;

4. the lorries were to be at the company's disposal at all times and to be used for no other purposes;

5. the drivers would carry out all reasonable instructions from company employees and pay all running costs.

On these facts the judge decided that it was a contract of carriage and not a contract of employment. The drivers were "lump" drivers.

This decision, however, was challenged in 1976 in *Ferguson* v. *Dawson & Partners Ltd*, where it was held that a contract to hire labour-only sub-contractors was a contract of employment, with the result that the sub-contractor was unable to rely on the previous decision in *Smith* v. *George Wimpey Ltd.* to the effect that, as the labour-only sub-contractor was self-employed, he had a statutory duty to take care of himself. This had the consequence that the sub-contractor was liable to the labour-only sub-contractor for a breach of reg. 28(1), Construction (Working Places) Regulations 1966, in that he had not provided a suitable guard rail. The fact that the contract was referred to as one of labour only, the workman providing his own equipment, did not detract from the reality that the labour-only sub-contractor was employed under a contract of service with the sub-contractor. The form of the contract is clearly irrelevant.

Two other cases worthy of examination are *Mersey Docks & Harbour Board* v. *Coggins & Griffiths Ltd.* in 1947 and *Garrard* v. *Southey & Co.* in 1952. In the former case M.D.H.B. had hired out to C. & G. the services of one of their skilled crane drivers, C. & G. being a firm of stevedores who wished to use the crane to load a ship in the Liverpool docks. In the contract of hire between the Board and the firm of stevedores, there was a clause embodying one of the Board's regulations to the effect that the driver so provided should be the servant of the applicants. When the crane driver was transferred to the firm of stevedores, the Harbour Board still paid him and retained the right to "hire and fire" him. The temporary employer could direct him as to what operations to perform with the crane but were in no position to tell him how to operate the crane. Whilst the loading of the ship was in progress a checker was injured when the crane driver negligently drove into him. The question arose as to who was vicariously liable for the negligence of the crane driver. In other words, who was his employer? The Harbour Board argued that the determinant was the contract, under the terms of which the firm of stevedores would be responsible, but the House of Lords thought differently. They thought that the mere form of a contract could not detract from the reality of the situation, namely who had the power and right to direct the crane driver as to how to perform his job? This depended on whose payroll he was on, and here the evidence showed that the Harbour Board, the permanent employer, retained the ultimate sanction of dismissal, i.e. the right to hire and fire, and so control over the employee had not passed to the temporary employer. The result was that, despite the wording of the contract, the Harbour Board had to pay for the crane driver's negligence as being vicariously liable.

The facts of *Garrard* v. *Southey & Co.* were similar, except that here the employees were electricians. Two electricians employed by electrical contractors were sent to wire up a factory. Their own firm, the electrical contracting company, continued to pay them, to pay their national insurance contributions and retained the right to dismiss them. Some of the equipment used by the electricians and most of the tools were supplied by the factory occupier except for special tools belonging to the electricians themselves. They were supervised, however, by a foreman employed by the factory occupiers, who was himself an electrician, and they followed the system of work operational in the factory. During the course of his work one of the electricians fell off a defective trestle on which he was working and which belonged to some builders who were also working in the factory. The question arose as to who was responsible as employer for providing the electricians with proper and safe plant and equipment and a safe place in which to work. Who was the employer owing the common law duty? It was held by the court that control over the way in which the electricians were to do their work, the time when they were to do it, etc., had passed to the temporary employer, the occupier of the factory, even down to the two operatives taking meals in the factory canteen with the other full-time employees at the factory.

Decisions like *Garrard* v. *Southey & Co.* need to be studied with care, in that there is an initial tendency to treat such persons as the electricians or anyone else from outside the permanent workforce who comes into the factory to do work as an independent contractor, with the result that the independent contractor may not be protected by certain provisions of the Factories Act 1961 (e.g. those applicable only to employed persons), and also that he will be liable for any negligent acts. It is easy to believe that the only protection the factory occupier owes to the oncoming contractor is that laid down in the Occupiers' Liability Act 1957, a lower standard of care which makes the factory occupier liable only for "unusual" dangers to the contractors, i.e. those not to be expected in the course of employment. On that basis, it is doubtful whether the defective trestle would have qualified as an unusual danger and so the factory occupier would not have been liable. In truth, however, certain sections of the Factories Act 1961, e.g. ss. 14(1), 28(1), 29(1), exist to protect independent contractors (*see* §**33**, p. 208).

Once, however, it is admitted that one of the principal criteria in determining the relationship of employer and employee is control, it is obvious that there are many situations in which the factory occupier assumes the role of an employer. The fact that the oncoming temporary workforce brings its own gear and tools does not prevent its members being employees, as we saw in *Ferguson* v. *Dawson & Partners Ltd*; the fact that the permanent employer retains the right to hire and fire and pay national insurance contributions and wages is not in itself sufficient to prevent the factory occupier in which the work is being done, or the temporary employer, from becoming liable to set up a safe system of work and provide proper and safe plant and equipment. It is submitted that the true test of an employer–employee relationship is whether, as Denning, L.J., said in *Stevenson, Jordan & Harrison* v. *Macdonald & Evans Ltd.*, the man is employed as part of the business and his work is done as an integral part of the business, as distinct from where, although the work is done for the business, it is not integrated into it but merely accessory to it. The inference that the man has become integrated into the workforce is much more easily drawn where he is not skilled or where equipment, particularly complicated equipment, is not transferred with him.

On most occasions when employees are working away their permanent employer will remain responsible for safe plant and equipment and a safe system of work, etc., whilst the party on whose premises the work is being done will be an occupier, owing a lower standard of care at common law to the oncoming workforce. When there is someone on the staff of the occupier or the party on whose premises the work is being done who is in a position to direct how the work be done and advise generally, liability for safe work systems and plant and employees rests with the new temporary employer. If not, the occupier may well, however, be subject to certain duties under the Factories Act 1961 (*see* §**33**, p. 206).

(iv) *Complaints about unsafe place of work, plant and equipment.* There have been cases where employees have repeatedly complained about the dangerous conditions in which they work. Indeed if later a civil action is brought against the employer, employees will often be asked whether they did not complain about the bad and unsafe working conditions. For this reason, it is necessary to find out the exact legal effect of making a formal complaint to the foreman. Several case histories throw more light on this problem, the result of which seems to indicate that if a formal complaint is made and is found to be justified, but nothing is done to rectify the position, the employer, if he can be taken to have notice of the complaint (which will not be the case if the complaint is made to a union official as distinct from a foreman or someone else associated with management), will be held to be negligent at common law. This sword, however, is two-edged, with the consequence that complaints made by employees may justify a finding that they have been contributorily negligent or have agreed to run the risk of the injury, particularly the former, these being defences to the employer (the latter, it seems, following *I.C.I. Ltd.* v. *Shatwell*, being restricted to a defence for breach of statutory duty). The first side of this two-edged sword can be illustrated by the 1970 decision in *Smith* v. *Vange Scaffolding & Engineering Ltd.* S was employed by V who, under a contract with J.B., did whatever scaffolding work J.B. required on a building site where J.B. were doing work for the owners. While going from the place of work to the changing hut at the end of the day, S injured himself when he tripped over a cable of a welding machine which had been put there by an employee of J.B. The fact that access to and from the scaffolders' place of work was unduly difficult because of obstructions was well known to V's representatives on the site, but they did not formally complain to J.B. It was held by MacKenna, J., that V was liable at common law to S because it was the duty of V's representatives on site to make complaints to J.B. about obstructions, and their failure to make them could be construed as a cause of S's accident that was sufficiently proximate to establish V's liability. And because of this V was liable under the terms of the building contract to indemnify J.B. because it was enough, for the purposes of indemnity, that S's injuries (as distinct from the liability of J.B.) should be connected with V's acts or omissions.

The other side of the coin is the decision in *Bowater* v. *Rowley Regis Corporation* where an employee of the corporation in the course of his employment as a dustcart operator had to take out a particular horse which he knew to be frisky and difficult to control. He complained on several occasions to his immediate employer about the wayward behaviour of the horse, but nothing was done and no substitute horse found. One day during the course of his employment the employee was injured as a result of the horse's unpredictable behaviour. When sued for damages for negligence at common law, the employers sought to argue that the employee had knowingly and therefore willingly taken out a horse which was dangerous and that, as a result, he could not complain about his injuries, or alternatively that his injuries were the result of his contributory negligence, which meant that his damages should be reduced to the extent of his own fault. Neither of these arguments impressed the court, which took the view that, as the employee had had no choice in the matter of which horse was to be used, coupled with the fact that he had complained on a variety of occasions, he had not voluntarily assumed the risk of injury involved nor could his conduct be construed as contributory negligence.

(*b*) *Safe plant and equipment for the job.* An employer has to provide and maintain safe plant and equipment and appliances for his employees to do the job. This duty is essentially twofold, one of provision and then one of subsequent maintenance.

(i) *Provision.* Whether safe plant and appliances have been *provided* is a question that has been the subject of much case law. The general view seems to be that equipment has been provided if, although not actually put into the physical possession of the employee, he knows where the appliance or equipment can be obtained and access to it is reasonable. This was the view taken in the leading case of *Norris* v. *Syndic Manufacturing Co. Ltd.* in 1952, where the court said that equipment was provided if it was "ready to hand in a proper place". Certainly, it would appear, the expression "provide" cannot be construed to mean that the employee be forced to wear the appliance or use the equipment. At least this was the position before H.A.S.A.W.A. became law, though certain regulations imposed exceptions where appliances had to be worn and not merely provided, e.g. the Protection of Eyes Regulations 1974. In *James* v. *Hepworth Grandage Ltd.* in 1968 it was held that spats had been provided as protection against metal splashes when the employers had posted a notice in a reasonably prominent position to the effect that spats were obtainable from the stores, even though the plaintiff, who had been injured through not using spats, was illiterate (whether this decision would be followed today is doubtful in the light of the employer's duty to supervise the wearing of safety appliances, as laid down in s. 2(2), H.A.S.A.W.A., coupled with the employee's duty to co-operate with his employer and take reasonable care for his own health and safety).

Presumably strong evidence that an employee has been supplied with the necessary plant and appliances is written evidence, at least, in the case of movable appliances obtained from the stores or obtained for the employee's personal safety. It was not necessary at common law (though this may well have altered in the light of ss. 2 and 7, H.A.S.A.W.A.) for employers through their supervision to ensure that experienced employees used the appliances supplied or provided. An instructive case is *Woods* v. *Durable Suites Ltd.* in 1953. D.S. employed W to spread synthetic glue in their veneer department. So as to minimise the risk of dermatitis, they put up a notice in or near the work department outlining the steps to be taken to avoid contracting dermatitis and instructing the men in the use of barrier cream soap and washing facilities which were provided. W was a very experienced workman and he was specifically instructed by management in the precautions necessary. Unknown to the employer, however, W did not take these precautions, as a result of which he contracted dermatitis. In an action against the employer for negligence at common law

W failed, the court saying that an employer is not expected to wet-nurse his most experienced employees.

This duty will not have been carried out if no plant is provided at all, as was made clear in the 1944 decision in *Lovell* v. *Blundells & Crompton & Co.* L was employed by B. & C. to overhaul a ship's boiler tubes. He was not able to reach all the tubes and so he found some planks for himself and set up his own staging. This planking proved unsatisfactory and L was injured. Evidence suggested that the employer had not provided any planking nor set up a system of work. On these facts the judge had no hesitation in finding the employer liable. Moreover if the correct quantity of plant for the job and materials is not provided, the employer will also be liable. In *Machray* v. *Stewarts & Lloyds Ltd.* M was employed as a rigger. One day whilst working 70 ft. above ground a loop section of pipe, weighing over 10 cwt., had to be fixed. In order to do this it was necessary to move the loop section 14 ft. There were certain established methods of doing this, firstly by using the ground driver (who was not available), second by using chain block tackle. M tried to get hold of two sets of chain block but only one was available at that time. Earlier M had been instructed that the job was urgent by the chargehand, so he used a single chain block, as a result of which the loop swung out of control and injured him. He sued his employer for common law negligence, to which the employer argued that M had been contributorily negligent. It was held by the court that, although the way of doing the job chosen by M was clearly less safe than if two sets of chain blocks had been used, he was not guilty of contributory negligence. During the course of his judgment the judge explained the law as follows: "... when I find a workman ... adopting a course of conduct not for the sake of saving himself trouble, but in order to get on with his employer's business, and when I find that he has been prevented from doing the work in the way in which he would have preferred to do by the employer's breach in not providing him with proper tackle, I am very slow to put any blame on him...."

The common law does not prescribe that the employer provide brand new plant and equipment, a fact which came to light in the decision in *Parkes* v. *Smethwick Corporation* in 1957. Nevertheless older equipment should be inspected regularly and have the requisite repair and maintenance work done on it. The duty at common law is to provide equipment which will, or will when modified (e.g. guards on a machine) save the employee from foreseeable injury. It is a duty to take reasonable care. This really is another way of enjoining the employer to do what is reasonably practicable in the provision and maintenance of plant and systems. This involves balancing the likelihood of risk of injury against the cost and inconvenience of introducing safety devices. As Asquith, L.J., said in the leading case of *Edwards* v. *N.C.B.*: "... if it be shown that there is a gross disproportion between them—the risk being insignificant in relation to the sacrifice—the defendants discharge the onus on them." In other words if the likelihood of the injury occurring from failure to install the safety device is remote or minimal, the employer can take that risk.

(ii) *Maintenance.* The plant and appliances have also got to be *maintained* on a regular basis. Four situations can arise here which may involve the employer in liability:

1. he neglects to rectify known defects;
2. he neglects to rectify unknown but discoverable defects;
3. he fails to select suitable plant which a maintenance system would suggest as necessary;
4. he fails to implement a defect-reporting system.

1. *Systematic reporting of defects.* Maintenance means a regular system of inspection and testing of machinery and equipment by safety engineers. This duty delegated by the employer to the safety engineer involves the monitoring of defects in the company plant and the implementation of defects' reporting, following inspection and testing. Failure to implement a defects' reporting system can involve the employer in negligence, even though the system of inspection and testing is first class. This is exactly what happened in the classic case of *Barkway* v. *South Wales Transport Co. Ltd.* in 1950. As a result of a tyre bursting on a bus belonging to S.W.T.C., the bus was involved in a collision and B's husband, a passenger on the bus, was killed. Evidence showed that the system of inspection and testing by the bus company was excellent. Nevertheless the House of Lords held that the bus company was liable to the widow, because the company did not require its drivers to report incidents which could have led to hidden defects that might cause a tyre to burst. Hence the safety officer and his team should systematically monitor defects and keep defects' reports. Line management should also keep the safety personnel informed of defects as they arise so that repairs or part replacements can be carried out with the utmost expedition. If, when a defect arises which could prove dangerous, an employee merely informs his shop steward, management cannot be deemed to have notice of the defect, though, if the defect could have been uncovered by the use of reasonable care and skill, the employer will be liable if injury occurs because he ought to have known about its existence. When defects arise they should be communicated to both the safety officer and the shop steward, and once the regulations relating to Safety Representatives and Safety Committees have become operational after October 1978, unions will be kept informed of safety problems and defects in plant

and equipment and machinery in use as they arise on the shop floor, through the union-appointed safety representative. This should incidentally relieve the inspectorate of some of the present donkey work involved in plant safety, and joint inspections by management and safety representatives may well have the same effect as improvement and prohibition notices served by inspectors pursuant to their powers under H.A.S.A.W.A. In this way plant which is, or parts of plant which are, dangerous will be put out of use and clearly labelled as such until such time as the defect is rectified, unless the plant defect can be rectified while the machinery is in use.

2. *Neglect to rectify known defects.* If a defect has been reported to the safety engineer, the employer is normally deemed to have notice of it, and failure to rectify the problem resulting in injury to an employee will involve the employer in liability for negligence. The case usually cited in this context is *Clarke* v. *Holmes* in 1862. C was employed to oil a machine. One day the fencing broke and he reported this to his employer who agreed to take remedial action. Nothing was done and C was later injured. It was held that his employer was liable.

3. *Failure to rectify unknown defects.* Negligence is failure to do what a reasonable man would do or doing what a reasonable man would not do. This famous dictum was first expressed by Baron Alderson in *Blyth* v. *Birmingham Waterworks Co.* in 1856. Even though an employer did not know of the existence of a defect, but would have done had he been using reasonable care, he will be deemed to know about it and, if injury occurs to a workman, he will be liable. The sort of defect referred to here is a defect *discoverable through normal testing and inspections.* The leading case is *Webb* v. *Rennie* in 1865, where some scaffolding had been erected for shipbuilding purposes and the poles had been left in the ground for two years. W was instructed to dismantle the scaffolding and whilst he was doing this one of the poles snapped at the base and fell on him, injuring him. It was argued for the employer that he was not liable because he did not actually know of the existence of the defect. Nevertheless the employer was held liable to the workman and the court delivered a statement of law which has long been recognised as authoritative in such cases: "... and although in general an employer was not liable unless he knew of the danger, yet it was his business to know if, by reasonable care and precaution, he could ascertain whether the apparatus or machinery was in a fit state or not. ... Thus in the case of a manufacturer employing machinery which might be attended with danger to the persons employed about it ... the master must either ascertain the state of the machinery or apparatus himself; or employ some competent person to do so."

4. *Failure to select safe plant.* Once inspection, testing or other examination suggests that the plant or part should be replaced or the matter is brought to light in a report, the employer should select his new plant with care. The plant so selected must be safe and this means that the employer should delegate this duty, or rather performance of this duty, to a reputedly competent and respectable manufacturer. A question of considerable practical significance arising here is: does the delegation of the performance of this common law duty to a manufacturer absolve the employer from liability for negligence for injuries suffered by an employee of his whilst using plant or equipment in the way in which it was intended that it be used?

The imposition of new duties on manufacturers by s. 6, H.A.S.A.W.A., may absolve the employer from liability where, as in *Davie* v. *New Merton Board Mills Ltd.*, he has selected the manufacturer with care and has inspected the equipment for discoverable defects. At least, if such liability is not exclusively that of a manufacturer in such circumstances, it is arguable that the liability for injury caused to an employee in circumstances analogous with those in *Davie's* case should be joint as between employer and manufacturer. There is, however, a school of thought that maintains that the fact of delegation cannot operate to relieve the person delegating the performance of the duty, as here the employer, of the original duty cast upon him by statute or common law. This point has been made clear by the House of Lords in *Padbury* v. *Holliday & Greenwood Ltd.* where it was said that a main contractor would be liable for the negligence of a sub-contractor to whom part of the original contract work had been delegated or sub-let, unless the negligence happened to be ancillary to the performance of the contract with the main contractor. In *Padbury* sub-contractors were employed to put in window casements. Owing to the negligence of an employee of the sub-contractors a tool fell off a casement and struck a passer-by, causing him injury. It was held that the main contractor was not liable for this act of negligence which was personal to the workforce of the sub-contractor and the latter should as an employer have had control over it. So although an employer can delegate the performance of a duty to an independent contractor such as a manufacturer he cannot by so doing absolve himself of the duty. How far it is possible to reconcile this decision and its implications with the approach outlined in *Davie*'s case and apparently furthered by s. 6, H.A.S.A.W.A., is difficult to be sure. It seems better to follow *Davie*'s case as this is backed up by the imposition of new statutory duties upon manufacturers by s. 6, H.A.S.A.W.A., which by virtue of s. 47(2) makes it clear that civil liability may follow from the imposition of any regulations placed on manufacturers in pursuance of s. 6. Presumably it will be impossible for manufacturers to contract out of this liability by virtue of the Unfair Contract Terms Act 1977.

(*c*) *Delegation of statutory duties to competent person.* There is one set of circumstances in which the employer can absolve himself from liability by delegation, at least so it would seem from a succession of cases on the point. As Cockburn, C.J., observed in *Webb* v. *Rennie*, if the employer cannot personally carry out duties imposed on him for the benefit of his workforce, he must entrust those duties and their performance to those who can. If the employer delegates the performance of duties associated with the safety and health of his workforce to competent safety personnel, it seems from cases like *Johnson* v. *Croggan & Co.* and *Wilson* v. *Tyneside Window Cleaning Co.* that the safety officer would not be able to sue his employer for negligence in cases where the injury arose through the negligence of the safety officer. Moreover this approach may not be restricted to safety officers but may include all skilled and experienced personnel (as witness *Wilson* v. *Tyneside Window Cleaning Co.*, assuming that they have been trained and instructed in their safety obligations and have received the necessary supervision) who are injured through their own negligence. The case of *I.C.I. Ltd.* v. *Shatwell* shows that where a duty is imposed upon an employee for his own safety and benefit and he is negligent in discharging this duty, the courts will treat his injury as having been caused exclusively by his negligence (for the facts of *Shatwell*, *see* § **22,** p. 127). Though clearly the fact that an employer has nominated a safety adviser to oversee safety does not absolve the employer from liability for negligence in cases where injury is suffered by an employee. This would be tantamount to a flat denial of the existence of vicarious liability, the safety adviser after all being another employee, for whose negligence the employer will be liable. This raises the question whether the safety adviser and other skilled personnel should not have a term written into their contracts of employment to the effect that they should be entitled to a safe place of work.

(*d*) *Safe system of work.* This is the most comprehensive of the duties laid upon an employer. It ranges from co-ordination of various work departments to the general conditions of work, such as heating, ventilation, lighting and freedom from risk of disease. It includes an effective system of supervision and training in the hazards of the job and education in the use of protective clothing and appliances. In-plant training as part of job instruction in safety hazards should be encouraged as part of the system of work. Job layout, permit to work systems, and planning for specific tasks are all covered under this heading.

(i) *Co-ordination of various work departments.* This was the problem in *Wilsons & Clyde Coal Co. Ltd.* v. *English* where E, who was employed in W.C.C.'s coal mine, was injured whilst he was passing along one of the underground haulage roads at the end of the day shift at the same time as some haulage plant was set in motion. Obviously a safer and proper system would have required that the haulage plant be stopped simultaneously with shifts coming to an end and become co-ordinated into the shift system.

(ii) *Training and instruction in the hazards of the job*. It has generally been accepted since the decision of Lord Wensleydale in *Weems* v. *Mathieson* in 1861 that "A master is no less responsible to his workmen for personal injuries caused by a defective system of using the machinery than for injuries caused by a defect in the machinery itself". Thus to minimise liability for negligence at common law (and now to avoid criminal liability under H.A.S.A.W.A.) an employer must see to it that his employees are instructed and trained in the intrinsic hazards of the job. Impetus was given to this approach by the Industrial Training Act, passed in 1964 to place levies upon employers for the training of their employees. This Act also set up Industrial Training Boards to determine whether a particular employer belonged to a particular industry and to grant exemption certificates to companies having what they considered to be satisfactory training arrangements. Levies are made at the order of the Secretary of State for Employment, but they must not exceed 1 per cent of the company's payroll and the company is then reimbursed when the training has been carried out. Whether exemption certificates are issued will depend on criteria approved by Manpower Services Commission and the Secretary of State for Employment. The law relating to training in industry is to be found at present in the Employment and Training Act 1973, which set up the Manpower Services Comission and in part modified the Industrial Training Act. In addition the Employment Protection Act 1975 and the regulations relating to Safety Representatives and Safety Committees provide specifically for union officials to have time off with pay for approved training, "approved training" referring to training approved by Manpower Services Commission consisting of representatives from the C.B.I., T.U.C. and educational establishments. Section 57 of the Employment Protection Act 1975, which lays these duties on employers, means that employers are going to have to pay for the time off taken by union officials and safety representatives for training. The actual training will be paid for by the union. Safety training is largely enforced through the inspectorate system. It will be seen in §**28** that, by virtue of s. 2(2), H.A.S.A.W.A., employers who fail to train employees in the hazards of the job are liable to prosecution by the inspectorate, and this is already happening.

Moreover even prior to the passing of H.A.S.A.W.A. in 1974 there were regulations prescribing that certain classes of employees had to be trained and certificated before being adjudged competent to perform their job properly and safely, e.g. the Power Presses Regulations 1965 and the Abrasive Wheels Regulations 1970. These regulations were and still are enforced by the inspectorate under powers given to them in the Factories Act 1961 and now in H.A.S.A.W.A. It would appear from the decision in *Tesco Ltd.* v. *Nattrass* (*see* §**30,** p. 199 in relation to s. 36, H.A.S.A.W.A.) that the implementation of training schemes by companies will help to ensure that in part, if not wholly, employers and directors can offload some of the statutory criminal responsibility. Though how far this applies to H.A.S.A.W.A., as distinct from the Trades Descriptions Act 1968, is doubtful.

Thus with the passing of s. 2(2), H.A.S.A.W.A., more and more employees are going to be trained in industry, with safety being built into the training requirement for the job. This last is important; s. 57, Employment Protection Act 1975 (duty to allow union officials to be trained), plus the requirements of the Employment and Training Act 1973, both consistent with the United Kingdom's obligations under Art. 10 of the European Social Charter, mean that training in industry, including safety and the law relating to safety is now becoming established, and the requirement for safety training under s. 2(2), H.A.S.A.W.A., is merely part of this training, though a crucial part. Management training will normally be carried out at management training centres, whilst the training of union officials and union members will take place at W.E.A. centres, polytechnics and technical colleges. This applies also to safety training, though the facilities of Ro.S.P.A. and the British Safety Council may also be used.

(iii) *Effective supervision.* If the system of work is to be safe, employees must be supervised effectively and competently. This means that the supervisor has got to be trained in his role and responsibilities and supervisory training must be therefore regarded as of the utmost importance in the future. The key to the operation of a safe system of work revolves round good supervision; it will be the supervisor's job to see that employees do their job safely and to instruct the workforce in their obligations under H.A.S.A.W.A. and to see that the safety regulations are complied with, and films and teach-ins by the safety personnel in this respect would be very helpful. Above all it is the role of supervision to see that the law is enforced throughout the workforce, to see that employees wear the necessary safety clothing or eye protection and to see that employees are made aware of the availability of protective clothing and the way in which the law and changes in the law affect them. The effect of not having supervision trained under H.A.S.A.W.A. is prosecution and the effect of not doing so at common law can be even more tragic and costly to the employer.

The leading case, in which many previous authorities were cited is *Bux* v. *Slough Metals Ltd.*, a decision of the Court of Appeal in late 1973, and one of the most important cases of its kind in recent industrial injuries litigation. B, a die-caster, was employed by S.M. on piece-work in their die-casting foundry. His work consisted of melting ingots of aluminium alloy in a furnace, lifting out the molten metal with a ladle and pouring it into a die. No goggles had been provided during his training or for some two months after he started work. They were then supplied and B was told to wear them. He found that they misted up, told his superintendent that they were useless and stopped wearing them. No member of the management sought to induce him to wear them. He had been so employed for a year when some molten metal splashed up and, as a result, he lost the sight of one eye and, to a large extent, the other as well. He claimed damages for personal injuries, alleging firstly breach of reg. 13(1)(*c*), Non-Ferrous Metals (Melting and Founding) Regulations 1962, and second of the common law duty of care by failure on the part of the employer to instruct him to wear goggles and ensure that they were worn. The trial judge ruled that S.M. had "provided ... suitable goggles", within reg. 13(1)(*c*), but that the employer had been negligent at common law. He found that B would have followed instructions and persistent advice to wear the goggles, and that he was guilty of a mere technical breach of reg. 13(4) and assessed his contributory negligence at 20 per cent, awarding the employee £24,000. The employer appealed on the ground that, having complied with reg. 13(1)(*c*), he owed the employee no separate duty at common law to require or persuade him to wear goggles. It was held by the Court of Appeal that:

1. regulations in no way superseded an employer's ordinary common law obligations to care for the safety of his employees and the regulations ought not to be expected to cover every possible kind of danger; accordingly, although the employer had complied with reg. 13(1)(*c*) in providing the employee with suitable goggles, the employee was still entitled to pursue his claim at common law for negligence; and

2. the question whether or not an employer's common law duty of care extended to instructing, persuading or insisting on the use of protective equipment depended on the facts, and there was no reason to disturb the judge's finding that the employee's claim at common law succeeded; but the employee had clearly been in breach of reg. 13(4) by failing to make "full and proper use" of the goggles and his apportionment of blame should be 40 per cent (reg. 13(1) of the Non-Ferrous Metals (Melting and Founding) Regulations 1962, which was replaced on 10th April 1975 by the Protection of Eyes Regulations 1974, provided "There shall be provided and main-

tained suitable protective equipment of the types and for the operations or work respectively hereafter in this regulation specified for the protection of persons employed in any such operations or work, that is to say— . . . (*c*) suitable goggles or other suitable eye protection for persons employed in—(i) work at a furnace where there is risk to the eyes from molten metal; or (ii) pouring or skimming; or (iii) work involving risk to the eyes from hot sand being thrown off").

During the course of giving judgment Edmund Davies, L.J., referred to the judgment of Lord Reid in *Gill* v. *Donald Humberstone & Co. Ltd.* Lord Reid, speaking of the Building (Safety, Health and Welfare) Regulations 1948, said:

"I find it necessary to make some general observations about the interpretation of regulations of this kind. They are addressed to practical people skilled in the particular trade or industry, and their primary purpose is to prevent accidents by prescribing appropriate precautions. Any failure to take prescribed precautions is a criminal offence. The right to compensation which arises when an accident is caused by a breach is a secondary matter. The regulations supplement but in no way supersede the ordinary common law obligations of an employer to care for the safety of his men, and they ought not to be expected to cover every possible kind of danger . . ."

"In these circumstances it would [said the judge in *Bux* v. *Slough Metals Ltd.*], in my judgment, be idle for the employers to urge that they have fulfilled their duty of reasonable care for the safety of their workmen by simply providing them with goggles and then stand by while the men did dangerous work with their eyes wholly unprotected."

Stephenson L.J. in the Court of Appeal observed "The employer must try to make the law of the land a rule of the factory. Easier said than done, but it is the duty of an employer . . . to take reasonable steps to make it a rule. . . . And in this case they were not taken, nor was the plaintiff ever told: 'I am sorry; we have to provide you with suitable goggles or we commit a criminal offence, and what is more, if you don't wear the ones we provide, you commit an offence too and may be fined yourself.'"

(iv) *Planning for specific isolated tasks.* A safe system of work has got to be devised and laid out in the case of single isolated tasks just as in the case of tasks which have to be performed every day. Authority for this point is to be found in *Rees* v. *Cambrian Wagon Works Ltd.* in 1946. An employee who was employed with others in the workshop of C.W.W. started to move, in conjunction with others and under the supervision of the works manager, a heavy iron cog wheel for repair. This cog wheel weighed some 15 cwt., was 4 ft. 2 in. in diameter and had a width of 8 in. So that the cog wheel might be brought to the ground a sufficiently wide plank 2 in. thick was used, a sloping wedge being provided for placing under the end of the plank. This wedge was not wide enough and as a result the plank started to rock, causing the wheel to overbalance. The employee was killed. It was held that there had been a failure by the employers to provide proper plant in the form of a wedge of sufficient width, even though a very unusual and one-off operation was being undertaken; moreover the employers could not delegate this duty to the works manager.

(v) *Careful selection of competent employees.* Care should be taken in recruiting employees and in training and supervising them. Otherwise liability may follow if one employee injures another, and the employer will be vicariously liable. Also if an employee or employees deliberately inflict injury on another employee, this can be attributed to the lack of care of the employer in selecting his workforce. Two cases are frequently cited in this context, *Hudson* v. *Ridge Manufacturing Co. Ltd.* and *Smith* v. *Crossley Bros.* In the former case H was injured at work owing to a practical joke being played upon him by a co-employee about whom it was known by management that he was a practical joker and had been in the habit of indulging his weakness for the past four years. It was held that the employer was liable to H for his injury because, such conduct on the part of the co-employee having occurred before, it was reasonably foreseeable that it might happen again. In *Smith* v. *Crossley Bros.*, however, two apprentices injected compressed air into the anus of a third apprentice, causing him injury. It was held that his employer was not liable, as this sort of behaviour was out of character for the apprentices and had not occurred before: it was not reasonably foreseeable. Whether *Smith* v. *Crossley Bros.* would have been thus decided today is more doubtful, especially in the light of the employer's duty to instruct, train and make sure that there is effective supervision. It is submitted that on those facts today the employer would have been liable for negligence, following *Bux* v. *Slough Metals Ltd.*, for failing to have effective supervision.

(vi) *Ignoring safety warnings and devices.* The case law on this subject, which is extensive, suggests that the employer will have discharged his duty to take reasonable care for his men if he makes available safety equipment and informs his employees where it can be obtained. Moreover he is expected to post safety warnings but is not expected to stand over his experienced workmen exhorting and instructing them to observe the notices, obey the law and wear their safety gear. This at least was the position established by leading cases like *Woods* v. *Durable Suites Ltd.* (1953), *Qualcast Ltd.* v. *Haynes* (1959), *McWilliams* v. *Sir William Arrol Ltd.* (1962) and *James* v. *Hepworth Grandage Ltd.* (1968). There are, however, two leading cases, *Clifford* v. *Challen &*

Sons Ltd. (1951) and *Bux* v. *Slough Metals Ltd.* (1973), which go the other way and suggest that exhortation and instruction are necessary and that the problem may well be related to ineffective supervision, whilst *Boyle* v. *Kodak Ltd.* decides that, although the employee may be in breach of the regulations, if it can be established that the breach was in any way due to the failure of the employer to exhort the employee to follow the regulations and that supervision was ineffective, the damages will be apportioned as in contributory negligence cases. The story starts with *Clifford* v. *Challen & Sons Ltd.* in 1951. An employee suffered dermatitis because water was not available to wash synthetic glue from his hands. It was held that there was no breach of what is now s. 58, Factories Act 1961, since this is essentially a welfare section, but that the employers were liable at common law for negligence for failing to insist on the use of protective barrier cream. Then follows a quartet of cases which suggest that the important thing is to establish that the employer's breach was the real cause of the employee's injury or disease, in other words that the issue is really one of causation. This had been the approach adopted in *Woods* v. *Durable Suites Ltd.*, *Qualcast Ltd.* v. *Haynes*, *McWilliams* v. *Sir William Arrol Ltd.* and *James* v. *Hepworth Grandage Ltd.* In *Woods* v. *Durable Suites Ltd.* in 1953 D.S. employed workmen to spread synthetic glue in their veneer department. To prevent the risk of dermatitis they put up a notice near the workplace outlining the precautions to be taken to avoid dermatitis and instructing the men in the use of barrier cream soap and washing facilities, all of which had been supplied. An experienced workman was specifically instructed by the manager in the protective measures, but unknown to management did not observe the precautions and caught dermatitis. He sued D.S. for not having set up a safe system of work. His action failed, the Court of Appeal taking the view that an employer is not expected to keep a constant eye on mature and experienced employees in order to make sure that they follow safety instructions. Similarly in *Qualcast Ltd.* v. *Haynes* in 1959 an employee of Q was a very experienced moulder. One day whilst casting at the moulding boxes the ladle of molten metal which he was holding slipped, and some of the metal splashed on to his left foot, causing him injury. He was not wearing spats or protective boots though he knew that these were available, spats free of charge and boots at a special price. It was held that the employer was not negligent in failing to operate a safe system of work. The fact was that the employee was an experienced moulder and by making protective clothing available to him the employer had taken reasonable care for his workman's safety, despite the fact that no pressure had been put on the employees to make them wear the spats.

Are these cases examples of causation in negligence? Are they tantamount to the courts saying that the real cause of the workman's injury was not the employer's breach, in that he provided the safety equipment necessary and made this fact known to employees, but the failure of the workman to use the equipment and observe the regulations? Particularly strong is this inference in cases where, as in *Qualcast* and *McWilliams*, the employee is a skilled operative. Certainly the decision in *McWilliams*' case in 1962 leaves little doubt that the issue was causation. These cases are saying that *but for* the employee's failure to look after himself, the injury would not have occurred. *McWilliams*' case goes further and argues that, even where the employer had not provided the necessary safety gear, as required by statute, if the employee would not have worn it, even if it had been provided, the employer is not guilty of negligence. *James* v. *Hepworth Grandage Ltd.* in 1968 reaffirms a similar principle, that the duty at common law is to take reasonable care. Employees were told on notices posted throughout the factory that spats were to hand for protection against metal splashes. It was held that the employer had carried out his common law duty of care to J, even though J, being illiterate, could not read the notice.

Two other cases which are relevant here are *Ross* v. *Associated Portland Cement Manufacturers Ltd.* and *Boyle* v. *Kodak Ltd.* In *Ross's* case in 1964 the relevant facts were that R's deceased husband had been a chargehand steel erector. In conjunction with two other men he was told to repair a wire safety net suspended below an aerial ropeway in the factory of A.P.C.M. This netting which was in a bad state of repair was 22 ft. above ground and supported by brackets held by cables attached to two pylons some distance apart. The steel erector was left to get on with the job in his own way without any instructions and without any proper equipment (which would have been a movable platform). The deceased steel erector was entitled to use any equipment he could find in the factory and he could have asked A.P.C.M.'s chief engineer for instructions or equipment, but he was not encouraged to do so even though he was new to the job and it was an unusual one for a steel erector. At one stage in the work the men decided that the only suitable equipment available so as to be able to carry on was a long ladder in the factory stores. While standing on top of this ladder, which was resting on the wire netting, R was thrown to the ground when the ladder moved forwards and backwards because of the partial collapse of the wire netting. He was killed. R's widow sued the factory occupier under s. 29(1), Factories Act 1961, to which the factory occupier responded that the sole cause of the accident was the steel erector's own negligence in using the ladder. The House of Lords, however, ruled that the factory occupier was two-thirds liable because his failure to provide proper equipment and keep the steel erector's place of work as safe as was reasonably practicable was a dominant cause of the accident. In contrast with the case of *Ginty* v. *Belmont Building Supplies Ltd.* (*see* §**22,** p. 80), suitable alternative appliances were not provided. Section 29(1), Factories Act 1961, provides that there shall, so far as is reasonably practicable, be provided and maintained safe means of access to every place at which any person has at any time to work, and every such place shall, so far as is reasonably practicable, be made and kept safe for any person working there. The expression "any person" has the effect of extending the scope of this duty to persons other than employees, i.e. to independent outside contractors. The dominant cause of this accident, in the opinion of the House of Lords, was the fact that the factory occupier had not carried out his statutory duty under s. 29(1), Factories Act 1961.

This approach was followed in *Boyle* v. *Kodak Ltd.* in 1969, where the relevant facts concerned reg. 29, Building (Safety, Health and Welfare) Regulations 1948. This regulation provided "every ladder shall, so far as practicable, be securely fixed, so that it can move neither from its top nor from its bottom point of rest. If it cannot be so securely fixed, it shall where practicable be securely fixed at the base or if such fixing ... is impracticable, a person shall be stationed at the base of the ladder to prevent slipping". This provision is applicable both to employers and employees. B, who was an experienced painter, placed his ladder against the top of a storage tank in order to start painting. He started up the ladder intent on tying the top of the ladder to the tank. On his way up the ladder collapsed and he fell, injuring himself. Evidence showed that he could have used a stairway made of iron which was attached to the tank to go up and secure the ladder. The House of Lords decided that on the facts both employer and employee were equally liable, since both were equally in breach of statutory duty and certain observations of the court in this case typify an attitude which is much more characteristic of the way in which the courts might deal with such a problem today. The approach of *Woods* v. *Durable Suites Ltd.*, *Qualcast Ltd.* v. *Haynes* and *McWilliams* v. *Sir William Arrol Ltd.* may well now have been superseded by *Ross* and *Boyle*, with the result firstly that both employer and employee can be in breach of duty leading to contributory negligence when the employee sues in respect of his injuries, and second that, despite the fact that the employee is highly skilled and experienced, the employer is not entitled to assume that the employee is aware of the law and the relevant regulations: it is incumbent upon the employer to instruct the operative in his duties and have the workman acquaint himself with the regulations. In *Boyle* the employer was not able to show that he had acquainted B with the regulations, therefore both employer and employee were found equally liable. B got damages, but they were reduced by 50 per cent to account for his own contributory negligence. Lord Diplock put the matter succinctly as follows: "The plaintiff establishes a prima facie cause of action ... by proving the fact of non-compliance with a ... regulation and ... injury as a result. He need prove no more. No burden lies on him to prove what steps should have been taken.... But if the employer can prove that the only act or default of anyone which caused or contributed to the non-compliance was the act or default of the plaintiff himself, he establishes a good defence."

The decisions in *Boyle* v. *Kodak Ltd.* and *Bux* v. *Slough Metals Ltd.* pave the way for the position adopted by H.A.S.A.W.A. and civil actions that may revolve round the interaction of ss. 2 and 7 of that Act, especially the duty of enforcing factory law and regulations through supervision, and the *Bux* case is probably closer to the position that will obtain under H.A.S.A.W.A. than is *Boyle*. The remarks of Stephenson, L.J., in *Bux* reflect the correct legal position at common law and under H.A.S.A.W.A. when he said "The employer must try to make the law of the land a rule of the factory. Easier said than done, but it is the duty of an employer ... to take reasonable steps to make it a rule. And in this case, they were not taken". Echoing the judge's sentiments in that case, it is recommended that if employers wish to keep within the law and keep their employees within it as well the employer should say to his employees, whether skilled or not, "I am sorry but the law says that you must take these measures for your own protection. If you do not do so, you commit a criminal offence and you may lose compensation in the event of your being injured through not complying with the law and wearing the equipment provided. Moreover, if you persist in breaking the law, I shall have no alternative but to dismiss you."

H.A.S.A.W.A. probably lays greater duties upon employers than either *Boyle* or *Bux*, in that s. 2(2) prescribes that all employers instruct, train and supervise their employees in the hazards of the job. This has the practical consequence that, if an employer is in breach of this sub-section, it is difficult to conceive of an employee being guilty of a criminal offence by virtue of a breach of s. 7. There would not have been a breach of s. 7 in those circumstances. Conversely, if the employee had been instructed in the way Stephenson, L.J. mentioned in *Bux*, and if he had been trained in the hazards of his job, and if this had been backed up effectively by supervision and an employee elected to break the law, e.g. by not wearing eye protection, it is submitted he can and should be punished and might lose part of his compensation. Furthermore following the 1977 decision of a Birmingham Industrial Tribunal in *Marsh* v. *Judge International House-*

wares Ltd. the employer would be justified in telling the employee that, if he continues to constitute a health or safety hazard, he will be dismissed from his job, and the dismissal will be lawful within the meaning of the Trade Union and Labour Relations Act 1974. Weight is lent to this approach by observations of Streatfield, J., in *Hudson* v. *Ridge Manufacturing Co. Ltd.*: when he said "If ... a fellow-workman ... by his habitual conduct is likely to prove a source of danger to his fellow-employees, a duty lies fairly and squarely on the employers to remove the source of danger." Since most safety problems in unionised firms can become industrial relations problems, it is to be hoped that the action of safety representatives will encourage their own union members to carry out the law and follow the regulations made for their own safety, without having to face the ultimate sanction of dismissal—the concerted action of safety representatives should serve to make supervision as effective as the court stipulated in *Bux* (*see* p. 97).

(vii) *Apprentices.* Obviously apprentices are less skilled and by implication less safety-conscious than trained and experienced employees. This fact was uppermost in the mind of Lord Denning in the 1969 Court of Appeal decision in *Kerry* v. *Carter*, where his Lordship stated that the employer owes a greater duty of care to apprentices in his workforce than to trained employees. Giving judgment in this case in favour of an injured apprentice Lord Denning reminded the employer that he should, had he been exercising reasonable care, have inquired into the experience of an apprentice in the working of circular saws and should have had him supervised more carefully.

(viii) *Noise.* The first reported case on noise where an employee received compensation for his employer's negligence at common law was *Berry* v. *Stone Manganese Marine Ltd.* in 1972. B became deaf as a result of working over a considerable period of time with pneumatic hammers. These hammers emitted a noise level of between 115 and 120 decibels, the tolerable limit being about 90 decibels. The company provided earplugs, but, as in *Bux*, effective supervision was not available to secure that the earplugs were worn. Nor was the nature of the danger outlined to the workmen. As a result earmuffs were brought in but were straightway rejected by B in conjunction with other employees as being uncomfortable and preventing him from hearing the other noises in the factory. As in *Bux*, the employer did not through supervision inform B and the others of the risks they were running in not wearing the earmuffs. B went completely deaf and the court found the employer liable. More particularly the court felt that, whereas with certain jobs the risk of danger was obvious, e.g. risk of danger to the eyes from molten metal, this was not so with deafness, which was a hidden and insidious condition that came upon an employee slowly over a period of time. For this reason the employers should have gone to much greater lengths than in the case of obvious dangers to apprise the employee of the risk he was running, by circulating propaganda and other didactic literature.

Following this case the Industrial Health Advisory Committee's Sub-Committee on Noise published a report, *Framing Noise Legislation*, based on its examination of the voluntary Code of Practice for Reducing the Exposure of Employed Persons to Noise 1972. This report recommends that H.S.C. propose legislation on the subject of industrial noise, and H.S.C. has already suggested that legislation to limit exposure to occupational noise should be made under H.A.S.A.W.A. to cover everyone at work except domestic servants.

Regulations are to be made under s. 15, H.A.S.A.W.A., consisting of the following six basic provisions:

1. Employers are to carry out surveys if the total amount of sound to which persons are exposed is likely to exceed the equivalent of eight hours' continuous exposure at 90 decibels;

2. where the surveys show that these limits are exceeded or are likely to be exceeded, employers should take all reasonably practicable steps to reduce the noise or the length of exposure or both. If exposure does not fall below the limits, it should be reduced to the greatest reasonably practicable extent;

3. employers should identify and mark all areas where the limits are exceeded or are likely to be exceeded;

4. employers should provide all persons entering or working in these areas with suitable ear protection which would reduce noise by at least the amount by which the limits are exceeded. People should not in general be allowed to enter or work in these areas without wearing the ear protection provided. The regulations should include an approval scheme for ear protectors, to be worked out in consultation with manufacturers and others;

5. employees should wear the ear protection provided when in these areas, co-operate with employers in the application of protective measures and not wilfully interfere with or misuse noise control devices and equipment;

6. machinery designers, manufacturers, importers and suppliers should warn purchasers if their machines are likely to produce hazardous noise when properly used. The exact nature of this requirement should be worked out in consultation with manufacturers and others.

The 1972 voluntary code of practice is also to be approved by H.S.C. under s. 16, H.A.S.A.W.A. This would give quasi-legal status to its guidance on the practical implementation of the regulations and should facilitate the effective enforcement of the new statutory obligations. Moreover, as technology improves in this field, more subordinate legislation or regulations or more approved or voluntary codes, are likely to appear dealing with noise control measures or test standards for particular machines or particular types of machines.

(ix) *Employer's liability*

1. *Residual liability*. Reference has been made to the residual nature of common law, especially the duty to take reasonable care which is the basis of negligence. The effect of this residuum of liability can be like a sting in the tail to an employer who has done his best to carry out the statutory requirements laid upon him, as cases like *Kilgollan* v. *Cooke & Co. Ltd.* and *Bux* v. *Slough Metals Ltd.* show. Indeed, a fair proportion of the statutory offences created by the Factories Act 1961 are consistent with the transition from common law liability to statutory format backed up by criminal sanctions. Matters like lighting, heating, access to and egress from work, freedom from work-induced injuries and/or diseases as well as not so obvious dangers such as attack from criminals and undue exposure to the elements, all qualify as aspects of the general duty to take reasonable care, which every employer owes to his employees at common law. Employees injured whilst, and as a result of, carrying out their employment are almost bound to be compensated, assuming that the employer is insured and that there are no valid defences.

2. *Indirect liability*. Equally important are those situations where an employer is indirectly or vicariously liable for torts and other wrongful acts committed by his employees and outside contractors. Though the number of occasions when an employee will be vicariously liable for the negligence of outside contractors will be fewer and limited to cases of strict liability.

The meaning of vicarious liability

The doctrine of vicarious liability, whereby an employer becomes liable in damages for the torts of his employees committed during and arising out of their employment, was a concession on the part of the judges to practical necessity. The origins of the concept are traceable back to the beginning of the nineteenth century, in other words to the time when the initial effects of the industrial revolution were beginning to create their inventories of human tragedy and personal suffering.

(*a*) *Features of the doctrine*. The important feature of the doctrine is that if one employee injures another employee or some other third party, e.g. an outside contractor or an innocent member of the public, and the injury is committed during the employment (or reasonably incidental thereto) and arises out of the employment, the employer will be vicariously liable for the injury caused by his employee (nor, since *Lister* v. *Romford Ice & Cold Storage Co. Ltd.*, will the insurer be able to exercise a right of indemnity against the employee, *see* p. 70). Vicarious liability is an example of strict liability not joint liability. The injured party may sue both the employer and the negligent employee but the employer is liable because of his status as an employer: the fault of the employee is transferred to the employer. In practice employees are not sued, for the obvious reason that they are not normally in a position to pay the damages required by the law in respect of the injury. Moreover there seems little doubt that the rapid growth of liability insurance, at first voluntary and now compulsory (Employers' Liability (Compulsory Insurance) Act 1969), has ensured the permanence of vicarious liability in the law of tort. Even as early as 1912 in the case of *Lloyd* v. *Grace Smith & Co.*, where the House of Lords made a solicitor liable for the fraudulent acts of his conveyancing clerk, cognisance was taken of the fact that the solicitor was covered by insurance. Although originally, because of the rule of common employment (whereby an employer was not liable for injury to one employee caused by the negligence of another employee), employers were generally only liable for the negligent acts of their employees causing damage or injury to third parties, gradually injuries to fellow-employees as well fell under the umbrella of vicarious liability. The final stage in this transition was the Law Reform (Personal Injuries) Act 1948, which specifically permitted an employee to sue his employer for injuries sustained by a co-employee's negligence, just as if he were not an employee and it is illegal under the Act for employers to attempt, either directly in a contract of employment or apprenticeship or indirectly in a contract embodying a pension scheme, to contract out of this liability; such a clause is null and void.

(*b*) *The negligent act or omission must arise out of and in the course of the employment.*

(i) *Acts arising out of the employment.* Case law has established that there must be a causal connection between the employment and the injury caused to the co-employee or the third party, whether he be an outside contractor or a member of the public. The leading case here is the 1862 appeal case of *Limpus* v. *London General Omnibus Co.*, where L's bus was overturned when the driver of an L.G.O. bus drove across it so as to be first at a bus stop to take all the passengers. L.G.O.'s driver conceded that the act was done intentionally, as there was some animosity between the two drivers and L.G.O. had issued strict instructions to their drivers that they were not to obstruct other buses. It was held, perhaps at first sight surprisingly, that the employer was liable. The fact was that the driver was acting within the scope of his employment when the buses collided and the fact that the employer had issued strict orders against this did not alter the fact. It would seem from this case that the scope of employment cannot be defined by the simple expedient of issuing an express prohibition to an employee, a result which is to the advantage of the employee.

There have been occasions, however, when this approach has not found favour with the courts. A classic example occurred in *Twine* v. *Bean's Express Ltd.* in 1946. B.E. hired out vans to commercial undertakings. The van drivers were to remain employees of B.E. and were strictly prohibited from giving lifts to persons or from carrying passengers. There was a notice displayed in a prominent position in the cabs to this effect. The driver of one of the vans hired to the Post Office gave an employee of the Post Office a lift. Subsequently, owing to the driver's negligent driving, the van was involved in a collision and the employee was killed. His widow sued the employer and the trial judge dismissed the action. An appeal by his widow was also dismissed on the ground that the employer owed no duty of care to her husband. Giving judgment Lord Greene, M.R., observed:

"Was the driver, in giving a lift to the deceased man, acting within the scope of his employment? The answer is clearly, No. He was doing something which he had no right whatsoever to do, and qua the deceased man he was as much on a frolic of his own as if he had been driving somewhere on some amusement of his own quite unauthorised by his employers.... He was in fact doing two things at once. He was driving his van from one place to another by a route which he was properly taking when he ran into the omnibus, and in driving the van he was acting within the scope of his employment. The other thing which he was doing simultaneously was something totally outside the scope of his employment, namely, giving a lift to a person who had no right whatsoever to be there."

The principle in *Twine's* case was followed in *Conway* v. *George Wimpey Ltd.*, where a lift was given to an unauthorised person by an employee of Wimpey contrary to company instructions and the company was found not liable for the injury, though had the injured party been another employee riding in the same vehicle with the negligent employee, the company would have been liable. This much is clear from the decision in *Young* v. *Edward Box* in 1951.

In more recent times there has been a tendency to revive the approach in the *Limpus* case, with the courts being reluctant to allow employers to define the scope of their employee's relationship by means of express prohibitions. In *Century Insurance Co. Ltd.* v. *Northern Ireland Road Transport Board* in 1942 an employee of N.I.R.T.B. was employed to deliver petrol in tankers to garages. One day whilst delivering petrol to a garage in Belfast and whilst the petrol was being transferred from the tanker to an underground tank at the garage, the driver decided to have a smoke. Having lit a match he proceeded to throw it, whilst it was still alight, down by the underground tank. There was an explosion and considerable damage. It was held by the Court of Appeal that the employer was liable because the employee was doing what he was employed to do, deliver petrol. And although he was performing this job in a grossly negligent manner, this did not mean that he was not acting in the course of his employment. The same view was taken by the court in the 1976 decision in *Rose* v. *Plenty*. P was a milk roundsman employed by the Co-op. His employer had taken every conceivable precaution to stop the common but dangerous practice of roundsmen taking on young children and paying them for helping with the delivery of bottles of milk and the collection of empty bottles. One notice at the depot advised "Children must not in any circumstances be employed by you in the performance of your duties". Contrary to this instruction an employee did employ a boy. Whilst moving from one delivery point to another the boy rode on the milk float with one leg dangling over the side so that he could jump off quickly. Meanwhile the employee drove negligently with the result that the boy's leg was trapped and crushed between the wheel and the kerb. The trial judge ruled that the employee was 75 per cent and the boy 25 per cent to blame. On an appeal to the Court of Appeal to determine whether the employer was liable for the boy's injury, it was held that he was. Delivering judgment the court said that in disregarding the prohibition to employ anyone the employee was not stepping outside the ambit of his employment. The employee was doing what he was employed to do but was doing it negligently.

Decisions like these owe much to the existence of employers' and public liability insurance. Furthermore it is difficult to see how much difference s. 7, H.A.S.A.W.A., will make to these situations. One suspects that s. 7 will prove to have had no effect whatsoever on vicarious liability, at least as far as injuries to third parties (other than employees who injure themselves through their own carelessness) are concerned (*see* § **28,** p. 184). Another House of Lords' decision reaffirming this approach is *Westwood* v. *Post Office*, mentioned in § **19,** p. 47. Moreover, in the light of the definition of an employer–employee relationship by Denning, L.J., in *Stevenson, Jordan & Harrison* v. *Macdonald & Evans Ltd.*, the employer may find himself liable for the negligent acts or omissions of skilled employees over whose work he has no control. Thus in *Cassidy* v. *Minister of Health* in 1951 it was held by Lord Justice Denning that the Minister of Health was liable vicariously for the negligence in performing an operation of a full-time assistant medical officer of the hospital, even though the

Minister himself clearly had no control over how the doctor performed his job. The fact was that the doctor was part and parcel of the organisation and so was within the *Stevenson, Jordan & Harrison* rule, with the result that the Minister was ultimately liable. Similarly in *Stokes* v. *G.K.N. Ltd.* in 1968 the employer was held liable in circumstances where an employee died from scrotal cancer because the company doctor had been negligent in failing to advise on the proper safety precautions to be taken in the use and application of machine oil.

(ii) *Can an employee's crime be an act arising out of the course of the employment?* How can a crime committed by an employee occur within the course of a person's employment? It is easy to believe that the commission of a crime takes the matter outside the parameters of the employment contract, but this is not so. Rather the courts take the view that when an employee commits a crime, whether a technical one or not, the employee will be prosecuted, usually by the magistrates, whilst the employer pays for the loss. The case normally cited in this connection is *Morris* v. *Martin & Sons* in 1965, where a firm of fur cleaners was held liable for the loss of a stolen fur, whilst the employee was prosecuted and convicted of theft.

On other occasions statute or regulations provide for the prosecution of the employer in addition to the employee. An example of this occurs in the case of ss. 36 and 37, H.A.S.A.W.A.

(iii) *Acts arising during the course of the employment.* Acts causing injury occurring during working hours or reasonably incidental thereto all qualify for compensation. Reasonably incidental acts or omissions refer to injuries caused by one employee to another whilst changing clothes for work before starting or whilst having a meal in the factory canteen or whilst having a wash and brush up after work has finished. Sports' activities are not deemed to arise out of and in the course of the employment. This is the result of the decision in *R.* v. *National Insurance Commissioner*, ex parte *Michael*, where the National Insurance Commissioner ruled that a young policeman who was injured whilst playing football for his force's team could not qualify for industrial injuries' benefit, because he was not employed to play football even though he was encouraged by his superiors to do so. Although this case concerns state benefit and occurred in connection with the Social Security Act 1975, a similar interpretation would arise in a common law claim against the employer for negligence. Two cases illustrating the scope of the expression "acts arising in the course of the employment" for the purposes of vicarious liability are *Beard* v. *London General Omnibus Co.* in 1900 and *Iqbal* v. *London Transport Executive* in 1973. In *Beard's* case a bus conductor who turned a bus round when the driver was absent and injured B in the process was held by the

Appeal Court to be acting outside the scope of his employment. Similarly in *Iqbal's* case in 1973 a bus conductor drove a bus in the depot so as to get it back on the road and it was held that the employer was not liable, the court saying that it was not part of a bus conductor's job to drive a bus. If he elected to do so and injury occurred as in *Iqbal*, the employer would not be liable. In cases like this the employee would be faced with paying the damages in respect of the injury.

(*c*) *Quarrels at work*. The situations in which an employer is held vicariously liable for the conduct of employees concern acts of negligence on their part. Sometimes, however, the acts are provoked and intentional, e.g. assault. Here an employee who suffered a gratuitous assault at the hands of another employee would not be able to sue the employer for his injury, since the incident leading to the assault and the assault itself would be regarded as an act outside the course of employment. Though, in the light of social security cases involving industrial injury benefit, if the quarrel were associated with and arose out of the nature of the employment, e.g. an altercation about the feasibility of a project leading to assault, it is arguable that the injury arose out of and in the course of employment. A similar example might occur where a safety officer advised that an employee wear safety equipment and the employee took exception, this leading to a heated exchange followed by fisticuffs; it is suspected that the safety officer could sue his employer for damages. In *Warren* v. *Henlys Ltd.*, however, where an employee got into a heated argument with a customer over the price of petrol and assaulted him, it was held that the employer was not liable.

Employer's liability in tort for the acts of independent contractors

It is not general for an employer to be liable for the torts, particularly negligence, of independent contractors whom he appoints. Rather the outside contractor is liable for his own acts of negligence and for those of his employees. In situations where an employer could be liable for the negligence of or damage caused by an outside contractor, he should extract a contractual indemnity from the oncoming contractor. Five situations are normally relevant:

(*a*) where the employer ratifies the tort of the contractor, e.g. a manufacturer supplies safety equipment which does not meet the requirements of an improvement notice and the employer agrees to use the equipment with the result that an employee is injured;

(*b*) where the employer does not use reasonable care in selecting a competent contractor, for example he fails to make inquiries of the contractor or uses a contractor with a bad reputation when a contractor with a good reputation was available;

(*c*) where the liability for certain acts is strict and does not depend on the exercise of reasonable care, e.g. *Rylands* v. *Fletcher* (*see* § **54**, p. 259);

(*d*) where the work to be done by the contractor is intrinsically dangerous or is being performed on or near a highway. In the case of *Tarry* v. *Ashton* in 1876 an employer was liable for faulty work done by a contractor when a lamp overhanging the highway, which he was engaged to repair, fell and injured a passer-by on the highway. It was held that the employer not the contractor was liable for the injury. Nevertheless it is clear from the case of *Padbury* v. *Holliday & Greenwood Ltd.* that the employer would never be liable for *collateral* acts of negligence of the contractor or his employees, e.g. a tool being used in the repair of the lamp in *Tarry* v. *Ashton* falling and injuring a passer-by, would have been the responsibility of the contractor;

(*e*) where a statute makes an employer liable for the fault of an outside contractor, e.g. the Employers' Liability (Defective Equipment) Act 1969, or certain sections of the Factories Act 1961. In *Hosking* v. *de Havilland Ltd.* in 1949 a building contractor dug a ditch in the grounds of a factory belonging to the defendant and bridged the ditch with an unsound plank, as a result of which the plaintiff was injured whilst walking on the plank. It was held that the factory occupier was liable under what is now s. 29(1), Factories Act 1961, but was entitled to exercise his right of indemnity against the builder.

In all these cases where the employer/factory occupier is made liable for the torts of others, he should always endeavour to extract indemnity from the contractor and prevail upon the contractor to effect insurance for these eventualities.

Employers' Liability (Compulsory Insurance) Act 1969

Section 1 of this Act states that every employer, unless exempted, carrying on any business in Great Britain shall insure and maintain insurance under one or more approved policies with an authorised insurer or insurers against liability for bodily injury or disease sustained by his employees arising out of and in the course of their employment in Great Britain. The purpose of this insurance, which is compulsory, is to protect the employer against claims for negligence, both personal and vicarious, by his employees and to make sure that an injured employee receives compensation. This was considered most important by the 1966 Labour government responsible for the Act, since certain employers were unable to meet claims as damages increased and the scope of liability widened. Moreover the effect of devastating claims for personal injuries was to drive smaller employers into liquidation or bankruptcy. To ensure that personal injury claims were met in cases where employers were negligent, the government felt compelled to make this sort of insurance compulsory. Doubtless this has been a salutary measure welcomed by employers, employees and insurance companies. It came into force on 1st January 1972.

The Act states that most employers must have employer's liability insurance up to at least £2 m pounds so as to cover claims made against them. Certain of the big employers—local authorities, the police, the nationalised industries—are exempt from these provisions. Nor is insurance compulsory for domestic servants, independent contractors and certain closely related employees. Even so employers would do well to insure in respect of such persons, particularly contractors, under a public liability insurance scheme. The Act requires that every employer display a certificate of insurance for the benefit of employees. The cost of this insurance is often expressed as a percentage of the wages and salaries bill, the percentage varying with the cost of the risk involved in the particular type of business: businesses employing office and clerical staff would pay a lower premium than businesses employing factory workers operating machines. Failure to comply with the Act can lead to prosecution before the magistrates and, by virtue of the Criminal Law Act 1977 (Commencement No. 1) Order which came into force on 8th September 1977, the magistrates can order a maximum daily penalty of £500. It is seriously believed that the near universality of this insurance has led to the concept of reasonable care becoming synonymous in many work injury cases with strict liability and to damages being paid out in cases where, in the absence of such legislation, claims might not have been successful or the amount of damages might have been reduced, as in *Westwood* v. *Post Office* in 1974.

Employers' Liability (Defective Equipment) Act 1969

This is a short piece of legislation and was passed principally to repeal the decision in *Davie* v. *New Merton Board Mills Ltd.* in 1956. D was an employee of N.M.B.M. and during the course of his employment was injured when a drift broke and a piece flew off, striking him in the eye and putting it out. The drift was supplied by B, a reputable company who had obtained the supply of drifts from toolmakers in Sheffield. It was held by the House of Lords that the employer had discharged his common law duty to take reasonable care of his employees by having delegated the duty to provide safe plant and equipment and select the same with care, to a reputedly competent manufacturer or supplier.

Although many saw this decision as consistent with the general theory of fault in English law, the 1966 Labour government felt that unfortunate consequences might follow from this case. In particular it meant that the employee had no redress against his employer but had to proceed against the supplier or manufacturer and it was felt that this undermined the intimacy of the employer–employee relationship. Moreover the employee's task might be made that much more difficult if the manufacturer was based outside this country or if he had gone into liquidation or ceased trading. The Act therefore states in s. 1(1):

"Where ... (*a*) an employee suffers personal injury in the course of his employment in consequence of a defect in equipment provided by his employer for the purposes of the employer's business; and

(*b*) the defect is attributable wholly or partly to the fault of a third party (whether identified or not), the injury shall be deemed to be also attributable to negligence on the part of the employer (whether or not he is liable in respect of the injury apart from this sub-section), but without prejudice to the law relating to contributory negligence and to any remedy by way of contribution or in contract or otherwise which is available to the employer in respect of the injury."

In the Act "fault" means negligence; "personal injury" refers to the impairment of a person's physical or mental condition as well as any disease or loss of life. "Equipment" includes any plant and machinery, vehicle, aircraft and clothing. There is no distinction between machines and tools. Whether, however, products manufactured in the factory, as distinct from equipment used in the manufacturing process, is covered by the expression "equipment" in the Act is more doubtful. In *Parvin* v. *Morton Machines Co. Ltd.* in 1952 it was held that the fencing provisions of the Factories Act 1937 were not applicable to machines manufactured in the factory as distinct from machines used in the manufacturing process. If in a subsequent action by the employer against the manufacturer it is found that the defect was one of manufacture in respect of which the manufacturer was negligent, the Unfair Contract Terms Act 1977 would prevent the manufacturer from excluding or limiting his liability to the employer or user in the contract of sale or supply.

The denial of a product liability regime

In retrospect, the effect of the reversal of the *Davie* case by the Labour government's legislation of 1969 has been to put into cold storage the introduction of a system of product liability comparable to that in North America. This is not necessarily a bad thing, in that it is well known that the American system is beginning to bulge at the seams as a result of overburdening the manufacturer with liability. Serious commercial and underwriting problems have manifested themselves and some manufacturers are being forced to go out of business or reduce the number of commercial lines they operate (*see* § **46,** p. 246). In North America the position is the reverse of that in the U.K. with the manufacturer carrying the can for defects in his products whereas in England, owing to the Employers' Liability (Defective Equipment) Act 1969, the employer, as far as injuries to his employees are concerned, has had to carry the can for a defect in manufacture. Doubtless this imbalance will be corrected if the U.K. legal system implements the E.E.C. Draft Directive or the Strasbourg Convention on Product Liability, which appeared in final form in January 1977 (*see* § **47,** p. 246). But until this happens, employees injured by defective equipment should claim against the user/employer not against the manufacturer or supplier. Pearson has recommended the introduction of a system of strict liability for product defects causing injury.

The liability of occupiers

It is necessary to deal with occupiers and their liability in the context of negligence because the Occupiers' Liability Act 1957 is generally supposed to have amended the common law relating to occupiers. The liability of occupiers is based on negligence and insurance against claims is effected by means of a public liability policy, except in the case of employees. The term "occupier" is comprehensive and refers to most persons, individual and/or corporate, who are in possession or control of premises whether movable or not. Occupiers are often subject to further liability in addition to that imposed by the Occupiers' Liability Act 1957, which is often strict. A factory occupier has duties towards his employees and less often towards outside contractors which are strict (i.e. he is liable irrespective of fault); also under the Occupiers' Liability Act 1957 the factory occupier has a duty to take reasonable care to see that his visitors are safe from injury arising from structural defects in the premises occupied by him and from operations carried on on the premises. This Act assumed the existence of common law. At common law an occupier was expected to protect invitees (e.g. workmen going on to his premises) from *unusual* dangers. This would include protecting workmen from dangers not associated with their job, as *Indermaur* v. *Dames* decided in 1866. An employee gasfitter went to test some gas-burners in a sugar refinery and fell into a shaft that had not been fenced and was injured. The factory occupiers were held liable and the court issued the following warning to all factory occupiers and other occupiers as well: "With respect to such a visitor, we consider it settled law that he, using reasonable care on his part for his own safety, is entitled to expect that the occupier shall on his part use reasonable care to prevent damage from unusual danger, which he knows or ought to know. . . ." In other words it is the duty of the employer to instruct the employee in dangers associated with his work on the premises of the occupier, whereas it is the occupier's duty to warn and take steps to eliminate or minimise sources of danger which the oncoming workman would not be expected to know of and which the occupier either actually knew of or ought, with the exercise of reasonable care, to have known about. Being bound to exercise reasonable care and not being impressed with strict liability, an occupier is not deemed to have knowledge of hidden or latent dangers on his premises, unless he possessed particular or special knowledge in that particular area. For instance, an occupier would not be expected to know of a faulty electrical circuit or the intrinsic hazards of poor scaffolding, unless he was either a skilled electrician or a scaffolder.

(*a*) *Occupiers' Liability Act 1957*. Common law was divisive in its treatment of lawful visitors, observing the distinction between invitees and licensees. At common law occupiers were liable to invitees for unusual dangers of which they actually knew or ought reasonably to have known. In the case of licensees, they were only liable for dangers causing injury which they actually knew about. The distinction was justified by the argument that the invitee was invited on to the premises having an interest in common with the occupier, e.g. a business associate, whereas a licensee was on premises merely because common law or equity or statute permitted him to be there, e.g. a factory inspector or policeman. The 1957 Act swept away this distinction so that now occupiers owe the same duty to both categories of lawful visitors. The Act gave no protection to trespassers. As regards injuries suffered on the premises of another, their position is governed by common law, the exact state of which is uncertain, except that the occupier does not owe any duty to trespassers as regards the structural state of the premises—they take the premises as they find them; but as regards activities and operations being carried on on the premises, the occupiers must act in a humane way, as explained in *British Railways Board* v. *Herrington* (*see* p. 119).

(*b*) *Who is an occupier?* Occupation is based on control, which need not be exclusive to one occupier, as recent case histories have shown. In this way there can be several occupiers of one set of premises. In *Wheat* v. *Lacon* in 1966 the appellant's husband was killed when he fell down some steps at a public house managed by X and owned by the respondents. During the course of giving judgment the House of Lords observed that both X and the respondent were in occupation in that X was managing the premises and so had some say in the running of the premises, while the respondent who permitted X to take in paying guests in the private part of the premises defined the scope of X's managing activity. There existed between X and the respondent the relationship of licensee and licensor. It followed that both exercised some degree of control and that the respondent along with X had been correctly sued (on a technical point both were held not liable). Again in *Moloney* v. *Lambeth B.C.*, where a four-year-old child was injured by falling through a gap in the balustrade of a stairway in a block of council flats, the court ruled that, although not in actual physical occupation, the local authority was as much in control of the common staircase as the tenants in actual occupation.

Multiple occupation will have important repercussions for building and civil engineering and the law of health and safety relating thereto. *Wheat's* case has established that both licensor and licensee can be in joint control of the premises, e.g. a building site. *London Borough of Hounslow* v. *Twickenham Garden Developments Ltd.* has established that the relationship between building owner and main contractor is also that of licensor and licensee, and thus case law would seem to indicate that under s. 4, H.A.S.A.W.A. (laying duties on persons other than employers in control of premises), building owners and contractors, including main and sub-contractors, all being in control to some extent, can be prosecuted and can be liable for damages in the event of a lawful visitor, e.g. a factory inspector, being injured. In answer to those who allege that the building owner cannot be in occupation it is worth recalling that under most forms of building agreements he is the person entitled to call for indemnity, thus ensuring indirectly that the contractor and sub-contractor insure in respect of injury to persons and damage to property. This fact led Lord Atkin in *Smith* v. *Cammell Laird Ltd.* in 1940 to observe "the occupiers of a shipbuilding yard have complete control over the yard. . . . They can . . . terminate the licence to enter the premises . . . and can take any indemnities they think fit." It follows that a building owner who wishes to escape the consequences of a claim for damages along with the contractor should extract indemnity from the contractor, which is normally the position with respect to the standard form contracts. Moreover by virtue of subs. 4(3), H.A.S.A.W.A., he will be able to escape prosecution if he entered into a lease with another, a condition of which stated that the other should maintain or repair the premises and keep them free from hazards. This would apply in the case of a local authority letting out factory units for occupation. Building owners, main contractors and sub-contractors are subject to liability under the Occupiers' Liability Act 1957 in respect of injury to persons or damage to property and under H.A.S.A.W.A. they may be prosecuted if they have failed to take such measures as are reasonably practicable to make the premises safe.

(*c*) *What are premises?* It is clear from the Occupiers' Liability Act 1957 that the expression "premises" extends beyond real property and land and buildings thereon, though that must be regarded as its cardinal meaning. The Act states that movable structures, including vessels, vehicles or aircraft, are included. By virtue of this extended definition an employee of a sub-contractor who whilst giving a lift in a company van injures a local policeman investigating thefts on building sites renders the main contractor liable as occupier under the Occupiers' Liability Act 1957 to the injured policeman; and at the same time renders his immediate employer, the sub-contractor, liable for having failed to implement and supervise the implementation of a safe system of work. It is on account of this enlarged definition of "premises" that the giving of lifts by employees of the main or sub-contractor can involve the main contractor in liability as an occupier, since the vehicle is a structure of a movable nature covered by the Act, and the person to whom the lift is given can hardly be classed as a trespasser, having been invited into the vehicle. On this basis, *Twine* v. *Bean's Express Ltd.* and *Conway* v. *George Wimpey Ltd.* must be regarded as bad law.

(*d*) *The occupier owes a common duty of care to all lawful visitors.* Lawful visitors are invitees and licensees, the difference being that *invitees* are persons having an interest, often pecuniary or material, in common with the occupier. A foreign delegation on a visit to a British factory or an employee of a firm of suppliers delivering materials to a building site might well qualify as invitees. Clients visiting offices and customers visiting shops to purchase goods and equipment are all prima facie invitees and are entitled to the protection of the Occupiers' Liability Act 1957 and H.A.S.A.W.A. as being within the scope of the obligation laid upon occupiers of buildings as regards persons other than their employees, e.g. the general public. By virtue of subs. 4(2), H.A.S.A.W.A., occupiers of premises in which people work owe a duty to the general public to see that the premises are as safe as is reasonably practicable. In this way the lessee of a supermarket (and the lessor as well subject to s. 4(3) and any modification of the position which may have been agreed in the lease) becomes liable to prosecution under s. 33 if members of the public using the premises are injured as a result of the premises not being safe. In addition there will be liability at common law by virtue of s. 47, H.A.S.A.W.A., and under the Occupiers' Liability Act 1957.

Licensees are on the premises because they are permitted to be there by statute or regulation or bye-law, hence a factory inspector is a licensee and entitled to the protection of the Occupiers' Liability Act 1957 as a lawful visitor and H.A.S.A.W.A. in cases where persons are working in premises visited by him.

The Occupiers' Liability Act 1957 states that the occupier should take reasonable care to see that his lawful visitors will not be injured by the structural state of the premises and from any operations being carried on on the premises about which he knew or ought to have known (for the protection afforded by s. 4, H.A.S.A.W.A., *see* §**28**, p. 180). The duty of an occupier under the Occupiers' Liability Act 1957 is the common duty of care because it is owed equally to both invitees and licensees, whereas prior to the Act this was not so.

(*e*) *Can the occupier contract out?* Under the Occupiers' Liability Act 1957 it is permissible for an occupier to contract out of liability under the Act, though the Act states that a notice without more shall not have the effect of absolving him from his duties under the Act. Nevertheless it is well established that the display of a notice warning of dangers on the premises and erected in a prominent position, so that reasonable people would see and read it, may be construed as the defence of contributory negligence or *volenti non fit injuria*, the latter being tantamount to the injured person having agreed to run the risk of injury. The number of cases, however, in which the courts have shown themselves ready to regard a notice as absolving the occupier completely (*volenti non fit injuria* being a complete defence, unlike contributory negligence which is merely partial) are few, and the likelihood of their doing so where the visitor visits the premises during the course of his employment is even more remote. This is illustrated by the decision of the Court of Appeal in 1973 in *Burnett* v. *British Waterways Board*, where B, a lighterman employed by X, was injured by the admitted negligence of B.W.B.'s employees when they were moving his barge from the river into the dock belonging to B.W.B. When the accident happened the barge was still in the tidal way a small way from the dock gates; the rope supplied by B.W.B., which was being used to pull in the barge, was defective. The rope snapped, knocked over B and injured him. B.W.B. argued that, in spite of their admitted negligence, they were absolved from liability by a notice displayed outside B.W.B.'s office which said "Lightermen ... availing themselves of the facilities and assistance of the servants of the B.W.B. do so at their own risk and upon the understanding that no liability whatsoever shall attach to the B.W.B. or its servants for any loss, damage or injury from whatever cause arising to the craft or to any person on board...." During his long period of employment as a lighterman B had become aware of and had read the notice. On appeal to the Court of Appeal it was held that the occupier failed in his appeal and that the notice was not binding on the workman. Because the workman had been sent there by his employer he had been compelled to accept the "facilities". He had no freedom of choice in the matter and could not have agreed to the terms of the notice, thus the defence of *volenti* was not available to the occupier. As was pointed out in *Burnett's* case and also in *White* v. *Blackmore*, the negligent act may not have been committed or the negligent equipment or structure may not even be on the occupier's premises when the plaintiff reads the notice. The existence of the notice cannot therefore be regarded as absolving the occupier from future acts of negligence or from the existence of future dangerous structures. Indeed notices cannot any longer be used to exclude or limit liability of occupiers, since this is prohibited by the Unfair Contract Terms Act 1977.

(*f*) *Do occupiers owe duties to trespassers?* It is necessary to define a trespasser since the status of such a person is not always obvious. A person is a trespasser who

(i) goes on premises, whether movable or fixed, without being invited or permitted, the premises not belonging to him; or

(ii) remains on premises after the invitation or permission to be there has expired; or

(iii) deposits goods or articles on the land or premises without authorisation.

A person can be a trespasser right from the start or become a trespasser after being a lawful visitor. If a person is invited or permitted to enter premises for the the purposes of business or work and goes to a part of the premises which is private and which has no association with his business or work, he becomes a trespasser and must take the premises as he finds them. This at least was the position until fairly recently when two cases were decided which placed greater duties of care upon occupiers, including factory and office occupiers, towards trespassers, particularly where the trespasser was an employee. In *British Railways Board* v. *Herrington* in 1972 (a case not concerning an employee) B.R.B.'s electrified railway line ran close to a field where children often played and which was owned by the National Trust. Separating the field from the railway was a fence, but a part of the fence had been broken down and people had been in the habit of using the gap in the fence as a short cut across the railway line. B.R.B.'s employees were aware for some weeks beforehand that the fence needed to be repaired but nothing had been done. A six-year-old boy ran through the gap in the fence and was seriously injured when he came into contact with the live rail. It was held by the House of Lords that the boy was a trespasser but that the occupier owed him a duty to take steps to avoid injury to him which common sense and common humanity demanded. B.R.B. was in breach of that duty and was liable. During the course of giving judgment Lord Reid said:

"So the question whether an occupier is liable in respect of an accident to a trespasser on his land would depend on whether a conscientious humane man with his knowledge, skill and resources could reasonably have been expected to have done or refrained from doing before the accident something which would have avoided it. If he knew before the accident that there was a substantial probability that trespassers would come I think that most people would regard as culpable failure to give any thought to their safety. He might often reasonably think, weighing the seriousness of the danger and the degree of likelihood of trespassers coming against the burden he would have to incur in preventing their entry or making his premises safe, or curtailing his own activities on his land, that he could not fairly be expected to do anything. But if he could at small trouble and expense take some effective action, again I think that most people would think it inhumane and culpable not to do that.... It would follow that an impecunious occupier with little assistance at hand would often be excused from doing something which a large organisation with ample staff would be expected to do...."

The Pearson Commission has recommended that the law relating to occupiers should remain unchanged, save for the addition of protection to be extended to the "uninvited entrant".

This liberal attitude of the courts towards the safety of children who are technically trespassers, understandable no doubt, is also reflected in their treatment of employees who are injured when doing something which they have not been authorised to do; in other words when acting outside the scope of their employment. A case illustrating this point is *Uddin* v. *Associated Portland Cement Manufacturers Ltd.* A Pakistani was employed as a machine minder by A.P.C.M. One day in an effort to retrieve a pigeon he leaned across a revolving shaft in a place where he was not supposed to be and was injured, losing the use of an arm. His employment related solely to the cement packing plant, but his injury occurred as a result of coming into contact with the shaft which was part of the dust extracting plant. He brought an action for damages under s. 14(1), Factories Act 1937 (now s. 14(1), Factories Act 1961), alleging that the shaft was a dangerous part of machinery and required to be fenced. It was held by the Court of Appeal that the employer was liable, and although the employee was doing something which was wholly unauthorised in a place where he should not have been, he was not entirely precluded from recovering damages. The employee lost 80 per cent of his damages.

Employers and factory occupiers, already smarting from the effect of decisions like *Uddin* in 1965, must have been mortified by the decision of the House of Lords in *Westwood* v. *Post Office* in 1974 to award damages to the widow of a deceased employee killed at work but whilst patently trespassing (*see* § **19**, p. 47 for the facts of *Westwood*'s case). The likelihood of the *Westwood* principle being resurrected in the wake of H.A.S.A.W.A. must, however, be regarded as remote, in that s. 2(2) of that Act now requires employers to train their employees in the hazards of their job and this must include the hazards of trespassing. If, therefore, facts similar to those in *Westwood* recurred today, one would expect to see that the employee had been made aware of the job hazards and that if he had been so informed and chose to ignore the instruction and training, he might lose part of his compensation. This could certainly be the effect of s. 7, H.A.S.A.W.A., in the light of *I.C.I. Ltd.* v. *Shatwell* (discussed in § **22,** p. 127).

22. Fatal injuries at work

The law governing recovery of compensation by dependants for fatal injuries suffered at work by the breadwinner is dealt with by two Acts of Parliament, the Fatal Accidents Act 1976 and the Law Reform (Miscellaneous Provisions) Act 1934. The spheres of operation of these two Acts is different, yet at one time no civil action could be brought in respect of the death of a person whether happening at work or elsewhere. Death itself did not give rise to a cause of action nor did an action which the deceased person might have had in respect of his injuries, e.g. negligence or breach of statutory duty, survive his death. The first principle was made clear in *Baker* v. *Bolton* in 1818. This led to a belief that it was cheaper to kill than to maim. Gradually, however, the law relating to compensation for fatal injuries began to assume a more charitable complexion and the first Act of Parliament making provision for compensation in such cases was passed in 1846, Lord Campbell's Fatal Accidents Act.

Death creates two problems in law:

(*a*) does death itself give dependants a right of action for premature deprivation? and

(*b*) do actions which the deceased employee might have had survive for the benefit of others?

Parliament addressed itself first to death itself as a cause of action and whether a dependant, usually a widow, should be compensated for the death at work of the breadwinner. The various Fatal Accidents Acts passed since 1846 and culminating in the Act of 1976 have allowed dependents to claim in respect of the loss of the breadwinner caused by a fatal injury, often at work.

The second problem related to death at work (and elsewhere), whether others could inherit actions the deceased would or might have brought had he lived, was not dealt with by Parliament until 1934 when it passed the Law Reform (Miscellaneous Provisions) Act.

Fatal Accidents Act 1976

This Act came into force on 1st September 1976. Section 1(1) states that if death is caused by any wrongful act, neglect or default which is such as would, if death had not ensued, have entitled the person injured to maintain an action and recover damages in respect thereof, the person who would have been liable if death had not ensued shall be liable to an action for damages, notwithstanding the death of the person injured. Section 1 (2) states that every such action shall be for the benefit of the dependants of the person whose death has been so caused. And by virtue of s. 1 (3) a "dependant" refers to either a wife, husband, parent, grandparent, child, grandchild, brother, sister, uncle, aunt, or issue of any brother, sister or uncle or aunt. Reference in the Act to "injury" includes disease (e.g. scrotal cancer) and impairment of a person's physical or mental condition. Thus this provision coupled with the Limitation Act 1975 confirms the view that, where employees contract illnesses through their work which later prove fatal, the fatal nature of their condition not manifesting itself until years later, e.g. mesothelioma, dependants are entitled to recover. As with all the previous Fatal Accidents Acts the basis of the action for recovery under the 1976 Act is a purely financial one reflecting the loss caused to the dependant by the fact that the breadwinner is dead and can work no more. There is no room for sentiment. It is an action for cash, and feelings of grief and despondency which the bereaved dependant may have are not recoverable. It is not solace compensation. The amount of compensation the dependant will get is commensurate with the amount the breadwinner would have contributed by way of maintenance of the family with deductions for tax.

It is evident that if the defendant is the deceased workman's employer, the employer may raise any defence which he could have raised against the deceased, e.g. contributory negligence or possibly *volenti*. An important addition was made to this branch of the law in favour of the bereaved widow by s. 3 of the 1976 Act, re-enacting s. 4, Law Reform (Miscellaneous Provisions) Act 1971, which stated in subs. 2 that in assessing damages payable to a widow in respect of the death of her husband in an action under this Act, there shall *not* be taken into account the remarriage of the widow or her remarriage prospects. This had always happened before 1971. In this way a young attractive widow could by remarrying a wealthy man become very wealthy. On the other hand in similar circumstances, save that there is no wrongful act, another woman bereaved will only be entitled to state benefit and, if under 40, will not be entitled to any pension at all.

Law Reform (Miscellaneous Provisions) Act 1934

This Act provides that on the death of a person all causes of action vested in him survive for the benefit of his estate. Thus this legislation does not create any new causes of action where none existed before but rather allows a deceased person's estate to bring an action in the case of a workman injured at work who has died from his injuries. Normally dependants would benefit from this Act, but the basis of the claim is not dependancy, as in the Fatal Accidents Act 1976, but rather the action survives for the benefit of the deceased workman's estate. In this way the person to whom he had willed his estate, e.g. a mistress, would benefit from the Act. If the persons are the same in both cases, e.g. a widow, and she elects to sue the deceased's employer under both Acts, as she may, any damages under one must be taken into account when damages are awarded under the other. If the nature of the action which the deceased workman had against his employer was for negligence or breach of statutory duty or both, the employer can raise the usual defences such as contributory negligence and possibly *volenti*. Since negligence is a compound of three inter-related ingredients—duty of care, breach and damage—it may happen that the damage does not manifest itself until after the death of the employer. Where this is so, the injured workman's dependants can still sue the representatives of the deceased employer, though they must do this within six months of probate being granted or letters of administration being taken out. If the damage did not manifest itself until after the expiry of six months of probate or letters of administration, the action becomes time-barred. It is thought that damages might be greater under the 1934 Act, as here they are recoverable for pain and suffering and medical expenses coupled with loss of earnings up to the date of death. Normally a widow who becomes an executrix or administratrix on her husband's death lodges an action under both Acts, one in respect of the estate under the 1934 Act and the other in respect of herself and the family under the 1976 Act. If the deceased had started to bring an action before it became time-barred but then died from his injuries or disease, his dependants or personal representatives have a further three years in which to sue.

Congenital Disabilities (Civil Liability) Act 1976

This Bill received the Royal Assent on 22nd July 1976. It is an interim measure to afford compensation to persons who would not otherwise be compensated pending the implementation of the recommendations of the Royal Commission on Civil Liability for Compensation and Personal Injury, the Pearson Commission (*see* § **79**, p. 288), and even as an interim measure the Act must be welcome. When the Bill began its passage through Parliament its object was to provide financial relief to disabled children who suffer damage before birth and the inspiration behind the Act was the case of the thalidomide children. In the course of time the parameters of the Act have widened and amplified the scope of an employer's liability. The Act says that a child who is born disabled as a result of any breach of duty to either of its parents can claim damages against the tortfeasor, e.g. the employer of the parents or one of them. The child stands in the shoes of the parent.

As far as employers are concerned, every duty existing at common law or under statute which is owed to an employee automatically devolves on the child of that employee. If that child is born disabled, even if the parent suffered no injury, and even if many years elapse before the child goes to law and even if the employer did not know or could not know of the child's existence, the employer can find himself liable to the child for actions which took place before the child was conceived. By virtue of the Limitation Act 1975 coupled with the fact of minority a period of 21 years may elapse before the claim is made against the employer.

(*a*) *The Act's provisions.* The effect of s. 1 (2) of the Act is that:

(i) if an occurrence takes place which affects

1. either parent of a child in his or her ability to have a normal healthy child; or

2. the mother during pregnancy or birth so that the child is born with disabilities that otherwise would not have been present; or

3. the child itself during the course of its birth;

a cause of action may arise on behalf of the child.

By virtue of s. 1 (3) the person answerable for the occurrence is liable to the child if he would have been answerable to the parent. By the same section it is no answer that the parent suffered no personal damage, once a tort has been established linking either parent with an employer.

Assuming that the child is born alive (whether or not the occurrence took place before or after conception is not important), if the claim arises out of an occurrence taking place before conception, it is a defence to the employer that either parent knew of the risk of their child being born disabled at the time of the occurrence and voluntarily assumed that risk. This is made clear in s. 1 (4). Section 1 (6) says that liability to the child may be excluded or modified by a contract made with the parent affected; in other words, such contracts will exclude or limit the child's claim. Finally s. 1 (7) indicates that the parent's own contributory negligence will have the effect of reducing the child's damages.

(*b*) *Reason for the Act.* Common law gave no unified answer to rights for children in respect of antenatal injuries. Nor did statutes designed to protect particular classes of persons, e.g. the Factories Act 1961 and regulations passed thereunder, e.g. the Ionising Radiations (Sealed Sources) Regulations 1969 (though protecting "employed persons" including pregnant women) give entitlement to damages to children born deformed and disabled as a result of breach of statute or regulation. To meet this situation the Law Commission drafted and passed a new statute to await the consideration of the Pearson

Commission, but in the light of the outcry regarding the thalidomide tragedy its birth was expedited.

(*c*) *Relevance of the act to health and safety*. The scope of this Act is enormous. In the case of H.A.S.A.W.A. and regulations made thereunder, any duty or regulation which relates to the health, safety and welfare of an employee brings with it the possibility of further action at a later date by an unborn child. This is particularly true in the case of a female employee. What the child will have to establish is that its mother (or more rarely the father) had a good action in negligence or for breach of statutory duty against the employer. The fact that the mother herself suffered no injury is irrevelant. Situations such as those involving vinyl chloride are clearly envisaged here where a female employee who is pregnant is exposed to such a substance but suffers no ill effects nor is off work; the employer may find that 20 years later he has an action on his hands brought by her disabled child. Certainly breaches by the factory occupier/employer of the safety provisions of the Factories Act 1961, and to a lesser extent the health and welfare provisions, may give rise to actions for prenatal injuries, assuming that damage has been caused by the breach. Furthermore actions for breach of the general duties of H.A.S.A.W.A. which are brought at common law and for breach of any regulations passed pursuant to H.A.S.A.W.A. (which will carry civil liability unless they state to the contrary: s. 47(2)) will give rise to actions by children who are born disabled in respect of prenatal injuries.

(*d*) *Time limit for bringing actions*. In view of the rule that a minor can claim up to 18 years after its birth coupled with the fact that under the Limitation Act 1975 the action need not be brought until the injury or disease manifests itself *and* the period of three years has expired since such time, employers may find that they are defending actions up to 21 years and more after the "occurrence" happened. They have inherited the doubtful accolade of being employers of an unseen workforce.

22A. The employer's defences

Against this sombre backcloth of ever-increasing liability it would hardly be surprising if most employers began to wonder whether it was worthwhile being in business at all. However, there is one redeeming feature in the defences employers may raise.

(*a*) *Synopsis of the history of employers' defences*. At one stage in the development of industrial law there were at least three well-entrenched defences that employers could raise when sued in respect of injuries suffered by their employees—common employment, *volenti non fit injuria*, and contributory negligence which was originally a complete defence. With the passage of time, however, this wall of defences has been systematically chipped away by both beneficent legislation and the courts, who have found all sorts of excuses for not applying the doctrines of common law which might deprive the workman of his compensation for injury suffered at work. Two historic landmarks in this respect were the Employers' Liability Act 1880 and the decision of the House of Lords in *Smith* v. *Baker Ltd.* in 1891, interring the defence of voluntary assumption of risk of injury in cases of employee injury. Moreover the evolution of vicarious liability in the hands of the judiciary has done much slowly but surely to transform the balance of power in industrial legislative history.

Today the pendulum has swung full circle, to the extent that not only is the law weighted in favour of the injured employee but in addition is working, with the aid of the European Social Charter and decisions of the European Court of Justice, towards the creation of a floor of statutory rights for workers and employees. As regards defences of employers, however, there are only two. Firstly, where the statute or regulation creating the duty and liability provides a defence. Second, where the employee has been contributorily negligent, in which case he will lose a proportion of his damages to the extent that he is adjudged at fault. In practice, the second is likely to prove the better in the light of the paucity of statutory defences in this field, and in recent times there also seems to be some evidence of a resuscitation of the old common law defence of *volenti non fit injuria* (voluntary assumption of risk of injury) where an employee is in breach of a statutory duty imposed either by statute or regulation. Evidence of this revival is afforded by *I.C.I. Ltd.* v. *Shatwell*, the result of which seems to be that if an employer carried out his statutory duties to instruct, train and supervise his workforce in job safety and an employee chooses not to put into practice the safety training, he may lose all his compensation for injury.

(*b*) *The defence of* volenti. The substance of this defence is that the employee, knowing of the danger involved in a particular process or job or having been made aware of the danger through both formal instruction and training, by his conduct agreed to run the risk of being injured in the way in which he was injured. If this is a defence to a workman's claim it operates as a complete defence with the result that there is no negligence in the employer, and so he (or rather his insurer) is relieved of the burden of paying any compensation. It is precisely for this reason, i.e. that an employee might lose all his compensation, that the likelihood of this defence succeeding is remote. Indeed until the revival of this defence with the decision of *I.C.I. Ltd.* v. *Shatwell* in 1964 it was assumed that after the decision in *Smith* v. *Baker Ltd.* in 1891 the defence had ceased to exist. In *Smith* v. *Baker Ltd.* S was an employee of B and was employed to drill holes in rock forming part of a railway cutting. Whilst this rock-cutting work was being done, a crane operated by another employee of B habitually swung heavy stones over S. Both S and B knew that there was a risk that the stones might fall and injure S. Nevertheless the crane continued to be operated without any warning being given when it was jibbing and swinging. One day because the load of the crane was negligently strapped, a stone fell from the crane and injured S, who later sued his employer for negligence. The action was successful and the defence of *volenti non fit injuria* (to one who is willing no harm is done), when raised by the employer, failed. The House of Lords took the view that the fact that the employee knew of the risk of injury was not tantamount to his having agreed to run the risk. In furtherance of this argument the fact that S was an employee negatived consent on his part, in that he was not really in a position to exercise freedom of choice as to whether he accepted or rejected the system of work.

The decision in this case stood effectively for 75 years supported by a string of decisions including *Bowater* v. *Rowley Regis Corporation*. Then in 1964 *I.C.I. Ltd.* v. *Shatwell* was decided in the House of Lords and it seemed that there might be a partial revival of the *volenti* doctrine where there was a statutory duty placed on an employee. GS and JS were certificated and highly skilled shot-firers employed by I.C.I. A statutory regulation placed an obligation on the two employees personally, not on I.C.I. the employer, to make sure that certain precautions were taken when shot-firing was about to commence and that proper cover had been taken by persons nearby. The risk of danger associated with the job had been inculcated into the two men. They were aware of the risks of premature explosion; they knew of the statutory prohibition but nevertheless they elected to test because a cable was too short to reach the shelter, rather than wait a few minutes for their workmate to go and get another cable. JS handed two wires to GS who applied them to the galvanometer terminals, in consequence of which an explosion occurred. On appeal by I.C.I. to the House of Lords this court, reversing the decision of the Court of Appeal, ruled that the employer was not liable for the injury to the two employees. The reasons they gave were as follows:

(i) the employers were not in breach of a statutory duty, it having been laid by regulation upon the employees;

(ii) the two Shatwell brothers were skilled employees and well aware of the risk, a fact which induced Lord Pearce to say "The defence should be available where the employer was not in himself in breach of a statutory duty and was not vicariously in breach of a statutory duty through the neglect of some person of superior rank to the plaintiff and whose commands the plaintiff was bound to obey";

(iii) the employer could plead the defence of *volenti non fit injuria* to an action for vicarious liability.

I.C.I. Ltd. v. *Shatwell* must be regarded as relevant now that s. 7, H.A.S.A.W.A., places a specific duty on all employees to take reasonable care for their own health and safety and that of others, but must be distinguished from the similar situation in *Ross* v. *Associated Portland Cement Manufacturers Ltd.* (§ **21,** p. 100), in that in the *Ross* type of case the employer is also in breach of his statutory or common law obligations or both, so that the cause of the employee's injury can be ascribed at least in part to the employer's breach of statutory obligation. In *Ross*' case the employer was found to be two-thirds liable. The *Shatwell* situation should also be distinguished from *Boyle* v. *Kodak Ltd.*, where employer and employee were in breach of their statutory obligations, with the result that damages were apportioned in accordance with the Law Reform (Contributory Negligence) Act 1945. The *Shatwell* type of case is also distinguishable from *Bux* v. *Slough Metals Ltd.* (*see* § **21,** p. 96) because, as Lord Pearce said in *Shatwell*, there was no-one who was skilled enough, or at least this is what the evidence suggested, effectively to supervise the two brothers. Had there been such a person it is clear that the employer would have been vicariously liable. The *Shatwell* situation is difficult, if not impossible, to distinguish from *Johnson* v. *Croggan & Co. Ltd.* (*see* § **21,** p. 81).

A question of the utmost practical importance is whether in the future the courts are going to adopt the approach in *Boyle* v. *Kodak Ltd.* or are more likely to follow *Shatwell*. In the light of the recommendations of the Pearson Commission in favour *inter alia* of retention of tort liability, both defences will probably survive in principle, with more emphasis in practice being laid on contributory negligence. In addition, disobedience to safety laws and regulations may lead to prosecution followed by fines and dismissal from the job, such dismissal often being lawful within the meaning of the Trade Union and Labour Relations Act 1974.

Future trends notwithstanding, for the present it is thought that *Boyle* v. *Kodak Ltd.* reflects a more realistic approach, in that many of the situations where an employee is injured relate to joint negligence on the part of employer and employee. Now, however, in view of s. 2(2), H.A.S.A.W.A., an employer has a duty to train, instruct and supervise in job safety, the inescapable inference being that if he has failed to implement this duty, and an employee gets injured, it is the negligence of the employer that will be the exclusive cause of the employee's injury. On the other hand, if the employer has carried out his statutory duty under s. 2(2) and has had his workforce trained and instructed and supervised in job safety, and the employee acts in disregard of this and is injured, the injury may well be exclusively his own fault and the *Shatwell ruling will apply*.

A case similar to *Shatwell* is *Ginty* v. *Belmont Building Supplies Ltd.*, a decision of the Court of Appeal in 1959. G, an employee of B.B.S., was working on an asbestos roof which he knew was defective. He also knew that he ought not to work on the roof without boards and that there was a statutory duty laid upon him to that effect under the Building (Safety, Health and Welfare) Regulations 1948. There was also a statutory duty placed on his employer by the same regulations to provide boards for such work. Evidence showed that the employer had provided boards but that the employee had not used them. As a result of failure to use the boards G was injured and sued his employer for damages. It was held that his action must fail, the court taking the approach that the relevant question to ask was: whose fault had caused the accident? Was it the fault of both the employer and the employee, as in *Boyle* v. *Kodak Ltd.*? Or was it the employee's conduct alone by which the accident was caused? The court found that it was the employee's conduct alone, his lack of care for his own safetv, that was the cause of his injury.

Rescue situations

Occasions will arise where an employee or some other third party goes to the assistance of an injured workman and is himself injured in consequence. If the situation discloses a genuine rescue attempt, a wealth of case law argues that the employer cannot raise the defence of *volenti* to debar the employee from recovering damages, even though it might appear that the employee had agreed to run the risk of injury involved. The reason is that the rescue is construed as a foreseeable consequence of the employer's negligence in allowing the first employee to be injured. In other words, in legal analysis rescue is really a question of remoteness of damage. Two cases advertise this fact very well. In *Haynes* v. *Harwood & Son* in 1935 Harwood's employee left a horse van in a busy street unattended. Later the horses bolted with it. Haynes was a policeman and he realised that if the horses were not stopped, some children playing in the street would be injured. At some risk to himself he succeeded in stopping the horses but was injured in the act. He sued Harwood for damages for negligence and the employer pleaded the defence of *volenti*. It was held by the Court of Appeal that the defence must fail: Haynes' conduct was a foreseeable consequence of the original act of negligence of Harwood's employee, nor was Haynes construed as being contributorily negligent. Similarly in *Baker* v. *Hopkins Ltd.* in 1959 another genuine rescue attempt arose out of an employer's negligence in not setting up a safe system of work. H was a building contractor and was engaged in cleaning out a well. Because it had proved difficult to clean out the well by means of hand-driven pumps, petrol-operated pumps were used. The exhaust from the engine on the pump caused a lethal concentration of carbon monoxide to form in the well. Two employees of H went down into the well to carry on the job of cleaning it out and were overpowered by the carbon monoxide. B, who was a doctor, after hearing of what had happened, also went down into the well, but soon realised that the two employees were dead. He then indicated to those at the top of the well that he was ready to come up, but unfortunately the rope which was being used to pull B up caught on a projection, as a result of which B himself was overpowered by the fumes and died. It was held that the employers were negligent in respect of their two employees in not setting up a safe system of work, the rescue action of B was a foreseeable consequence of this negligence and the defence of *volenti* was inapplicable.

The rescue attempt must be genuine. If the situation surrounding the rescue is such that no reasonable man would take the chance of attempting to rescue someone from the predicament which he is in, either the defence of *volenti* will apply so that the employee will lose his compensation or he will be found contributorily negligent. This might be the situation where an employee has got his foot trapped under a railway sleeper with dusk approaching and the London to Glasgow express is travelling through at 100 m.p.h.; here a rescue attempt by another employee is to take a chance that most reasonable men would decline, and if the employee lived to bring subsequent action it is likely that his action would fail.

The defence of contributory negligence

This defence for the employer and/or factory occupier is partial, as distinct from *volenti* which is complete. It is for that reason of greater practical value. It was seen that until *Shatwell* there was a general reluctance on the part of a judge to deprive an injured workman of all his compensation. But because contributory negligence is merely a partial defence and means that loss shall be apportioned between employer and employee, the courts were less reluctant to say that the employee had contributed to the injury which he suffered and to that extent his damages should be reduced. This has been the legal position since the Law Reform (Contributory Negligence) Act was passed in 1945. Prior to that Act contributory negligence had been a complete defence. Its survival is guaranteed by s. 47, H.A.S.A.W.A., to the effect that civil actions will be governed by common law criteria.

(*a*) *What is contributory negligence?* It is difficult to define contributory negligence although over the years many valid tests have been propounded. Perhaps the matter was best put in the early days of the emergence of the defence by Lord Ellenborough, C.J., in the classic case of *Butterfield* v. *Forrester* in 1808. F negligently allowed a pole to fall across and block the public highway. B, riding his horse violently at dusk, did not see the pole, collided with it and injured himself. It was admitted that F was negligent; nevertheless B's action against him failed because of his own contributory negligence. During the course of giving judgment Lord Ellenborough spoke as follows: "One person acting in fault will not dispense with another person taking reasonable care in respect of his own safety." Not to take reasonable care in respect of one's own safety is tantamount to contributory negligence. Thus s. 1(1), Law Reform (Contributory Negligence) Act 1945, states:

"Where any person suffers damage as the result partly of his own fault and partly of the fault of any other person or persons, a claim in respect of that damage shall not be defeated by reason of the fault of the person suffering the damage, but the damages recoverable in respect thereof shall be reduced to such extent as the court thinks just and equitable having regard to the claimant's share in the responsibility for the damage...."

Section 4 of the same Act further states, defining "fault":

"... fault means negligence, breach of statutory duty or other act or omission which gives rise to a liability in tort or would, apart from this Act, give rise to the defence of contributory negligence...."

(*b*) *Contributory negligence and alternative danger*. Where the negligent conduct of the defendant is such as to put the plaintiff in fear of his own life or limb, as a result of which he decides to take evasive action which results in his being injured, this is not contributory negligence. Old authority for this proposition exists in the form of *Jones* v. *Boyce*, where B's conduct in driving a stagecoach dangerously and negligently put J in fear of his life, as a result of which he opened the door of the coach and jumped out, causing himself serious injury. It was held that J's conduct did not amount to contributory negligence and he was entitled to damages. Recent examples of conduct which does constitute contributory negligence are to be found mainly in cases involving passengers in cars who fail to wear seat-belts. In the leading 1975 case of *Froom* v. *Butcher* Lord Denning, M.R., said in the course of giving judgment and contrasting masterly negligence (the action) and contributory negligence (the defence) "Negligence depends on a breach of duty whereas contributory negligence does not. Negligence is a man's carelessness in breach of duty to others. Contributory negligence is a man's carelessness in looking after his own safety...." In

Froom v. *Butcher* it was held by the Court of Appeal that where a defendant is liable in negligence for causing a road accident in which the plaintiff, a road user, is injured or the injuries are exacerbated by his failure to use the seat-belt, the plaintiff will be held to have been contributorily negligent. In the light of this decision employers should instruct their employees in the use and wearing of seat-belts and failure to observe these instructions on the part of employees who are passengers in the course of their employment will mean an apportionment and reduction in damages. Indeed because s. 7, H.A.S.A.W.A., lays a duty on employees to take reasonable care in respect of their own safety and because of the way in which this has been interpreted in *Shatwell's* case as giving rise to *volenti*, failure to observe instructions about the wearing of seat-belts by employees at work could mean loss of all compensation.

It is thought that instruction to employees about the wearing of seat-belts is also part and parcel of the setting up of a safe system of work by the employer which is now actionable as a breach of duty at common law by virtue of s. 47, H.A.S.A.W.A., and enforceable by fine and or imprisonment under s. 33. Similarly if through failure to instruct employees in these matters and related ones such as the giving of lifts to members of the public, an employee gives a lift to a member of the public, not having been instructed to the contrary, and the member of the public is injured as a result of the negligence of the employee in carrying out his employment, in addition he may be prosecuted for breach of s. 2(2), H.A.S.A.W.A. This interpretation would mean that cases like *Twine* v. *Bean's Express Ltd.* and *Conway* v. *George Wimpey Ltd.* would not be followed in the future.

Another way of referring to contributory negligence is in terms of the *last opportunity rule*. This argument was first put forward in *Davies* v. *Mann* in the last century, where the court took the view that if the plaintiff had had the last opportunity of avoiding the accident and failed to avail himself of it, he would be construed as being contributorily negligent. Also in view of the definition of fault in s. 4, Law Reform (Contributory Negligence) Act 1945, a breach of statutory duty on the part of the plaintiff and defendant can give rise to a case of contributory negligence, as we saw in *Boyle* v. *Kodak Ltd.* in 1969. Similarly where the employee trespasses at work, so long as there is a breach of either common law or statutory duty on the part of the employer (and it could be argued that a safe system of work is not being effectively enforced by supervision in such cases otherwise the employee would not be trespassing), there is a prevailing judicial tendency to treat such cases as contributory negligence, as evidenced by *Uddin* v. *Associated Portland Cement Manufacturers Ltd.* and *Westwood* v. *Post Office*. It is thought that contributory negligence will still continue to be as important as it formerly was. Indeed its future is specifically guaranteed by s. 47, H.A.S.A.W.A. Moreover the Pearson Commission has recommended the retention of fault liability in cases of injury at work in addition to an improved state industrial injuries scheme coupled with pensions calculated on the same basis as the new state pensions. It should be emphasised that *volenti* and contributory negligence are defences of the employer to a civil action brought by an employee; they have no relevance to a prosecution brought by an inspector or any other official.

SOCIAL SECURITY

23. The state's commitment to industrial injuries

The interest of the state in industrial injuries is recent compared with that of common law. The liability of an employer at common law goes back effectively to the early casualties in the mills and the factories of the industrial revolution, whereas the state's involvement in industrial injuries dates back to the Workmen's Compensation Act of 1897 and more recently the National Insurance (Industrial Injuries) Acts 1946 and 1965. The present law relating to state benefit for industrial injury is the Social Security Act 1975.

Entitlement to state benefit for injury at work is governed by different criteria from those of the tort system, owing to a different interpretation of the function of compensation traceable back to Lloyd George's National Insurance Act 1911. Whereas the tort system revolves round the concept of fault, the state system of industrial injuries' benefit is a feature of social insurance, in other words an extension of the welfare state. The former presupposes that entitlement to compensation for injury is consistent with someone having done wrong and so masquerades as a kind of retributive justice. The state system adopts the view that compensation for injury is one more facet of social need, with the result that entitlement depends on contribution to the state fund by the employer, employee and taxpayer, and not on proof that someone else has committed a wrong. The state system offers a kind of distributive justice whereby the loss is distributed throughout both sides of industry and the general body of taxpayers. The tort system is also gradually becoming a species of distributive justice, however, a device for loss distribution amongst the general body of premium payers of road users and employers, this trend having been accelerated by the growth of compulsory liability insurance. Moreover the view that the state seeks to offer equalisation compensation referable to the social criterion of need would appear to be hardly tenable in the case of industrial injuries where benefit has, since national insurance began, always been related to loss of individual earnings as distinct from being flat rate, as in the case of retirement pension, unemployment benefit and sickness benefit. Further comparison between social security and tort is to be found in § **60,** p. 267. Pearson has recommended that the industrial injuries' scheme provide the basis of improved provision for injury at work, with higher benefits for the first six months of incapacity, followed by pensions calculated in the same way as the new state pension.

Criteria of state industrial injuries benefit

The present law relating to state benefit for industrial injuries is to be found in the Social Security Act 1975. The state system provides for two types of benefit: firstly industrial injury benefit in ss. 50–55; and second, benefit for industrial diseases in s. 76. Just as employers' liability insurance benefits employees, so does the state system of benefit in the Social Security Act 1975; the self-employed are not entitled to industrial injury benefit but must rely on more expensive personal accident insurance against hazards associated with their job although this does not necessarily imply that company directors of public companies cannot benefit from the state system, in that most directors of public companies are employees of the company and pay towards their industrial injuries stamp. Moreover where a person sets up a small private company the director is regarded as an employee in his own company. The reason for this is that limited liability creates a division of legal personality between the company and the members who at any given time comprise the company. In this way those who set up and become directors in their own companies are both employers enjoying the advantages of limited liability (an advantage denied to non-corporate employers, including most of the professions for ethical reasons) and at the same time are entitled to state benefits as employees. The point was well made in the important appeal case of *Lee* v. *Lee's Air Farming Ltd.* in 1960. L's husband formed a crop-spraying company of which he and his wife were directors. Two years after the company had been formed L's husband was killed in a flying accident whilst piloting an aircraft for the purposes of crop spraying. His wife claimed compensation under the New Zealand Workers' Compensation Act 1922, and her entitlement turned on whether her deceased husband had been a "worker". Evidence showed that her husband owned 2,999 of the 3,000 shares of the company and was the governing director. It was legitimate, therefore, to ask whether he could be the employee of the company. It was held that her husband was an employee of the company and the principle in *Salomon's* case was applied. The Pearson Commission has recommended that the self-employed be covered by the state industrial injuries' scheme and entitled to flat-rate benefits.

(*a*) *What benefits are available?* Section 50(1), Social Security Act 1975, makes it clear that three types of benefit are available, and contribution conditions are irrelevant:

(i) industrial injury benefit (available in respect of accidents and diseases);

(ii) industrial disablement benefit (arising from an injury or a prescribed disease);

(iii) industrial death benefit (arising from an injury or a prescribed disease).

These are the three types of industrial injury benefit. In addition to this there is industrial disease benefit, provided for by s. 76. The difference between the two lies in the fact that an industrial injury arises from an accident or personal injury at work, while an industrial disease arises out of continued exposure to certain working conditions or industrial processes. All industrial diseases are prescribed in s. 76 Part 1 (*see* Appendix II). The list of fifty diseases is continually being augmented as new evidence uncovers still further hazards associated with differing working environments. Another distinctive feature is that industrial injuries tend to be associated with quick on-the-spot mishaps causing personal injury, whereas industrial diseases, e.g. pneumoconiosis, are slow and gradual in manifesting themselves. The former are generally redressed by injury benefit in the form of periodical payments, whereas the latter are more often redressed by benefit in the form of a pension (disablement pension). An industrial injury is also redressible by way of a pension if it lasts more than six months (i.e., the claimant is off work for more than six months). In practice, however, the great majority of industrial injuries are short-lived and the workman is generally back at work long before the expiry of six months. This is generally not so with industrial diseases which tend to be long-term and/or fatal. The Pearson Commission recommends that the industrial injuries scheme be the basis of compensation for injury at work and that industrial disablement benefit be retained.

(*b*) *Award of industrial injuries benefit.* Section 50(1), Social Security Act 1975, states that the sole criterion is that the injury "arise out of and in the course of the employment". To a limited extent this requirement reflects the causal connection of the common law and it is this causal link which has been responsible for the failure of a fair proportion of claims, as it has at common law. A classic example of this insistence on the two facets of the requirement that the injury "arise *out of and in the course of* the employment" is afforded by the case of *R.* v. *National Insurance Industrial Injuries Commissioner*, ex parte

Richardson, where a coloured bus conductor was attacked one night when carrying out his duties as a bus conductor. It was held by the High Court that the Commissioner had been right in upholding the decision of the insurance officer to refuse benefit on the grounds that although the injury arose in the course of the employment, it could not be said that the injury arose "out of" the employment. The inescapable (and for Richardson, lamentable) fact was that the injury was not inflicted upon him because he was wearing a bus conductor's uniform but rather because he was coloured. In situations like this claimants might like to avail themselves of s. 50(3), Social Security Act 1975, which, re-enacting s. 6, Industrial Injuries Act 1965, provides that an accident arising in the course of an employed earner's employment shall be deemed, in the absence of evidence to the contrary, also to have arisen out of that employment. This raises a presumption in favour of work-induced injury which it is incumbent upon the insurance officer to rebut. This he may or may not be able to do. *Richardson's* case is an example where the presumption was successfully rebutted. This sub-section is often invoked by the dependants of a deceased workman in cases where injuries have proved fatal and there is no evidence as to how the accident occurred. Pearson has recommended the retention of the requirement that the accident arise "out of and in the course of employment".

(*c*) *Playing sports*. Some of the Commissioner's decisions on playing games and sport are difficult to reconcile. Compare one case, where a fireman was injured while he was playing volley ball in his recreational period, he was encouraged but not obliged to play this game, and it was held that he was entitled to benefit, with another, where a laboratory technician was injured whilst playing football during the lunch hour as he had been encouraged to do by his superiors, as he was a member of the hospital team, it was held that he was not entitled to benefit. A 1977 decision of the High Court in *R.* v. *National Insurance Commissioner*, ex parte *Michael* seems to indicate that claimants who are injured whilst playing games are unlikely to qualify for benefit. In this case a young policeman was injured whilst playing football for his force's team and was refused industrial injury benefit. On an appeal to the High Court that court refused to set aside the ruling of the Commissioner. In spite of the fact that policemen were encouraged to play for the force team and improved their promotion chances by doing so, and were allowed eight hours of "duty time" per month in which to play, the Court of Appeal could not see its way clear to construe football matches as being "in the course of employment" (M appealed to the Court of Appeal after the High Court ruled against him). Participation in the factory or works football team, the court said, arose "out of" but not "in the course of employment", though the position would have been the reverse if the policemen had been ordered to take part in the game, e.g. as part of training.

This decision has the effect of limiting the recent trend towards construing liberally the expression "in the course of employment". It means that employees injured whilst playing sports will not be able to sue their employer for negligence or breach of statutory duty, and employers will not be vicariously liable for injuries committed by one employee upon another on the sports field. The employee will be able to claim both state benefit and common law damages if he is injured as part of training or if he is injured when playing a sport on the orders of his employer.

(*d*) *Acts incidental to the employment.* Games and sports apart, when an employee is injured doing an act or as the result of an omission which is not strictly speaking part of his employment, but is reasonably incidental thereto, it is normal for him to be held entitled to benefit, and the same position applies in tort at common law and in an action for breach of statutory duty. Thus where an employee was injured whilst having a wash and brush up before work or whilst putting on his coat after finishing work, he would normally be able to sue his employer. Similarly where an employee committed a tort in the same circumstances, e.g. allowing the washbasin to overflow with the result that the adjoining premises were flooded, the liability was that of the employer. At one time with regard to state industrial injury benefit, this was not the position and the expression "in the course of employment" was interpreted strictly. In *R.* v. *Industrial Injuries Commissioner*, ex parte *A.E.U.* (*No. 2*), a fitter who worked in a part of a factory where smoking was forbidden other than in special booths went to have a cigarette during a tea break. He saw that the booths were occupied and while rolling a cigarette outside one of the booths was injured by a passing fork-lift truck being driven negligently. The tea break lasted from 10.30 a.m. to 10.40 a.m. and the fitter was injured at 10.45 a.m. The fitter claimed industrial injury benefit, but his claim was disallowed by the insurance officer, the local tribunal and the Commissioner. The latter took the view that the extension of the tea break amounted to an interruption of employment and his claim must fail. On appeal to the Court of Appeal it was held that the Commissioner had taken a correct view of the law. In 1976, however, in *R.* v. *National Insurance Commissioner*, ex parte *East* the High Court, reversing the decision of the Commissioner, held that injury which an employee suffered in the factory canteen before clocking on was within the course of employment.

(*e*) *Employee acting in breach of regulations.* Section 52, Social Security Act 1975, provides that an accident shall be deemed to arise out of and in the course of employment, notwithstanding that the earner is at the time of the accident acting in contravention of any statutory or other regulations applicable to his employment, or of any orders given by or on behalf of his employers, or that he is acting without instructions from his employer, if

(i) the accident would have been deemed so to have arisen had the act not been done in contravention of any such regulations or orders, or without such instructions; and

(ii) the act is done for the purpose of and in connection with the employer's trade or business.

There are those who argue that s. 52 does not operate in the best interests of health and safety, in that it purports to provide industrial injury benefit where an employee is not observing or is acting in disobedience of safety codes, regulations and statutory duties. In this respect the provision would appear to have the same effect as the decision in *Westwood* v. *Post Office*, where the trespassing employee's widow got compensation. This argument, it is thought, fails to take account of the fact that the purpose of social security is to compensate and that fault has no role to play in the system. Moreover a negligent employee might always be punished for acting in breach of s. 7, H.A.S.A.W.A.

(*f*) *Travelling to work*. Section 53 states that an accident which happens while an employee is travelling to or from work will be deemed to arise out of and in the course of his employment if,

(i) the accident happened when the vehicle was being operated by or on behalf of his employer or by some other person by whom it is provided in arrangement with the employer; and

(ii) the vehicle was not being operated in the ordinary course of public transport.

In addition to this requirement the following conditions must be satisfied: firstly, the employee must be travelling by that method with the express or implied permission of his employer; second, he must be travelling as a passenger; and third, the accident would have been deemed to have so arisen even if the employee had been under no such obligation to travel in that way. Pearson has recommended by a majority that accidents occurring to and from work should be brought within the industrial injuries scheme.

(*g*) *Accidents happening in an emergency*. Section 54 extends the scope of accidents arising out of and in the course of the employment by providing that an accident happening to an employee in or about the premises where he is employed shall be deemed to arise out of and in the course of his employment if it happens when he is endeavouring, in an actual or supposed emergency, to rescue or give help to or to protect persons who are, or who are thought to be or possibly to be injured or endangered, or to avert or minimise serious damage to property. It appears that in the interpretation of this section (when it was s. 9, Industrial Injuries Act 1965) the tribunals and the Commissioner have not been as generous as the common law courts interpreting cases of negligence at common law. For instance in one case where a civil servant was visiting some private houses on the business of his employer, the National Assistance Board, and stopped a child coming down a hill on a bicycle out of control, it was held that he could not recover compensation in respect of the injury which he suffered. The Commissioner took the view that there was no obvious link with his employment here, although in another case, where a policeman standing on traffic duty outside property belonging to the Admiralty tried to stop a runaway pushchair with a child in it, as a result of which he was injured, it was held that he was entitled to recover in respect of his injury from the industrial injuries fund. But where was the link with employment here? A general duty on the part of the police to look after the welfare of the public including small children? The common law position in such cases was exemplified in 1963 in *Videan* v. *British Transport Commission.*

(*h*) *Accidents caused by another's misconduct.* Section 55 states that an accident shall be treated as arising out of employment if,

(i) the accident arises in the course of employment; and

(ii) the accident is caused by either

1. another person's misconduct, skylarking or negligence, or by action taken in consequence of any such misconduct, skylarking or negligence; or

2. the behaviour or presence of an animal (including a bird, fish or insect); or

3. the employee being struck by any object or by lightning; and

(iii) the employee did not directly or indirectly induce or contribute to the accident happening by virtue of any conduct on his part outside his employment or of any act not incidental to the employment.

The purpose of this provision is to offer further cover in respect of certain added perils of the employment which, in the absence of this section, would have been construed as arising outside the course of employment. Examples are where an employee is injured through the misconduct or skylarking of another employee, as where he is assaulted or injured by a prank of a practical joker. In the latter case of the prank, the injured employee may well be entitled to damages at common law, assuming that the prank is reasonably foreseeable as in *Hudson* v. *Ridge Manufacturing Co. Ltd.* in 1957. Where the employee is a victim of a gratuitous assault he will be entitled to state benefit if there is a link between the assault and the job, as where a foreman tells an employee off about poor workmanship or about wearing safety gear or rebukes him for drinking in the lunch hour, with the result that his work suffers, here there is a connection with the employment. If, however, the assault is merely coincidental to the employment, benefit is unlikely. This would occur where one employee was struck and injured by another employee as a result of the one having accused the other of sleeping with his wife. Nevertheless in the case of assault arising from bad feeling and quarrelling, even if there is merely a slight connection with the employment, the injured employee is likely to get industrial injury benefit and possibly damages at common law for failure on the part of his employer to set up and maintain a safe system of work through supervision.

Where an employee is injured through another's negligence, he can also recover benefit unless he "invited" the injury through an act outside his employment, e.g. a bus conductor becomes a policeman and goes to help an old lady who is being mugged. Nor will this sort of situation enable the employee to recover in tort either at common law or for breach of statute; he must keep strictly within the parameters of his employment. Nevertheless it is his employer's duty to see that his employees are given a safe system of work including safe premises and relative freedom from personal injury, both negligent and intentional. In the case of the bus conductor, he would be entitled to recover damages if his employer had not installed a signalling system for use on the buses late at night and at the weekends, as a result of which he is unable to get help when being assaulted. The same principle would also apply in the case of bank clerks or wages clerks at a factory who were not given the necessary protection from attack which is an intrinsic risk of their employment. Where the injury inflicted is of a criminal nature, as where there is a premeditated attack on a bank clerk, *ex gratia* compensation is available from the Criminal Injuries Compensation Board. Since, however, these payments are *ex gratia*, it may be better for the injured employee to proceed against his employer at common law and for breach of statute. Section 47, H.A.S.A.W.A., dealing with civil actions, states that the criteria of compensation for breach of the general duties of H.A.S.A.W.A. are those of common law, i.e. negligence. In addition, the injured bank clerk will be able to claim industrial injury benefit.

Where an employee is injured by an animal, including a bird, fish or insect, some connection with the employment must be shown, as where a farm worker is charged and injured by a bull, the employee not having provoked the beast, or where a zoo attendant is bitten by a snake. But common hazards will not qualify. If an employee is stung on the nose at work by a wasp, this injury cannot qualify for benefit unless his employment made him more vulnerable to this kind of injury, as where he was employed in a jam factory. Similarly where the postman is bitten by a ferocious dog, it is appropriate to ask whether the dog is a common hazard or a risk peculiar to the employment of a postman. It would appear, however, that postmen, milkmen and bakers can proceed against their employer for a breach of common law duty if they are particularly vulnerable to ferocious dogs and their employer has not instructed them how to handle this matter or provided them with some physical protection.

(*i*) *Common hazards*. Injury arising from a common hazard will not qualify for benefit. This applies also to travelling to and from work; if an employee is injured through travelling to work on a public transport vehicle, any injury suffered in consequence of the vehicle being in collision with another vehicle, will not entitle the employee to benefit, though if a vehicle normally operating as part of public transport is commissioned as a works bus and an injury is suffered by an employee whilst travelling in the works bus, he is entitled to benefit. Significantly the case of *R.* v. *Industrial Injury Benefits Tribunal*, *ex parte Fieldhouse* in 1974 established that the fact that a mileage allowance was given or travelling time paid for did not mean necessarily that the injury which arose was referable to the employment.

(*j*) *Plurality of systems of compensation*. Employees who are injured at work are well served by the law as compared with others who suffer accidents and disease. Firstly they can proceed against the employer at common law under the rule in *Wilsons & Clyde Coal Co. Ltd.* v. *English*; second, they can sue the employer for breach of statutory duty, such as the Factories Act 1961 or, by virtue of s. 47(2), H.A.S.A.W.A., any regulations which may be passed under the general duties of the Act. Third irrespective of contribution record they can claim industrial injuries benefit through the D.H.S.S. Fourth if employees are the victims of criminal attack, they can obtain *ex gratia* compensation from the Criminal Injuries Compensation Board (for more about these different systems, *see* § **25**, p. 151).

(*k*) *Prescribed industrial diseases*. Section 76, Social Security Act 1975, states that benefit is payable in respect of prescribed industrial diseases. Sub-section (2) makes it clear that a disease may be prescribed if it is a risk related to the employment of the employee and not common to all persons. Thus it is well established that pneumoconiosis is a condition associated with mining, whereas in the case of poliomyelitis it is more difficult to associate it with any particular kind of employment, in that it is a condition common to all persons. If, however, a nurse were to contract this condition it would be arguable that it was related to and arose out of her employment. Again the common cold is hardly employment-related. If, therefore, it results in an employee being off work for more than three days, it is unlikely that he can recover benefit. It is just conceivable that he may do so if he worked in freezing conditions. Part 1 of s. 76 enumerates fifty prescribed industrial diseases and no doubt with improvements in scientific and medical detection this number will increase (the list of prescribed industrial diseases and the occupations to which they are related will be found in Appendix II). It is important to remember that a person will not be entitled to industrial disease benefit unless:

(i) he was suffering from a disease which was prescribed, and

(ii) the disease must have arisen from the occupation prescribed as giving rise to that condition, e.g. mesothelioma must arise from working or handling or manufacturing or doing something in the course of one's employment with asbestos.

The Pearson Commission has recommended that the time taken to prescribe occupational diseases be reduced and that in addition to prescribed diseases there be an individual proof system whereby, although a disease was unlisted, if the claimant could show that his disease arose out of his occupation he would be entitled to benefit.

Suspension from work on medical grounds

There is a close connection between the operation of s. 76, Social Security Act 1975 and ss. 29–33, Employment Protection Act 1975, and the latter could involve the employer in paying out considerable sums of money. Section 29, Employment Protection Act 1975, states that an employer is required to pay an employee suspended from work on medical grounds a week's pay (s. 31) for each week of suspension until a period of 26 weeks has elapsed. Subject to certain other conditions being satisfied (*see* §**44**, p. 239) which are set out in s. 29, Employment Protection Act 1975, an employee is entitled to up to 26 weeks' pay where he has been suspended from his normal work because of medical risk but is not incapable of employment. In other words s. 29, Employment Protection Act 1975 is in no way a substitute for sick pay. There are at present fifteen grounds set out in Schedule 2 to the Employment Protection Act on which an employer must suspend an employee from work on pay, and one can only suppose that, since there are now fifty prescribed industrial diseases, the list of grounds set out in Schedule 2 for the purposes of suspension on medical grounds is likely to increase (as for the exact grounds on which employees are entitled to remuneration for suspension on medical grounds together with the potential scope of ss. 29–33, Employment Protection Act 1975, *see* §**44**, p. 239.

Training and the Social Security Act 1975

Recent decisions, particularly that of the Court of Appeal in *R.* v. *National Insurance Commissioners*, ex parte *Michael* (*see* p. 136), have established *obiter* that where an employee is injured whilst being trained (and sport may be classed as part of his training programme, e.g. a policeman or fireman), this injury is to be construed as arising out of and in the course of his employment. This trend is important; now that employers are statute-bound to train their employees in job safety, one can assume that if an employee is injured whilst participating in an approved course, i.e. I.T.B. approved, it is almost a foregone conclusion that he will be entitled to benefit under s. 50(1), Social Security Act 1975 (s. 2(2), H.A.S.A.W.A., consistent with the general obligation laid on employers to train under the Employment and Training Act 1973, requires all employers to train their workforce in job safety). Since this training could take place at a variety of appropriate venues, from hotels, formal training centres, polytechnics and technical colleges through to management centres and W.E.A. centres, the place where the injury occurs would appear to be irrelevant. In certain types of training (which were obligatory before the Employment and Training Act 1973 and H.A.S.A.W.A.), e.g. training in the use of abrasive wheels and power presses, one could foresee that the operation of the Social Security Act 1975, or rather that part concerning industrial injury benefit, was likely to be particularly appropriate. Now, since April 1975 (when H.A.S.A.W.A. came into force) all types of jobs require safety training to be built into the training of employees, with the result that the scope of s. 50, Social Security Act 1975, is widened. Employees injured whilst undergoing compulsory training should be entitled to industrial injury benefit.

(*a*) *Notification of accidents*. Section 88, Social Security Act 1975, states that regulations may be made requiring employers to make reports of accidents in respect of which industrial injury benefit is payable; in addition, to furnish the necessary persons with any information required for the determination of claims; and to take such steps as may be specified to facilitate the giving of notice of accidents, the making of claims and their determination. Such regulations have been made, in the form of the Social Security (Claims and Payments) Regulations 1975. Regulation 22 provides as follows:

"(1) Every employed earner who suffers personal injury by accident in respect of which benefit may be payable shall give notice of such accident, either in writing or orally, as soon as is practicable after the happening thereof: Provided that any such notice required to be given by an employed earner may be given by some other person acting on his behalf.

(2) Every such notice shall be given to the employer, or (if there is more than one employer) to one of such employers, or to any foreman or other official under whose supervision the employed earner is employed at the time of the accident or to any person designated for

the purpose by the employer, and shall give the appropriate particulars."

(*b*) *The appropriate particulars.* Schedule 4 of the regulations prescribes that these are as follows:

(i) full name and address and occupation of the injured employee;

(ii) date and time of accident;
(iii) place of accident;
(iv) cause and nature of injury;
(v) name, address and occupation of the person giving the notice, if other than the injured person.

(*c*) *Regulation 23: the obligations of employers.* Regulation 23 describes the obligations of employers as follows:

"(1) Every employer shall take reasonable steps to investigate the circumstances of every accident of which notice is given to him or to his servant or agent in accordance with the provisions of reg. 22 and, if there appears to him to be any discrepancy between the circumstances found by him as a result of his investigation and the circumstances appearing from the notice so given, he shall record the circumstances so found.

(2) Every employer who is required so to do by the Secretary of State shall furnish to an officer of the Department within such reasonable period as may be required, such information and particulars as may be required—

(*a*) of any accident or alleged accident in respect of which benefit may be payable to, or, in respect of the death of, a person employed by him at the time of the accident or the alleged accident; or

(*b*) of the nature of and other relevant circumstances relating to any occupation prescribed for the purposes of Chapter V of Part 2 of the Act (for industrial diseases purposes) in which any person to whom or in respect of whose death benefit may be payable ... was or is alleged to have been employed by him.

(3) Every owner or occupier (being an employer) of any mine or quarry or of any premises to which any of the provisions of the Factories Act 1961 applies and every employer by whom 10 or more persons are normally employed at the same time on or about the same premises in connection with a trade or business carried on by the employer shall,

(*a*) keep readily accessible a book or books in a form approved by the Secretary of State [an approved 'Accident Book'] in which the appropriate particulars of any accident causing personal injury to a person employed by the employer may be entered by that person or by some other person acting on his behalf (e.g. the safety officer); and (*b*) preserve every such book, when it is filled, for the period of three years beginning with the date of the last entry therein."

Criteria of industrial disease benefit

As distinct from an industrial accident, an industrial disease must be prescribed within the meaning of Part 1 or Part 2 (relating to pneumoconiosis) of s. 76, Social Security Act 1975, before a claimant is entitled to benefit. The list of prescribed industrial diseases is to be found in the Social Security (Industrial Injuries) (Prescribed Diseases) Regulations 1975 and the Social Security (Industrial Injuries) (Prescribed Diseases) Amendment Regulations 1975 and the Social Security (Industrial Injuries) (Prescribed Diseases) Amendment Regulations 1977. Section 76, Social Security Act 1975, provides that an employed earner can also claim industrial disease benefit when, even if he is not suffering from a prescribed disease, he suffers a personal injury not caused by an accident but due to the nature of his employment. An example of this would be where a man employed to lift flagstones injured himself in such a way as to give rise to a hiatus hernia. Here there is no accident in the traditional sense, as defined in *Fenton* v. *Thorley & Co.* in 1903 by the House of Lords: "... an accident is an unlooked for mishap or an untoward event which is not expected or designed. Any unintended and unexpected occurrence which produces hurt or loss...." Examples of this would be tripping over a ladder negligently left on the ground, or being hit by a falling chisel negligently placed on a window sill. This is different from the case of the man who suffers a hiatus hernia through lifting flagstones or contracts fibrositis whilst driving a van with the window down and who therefore suffers personal injury due to the nature of his employment.

Although normally an employee who contracts a prescribed disease must be in an occupation listed in Part 1 of s. 76, Social Security Act 1975, if he is unable to establish the link between prescribed disease and occupation, he will still be entitled to benefit if he can show that either (*a*) it was an accident arising out of and in the course of the employment, within the meaning of s. 50(1), or (*b*) that it was a personal injury due to the nature of the employment within the meaning of s. 76(1); an example of this might be where an employee suffers cramp of the hand or forearm due to repetitive movement. Section 76, Part 1, relates this to handwriting and typing employment activities, but it could arise where a man was working a pneumatic drill and in such a case he would be entitled to qualify for benefit under s. 76(1) as a personal injury due to the employment. It is also probable that, owing to the sympathetic attitude evinced by the authorities, a personal injury due to the nature of the employment within the meaning of s. 76(1) and giving rise to a disease (i.e. an accidental as distinct from prescribed disease) would probably qualify for benefit under s. 76(1), an example of this might occur where a man employed to lift flagstones injured himself as a result of not lifting the weight properly and in consequence suffered a heart attack. Bred of the same sympathetic attitude is the fact that an employee who suffers a disease after being criminally attacked during his employment and because of it will probably qualify for benefit under s. 76(1); this might occur where a rent collector was physically attacked and robbed by a gang of youths, and, as a result of the attack, suffered a nervous breakdown and/or heart attack. It should be remembered that under the industrial injuries scheme operating under the Social Security Act 1975, victims of intentional and violent injuries, in other words victims of criminal attacks, can claim industrial injuries benefit and industrial disease benefit; and if death follows, industrial death benefit will be payable as well. If, however, industrial injury or industrial disease benefit is paid under the Social Security Act 1975, no compensation can be claimed from the Criminal Injuries Compensation Board, though if the employer had failed to set up and enforce a safe system of work (as might be the case with a criminal attack), damages at common law would be payable as well. Where common law damages have been paid in respect of injury at work either from negligent or violent attack, s. 2(1), Law Reform (Personal Injuries) Act 1948, provides that one-half of the social security benefits, i.e. industrial injury benefit, industrial disablement benefit or sickness benefit, shall be deductible from any damages representing actual loss of earning and/or profits or loss of earnings or profits calculated to occur. This means that the damages representing pain and suffering are not reduced in any way because social security benefits, or rather one-half of the social security benefits, are not deductible from damages representing pain and suffering as distinct from loss of earnings or profits. Moreover if a disablement pension is increased by a constant attendant allowance, no deductions will be made from the damages.

The Pearson Commission has recommended that *all* social security benefits relating to injury be deducted.

Industrial injury and industrial disablement benefit compared

Unlike other social security benefits such as sickness benefit, unemployment benefit and retirement benefit, neither industrial injury benefit, nor industrial disablement benefit, nor industrial death benefit depends on contributions. It is clear also from s. 50, Social Security Act 1975, that the self-employed cannot benefit from either industrial injury benefit, industrial disablement or industrial death benefit (as far as the widow of a self-employed person is concerned);

although it will be recalled that a company director of his own company is an employee for these purposes, following the decision in *Lee v. Lee's Air Farming Ltd.*, mentioned on p. 134. Section 50 refers specifically to employed earners.

The bases of industrial injury benefit and industrial disablement benefit differ, in that with industrial injury benefit the criterion is loss of earnings. Industrial injury benefit exists in respect of accidents or diseases or accidental diseases arising out of and in the course of employment. Benefit lasts for up to six months from the date the evidence shows that the accident occurred or the disease manifested itself. But with industrial injury benefit claims, the employee must be able to show that he was precluded from working as a result of the accident or disease, whereas with industrial disablement benefit this is not necessary, since benefit is granted in respect of disability or loss of faculty rather than loss of earnings.

In order to qualify for industrial injury benefits the claimant must have been unfit for work for at least three days and no payment will be made in respect of the first three days of being off work, a fact which operates in a way like excess clauses in insurance contracts, e.g. the motorist paying the first £25 as regards third-party claims. If the injured employee is, although disabled, capable of doing other sedentary work or work of a lighter nature, he is not entitled to industrial injury benefit though he will be entitled to industrial disablement benefit plus certain other benefits such as a special hardship allowance (*see* below). Industrial injury benefit consists of a compound or flat-rate industrial injury benefit plus earnings-related supplement after 12 days.

(*a*) *Industrial disablement benefit.* Unlike industrial injury benefit the criterion of benefit here is not incapacity from work and consequent loss of earnings. Accidents are of two types, either (i) they incapacitate a man from work completely with consequent loss of earnings, or (ii) they deform, disable or mutilate a man but do not incapacitate him from work, even if he is forced to take different employment. Injuries of the first type are compensated by industrial injury benefit, but injuries of the second type, even if they occur during the first six months of industrial injury benefit period, entitle the injured employee afterwards to either an industrial disablement pension or in less serious cases to an industrial disablement gratuity. The difference lies in the gravity of the disability. If it is less than 20 per cent according to the tariff laid down in the Social Security (Benefits) Act 1975, the employee will become entitled to an industrial disablement gratuity; if over 20 per cent, to an industrial disablement pension. In addition, and this is one of the most important characteristics of industrial disablement benefit, it becomes payable automatically after the expiry of the six months' industrial injury benefit period either in the form of a pension or gratuity, irrespective of loss of earnings. If an employee, as a result of an accident or disease in respect of which he has been in receipt of industrial injury benefit, is at the end of the six-month period capable of going back to work, even if to a different job, he will be entitled still to industrial disablement pension or gratuity in respect of his partial or total, permanent or temporary disablement, disfigurement or loss of faculty. Since this benefit is in respect of loss of faculty, disfigurement and disablement, it is payable irrespective of loss of earnings.

(*b*) *Industrial disablement, invalidity and sickness benefit.* It may be that an employee not qualified for the higher rate of industrial injury benefit as a result of not having suffered an accident or disease at work is off work through sickness. Sickness benefit is payable at a lower rate than industrial injury benefit; its purpose is to compensate for loss of earnings, though benefit depends on contribution record, unlike industrial injury and industrial disablement benefit. Sickness benefit is flat rate, with an earnings-related supplement (first introduced in 1966) payable after 12 days. This applies only to employees, as the self-employed person is not entitled to the earnings-related supplement. This earnings-related supplement expires after six months and, originally on the expiry of this period, sickness benefit used to continue but at a flat rate, even after 1966 when earnings-related supplements were introduced into sickness benefit for the first six-month period. After the expiry of the six-month period, although payable for the rest of a sick person's life (assuming that three years' work had been done and 156 contributions paid), the earnings-related supplement would cease. Then in 1971 the National Insurance Act, in order to make up for the deficiency caused by the absence of the earnings-related supplement, introduced the *invalidity* benefit, and this is paid at a higher rate than flat-rate sickness benefit.

(*c*) *No sickness and industrial injury benefit but industrial disablement and sickness or invalidity benefit possible.* Because sickness benefit and industrial injury benefit and invalidity benefit are designed to compensate an employee for loss of earnings, it follows that an employee cannot be in receipt of sickness benefit and industrial injury benefit, or invalidity benefit and industrial injury benefit. But since industrial disablement benefit (either pension or gratuity) does not seek to compensate for loss of earnings but rather seeks to replace loss of faculty and disablement, an employee can be in receipt of both sickness benefit and disablement pension or gratuity, or alternatively invalidity benefit and disablement benefit. Once an employed earner has paid 156 contributions, or done three years' work, he can receive sickness benefit followed by invalidity benefit for the rest of his life, coupled with either disablement pension or gratuity, as long as the disablement lasts.

(*d*) *Unemployability supplement.* If an employee is not entitled to invalidity benefit, e.g. he lacks the requisite number of contributions, and it appears to the medical board that it is unlikely that he will ever work again as a result of accident or disease caused at work, or, alternatively, if he cannot earn more than £104 per annum, he is entitled to an unemployability supplement. This is where his disablement caused by either accident or disease at work is so severe that working is an impractical proposition ever again. This unemployability supplement has the effect of making the disablement pension or gratuity earnings-related.

Other social security benefits

(*a*) *Special hardship allowance.* This benefit is payable to an employee in receipt of disablement benefit, either during the first six months' industrial injury benefit period, or after the expiry of the six-month period when the disablement period follows the industrial injury benefit period, when,

(i) the employee cannot, and is likely never to be able to, follow his regular employment; and

(ii) he is not able to carry on with employment of an equivalent nature.

Thus this benefit differs from invalidity benefit in that the latter presupposes that the employee is incapable of work, whereas the special hardship allowance presupposes that the employee has suffered a drop in earnings as a result of having to do a different job.

(*b*) *Allowance for wife and dependent children.* An employee will also be entitled to allowances in respect of his wife and dependent children, but only where he is in receipt of an unemployability supplement, or alternatively is an in-patient at a hospital.

(*c*) *Hospital treatment allowance.* This is an allowance paid to an employee in receipt of a disablement pension which is less than 100 per cent according to the tariffs set out in the Social Security (Benefits) Act 1975. Whilst such employee is an in-patient at a hospital he will receive a 100 per cent disablement pension. This is to encourage disabled persons to go into hospital.

(*d*) *Constant attendance allowance and exceptionally severe disablement allowance.* The constant attendance allowance is paid in cases of industrial and non-industrial injury victims. There are no contribution conditions, but the circumstances in which it will be paid are rare and refer to a person so severely disabled, with 100 per cent disability, that he is completely helpless and needs constant attention from a nurse or someone acting in that capacity both day and night. If this appears to the medical authorities to be a permanent condition, i.e. the person or child is a permanent invalid, the person looking after him, e.g. a parent, is entitled to an exceptionally severe disablement allowance, which is double in value the constant attendance allowance.

(*e*) *Industrial death benefit.* By virtue of s. 67, Social Security Act 1975, a widow of an employee fatally injured at work either by an accident or by a prescribed disease or by an accidental disease is entitled to industrial death benefit, which consists of flat-rate injury benefit plus earnings-related supplement payable after 12 days. In exceptional circumstances where a husband is being maintained by a wife, he can in respect of her death at work claim death benefit and so can the parents of a deceased employee. But before a widow can claim, she must have lived with her husband or have been maintained by him. Like industrial injury and industrial disablement benefits there are no contribution conditions; a person can be killed on the first day at work and these benefits will accrue to the specified persons. Industrial death benefit consists of a pension which lasts for six months after the death. Upon the expiry of the six-month period the pension reduces and it will stop altogether if the widow remarries or goes to live with another man. On remarriage, however, she will receive a gratuity which is the equivalent of one year's pension. When upon the expiry of the six-month period the pension reduces, the widow can receive a further pension at one of several rates, depending on her age and whether she has any dependent children.

24. Common law damages and social security benefits

It seems that the social security system does not subrogate itself to the tort system of compensation; the result is that an injured employee can be doubly indemnified or doubly compensated in respect of loss of earnings and disablement. The degree to which this in practice happens, assuming that no personal accident insurance has been bought (in which case the decision in *Bradburn* v. *G.W.R.* in 1874, which allows for double indemnity, applies), is limited by virtue of s. 2(1), Law Reform (Personal Injuries) Act 1948. This provides that in any action for damages for personal injuries, there shall, in the assessment of those damages, be taken into account, as against any *loss of earnings or profits* which have accrued or are likely to accrue to the injured person, *one-half* of the value of any industrial injury and/or industrial disablement or sickness benefits for *five* years beginning with the time when the cause of the action accrued, i.e. when the damage manifested itself. Social security benefits are offset as against damages awarded for loss of earnings or profits (made or probable), not against damages awarded for pain and suffering. Moreover no deduction is made in the assessment of damages when a disablement pension is supplemented by a constant attendance allowance or an exceptionally severe disablement allowance (*see* p. 149).

25. Criminal injuries

There are innumerable situations where criminal liability and liability in tort overlap. Where employees are the victims of intentional and premeditated acts during their employment, e.g. assault, they can sue their attackers in tort in the hope of being awarded damages; alternatively, if the assault discloses a faulty system of work, the victim of the assault can sue an employer for failure to take reasonable care at common law (*see* § **21,** p. 93), as where a rent collector employed by a firm of estate agents or by a local authority is assaulted and robbed of rent moneys collected, or where a bank clerk is attacked and robbed during the course of his employment. But apart from being torts, these types of conduct, acts of assault and robbery committed against persons in the course of their employment, are also criminal matters, and if the offenders are apprehended, prosecution will follow in the Magistrates' Court or the Crown Court.

The purpose of criminal law

Although the relationship between crimes and torts has been mentioned in § **8,** p. 15, it might be worth recalling that one of the functions of criminal law is to punish offenders either by imprisonment or fine. Originally, therefore, compensation had no role to play in criminal actions. Moreover when a fine was imposed in criminal proceedings it bore no resemblance to the injury suffered by the victim, especially in cases of violence. It was to deal specifically with crimes of violence that the Criminal Justice Act 1972, s. 1, allowed Crown Courts and the Magistrates' Court to make awards of compensation in favour of victims of crimes involving personal injuries or damage to property, up to a maximum of £400. This power is now to be found in the Powers of Criminal Courts Act 1973, ss. 35–38. The substance of these provisions is as follows.

(*a*) *Section 35(1)* states that a court before which a person is convicted of an offence, in addition to dealing with him in any other way, may make an order (a compensation order) requiring him to pay compensation for any personal injury, loss or damage resulting from that offence or any other offence. This general power contained in subs. (1) is modified by subs. (3), which states that no compensation order shall be made in respect of loss suffered by the dependants of a person in consequence of his death and no such order shall be made in respect of injury, loss or damage due to an accident arising out of the presence of a motor vehicle on a road (with certain exceptions). Sub-section (4) then goes on to say that in determining whether to make a compensation order and in determining the amount to be paid by any person, the court shall have regard to his means. Sub-section (5) limits the amount of compensation that may be awarded to £400.

(*b*) *Section 38* deals with the effect of compensation orders on a subsequent award of damages in a civil court. Sub-section (2) of this section provides that the damages shall be assessed without regard to the order; but where the whole or part of the amount awarded by the order has been paid, the damages awarded in the civil proceedings shall not exceed the amount (if any) by which, as so assessed, they exceed the amount paid under the order. This means that if a compensation order has been made either by the Crown Court or by the Magistrates' Court against an offender to the extent of £300 and he has already paid half of this, the damages awarded in a subsequent civil action cannot be more than £150. This serves to ensure that the extent of damages awarded in civil cases is not greater than compensation orders made by criminal courts. Sub-section (3) of the above section then says that where the whole or part of the amount awarded by the order remains unpaid and damages are awarded in later civil proceedings, the court shall direct that the judgment—

(i) if it is for an amount not exceeding the amount unpaid under the order, shall not be enforced; or

(ii) if it is for an amount exceeding the amount unpaid under the order, shall not be enforced as to a corresponding amount; without the leave of the court.

In other words, the purpose of subs. (3) is to see that the victim of the offence who later brings civil proceedings is not compensated twice. It is regarded as sufficient if the victim is compensated as from the compensation order made by the criminal court, and only with leave of the civil court can the damages be for a sum more than the amount unpaid on the compensation order.

In addition to these powers given to criminal courts victims of criminal offences who suffer injury or die whilst carrying out their employment, or their dependants, can benefit under the provisions of s. 50(1), Social Security Act 1975, or even from a personal accident insurance policy, which would normally have been taken out by a self-employed person as he is not entitled to benefit under the provisions of the Social Security Act 1975; in this way, though it is not compulsory, a self-employed person ensures that he is compensated if he is the victim of a crime of violence, or that in the event of his death his wife and family are compensated by virtue of the cover afforded by a life assurance policy.

The Criminal Injuries Compensation Board (*C.I.C.B.*)

Social security apart, assuming that the injury which has been suffered by the victim of the crime during the course of his employment is not fatal and assuming that a personal accident insurance policy has not been taken out, if the means of the offender are small and/or he is without employment or is in and out of employment, the likelihood of the victim being compensated by the offender is remote, unless a tort can be proved against the victim's employer for having failed to set up a safe system of work (which is not impossible). It was for this reason that the criminal injuries compensation scheme was set up in 1964, so that compensation might be made available on an *ex gratia* basis to victims of criminal offences, whether injured during the course of their employment or not. And, it seems, the Criminal Injuries Compensation Board has justified its existence, so much so that it is now proposed that it be put on a statutory footing, though even before this appeals lay to the High Court as to whether the Board was acting within the power given to it.

(*a*) *When does the C.I.C.B. award compensation?* The Criminal Injuries Compensation Scheme purports to make compensation available to persons (not just employees) who have suffered personal injury as a result of a crime of violence or as a result of arresting or attempting to arrest an offender or suspected offender (consistent with the power of citizen's arrest given in the Criminal Justice Act 1972) or arising from the prevention or attempted prevention of an offence or from the rendering of assistance to any constable who is arresting or attempting to arrest an offender or a suspected offender or who is preventing or attempting to prevent the commission of an offence.

It is thus clear that the C.I.C.B. does not have the jurisdiction to make awards of compensation where personal injury arises as a result of offences being committed in factories and other places of employment, e.g. as a result of a breach of s. 14(1), Factories Act 1961, or s. 2, H.A.S.A.W.A. Injuries arising from traffic offences are also outside the province of the C.I.C.B.'s jurisdiction. That is not to say, however, that all occupational hazards are not compensable under the criminal injuries compensation scheme. If a person's job renders him particularly vulnerable to criminal attack, e.g. a bank clerk or a petrol attendant or a policeman, it is clear from past precedent that compensation is payable by the C.I.C.B.

The C.I.C.B. also make compensation awards in favour of the dependants of victims, since deceased, of crimes leading to personal injury. This has occurred in cases where a policeman has been killed whilst attempting to arrest a dangerous criminal or where a fireman was killed whilst fighting a fire caused by someone convicted of arson. As with common law actions and actions for breach of statutory duty against employers, the C.I.C.B. will consider carefully whether the injury sustained by the victim was caused by the crime; causation is a relevant factor. If whilst chasing a thief round the zoo, a security officer is bitten by a snake or has a brick thrown at him by an angry boy whom he knocks off his bicycle, it is doubtful whether he could recover compensation from the C.I.C.B. Moreover most C.I.C.B. cases show that contributory negligence is a factor which the C.I.C.B. take into account in awarding compensation. If whilst carrying out his employment a dance-hall "bouncer" was badly beaten up and razored by a gang of thugs, and it was later established that the bouncer was prone to get involved in this sort of affray, i.e. was a trouble-maker always looking for a fight, the C.I.C.B. might refuse to make an award or reduce the amount of compensation he would receive.

(*b*) *How much compensation will the C.I.C.B. award?* C.I.C.B. assesses compensation in a way similar to the common law courts, but no compensation of less than £50 can be awarded and loss of earnings which exceed twice the industrial average earnings will not be awarded except where victims of criminal attack are turned into vegetables. Here very high awards may be made, especially if the victim is young with good career prospects. Compensation to dependants may also be awarded where death results from the attack, in a way similar to that in which it is awarded under the Fatal Accidents Act 1976.

(*c*) *Who administers the C.I.C.B.?* The Criminal injuries compensation scheme is administered by a Board assisted by lawyers. This Board will often make interim awards pending the grant of an industrial disablement pension or gratuity based on a medical report. Social security benefits are deductible from awards of compensation made by the C.I.C.B. Legal aid is not available, but many industrial victims are often represented by a company solicitor or by a trade union official. The Board is under an obligation to deduct from the amount of compensation it would have awarded any damages awarded in a tort action or through an insurance company in the form of insurance moneys, any social security benefits and any pension rights accruing from the injury, if permanent or fatal.

Chapter 3

The Health and Safety at Work Act 1974, the Factories Act 1961 and the Contractor's Liability

THE HEALTH AND SAFETY AT WORK ACT 1974

26. Purpose of the Act

H.A.S.A.W.A. came about through the recommendations to Parliament of the Committee on Safety and Health at Work, which sat under the chairmanship of Lord Robens, in the years 1970–1972. This Committee was not impressed with the record of previous protective industrial legislation: "It is a melancholy fact that much of our preventive legislation was introduced following events that were both unforeseen and disastrous." In the next paragraph (para. 20, Cmnd. 5034) the Report goes on to say: "The present corpus of occupational safety and health legislation has its roots in earlier legislation introduced in a quite different social context. Attitudes and expectations about authority and decision-making have changed. Increasingly, employers and workers expect to take an active part in the making and application of legislation of this kind." The Report observes that the traditional approach to the control of health and safety problems had always been based on an extensive system of detailed statutory provisions administered and enforced by government departments and local authorities. Might there not, in the light of changing attitudes, be room for improvement with legislation itself indorsing a different approach?

Another criticism levelled at the previous state of the law was that, each successive statute being aimed at the remedying of a single ascertained evil or the problems of a particular industry rather than industry in general, there was a surplus of law. As the Report urged: "The first and perhaps most fundamental defect of the statutory system is simply that there is too much law.... Our present system encourages rather too much reliance on state regulation, and rather too little on personal responsibility and voluntary, self-generating effort.

This imbalance must be redressed. A start should be made by reducing the sheer weight of the legislation. There is a role in this field for regulatory law and a role for government action. But these roles should be predominantly concerned not with detailed prescriptions for innumerable day-to-day circumstances but with influencing attitudes and with creating a framework for better safety and health organisation and action by industry itself."

The tangle of jurisdictions

The Robens Report made it clear that previous industrial safety legislation was badly structured, with obsolescence being a chronic disease. "What disturbs us about industrial safety and health administration is that there are too many demarcation lines, and that they appear to have emerged more by historical accident than by design. The pattern of control is one of bewildering complexity" (para. 32). This point effectively made, the Report went on to castigate the existence of nine separate health and safety statutes which dealt with factories, commercial premises, mines and quarries, agriculture, explosives, petroleum, nuclear installations, radioactive waste disposal and alkali emissions. This accounted for the existence of seven separate inspectorates (factories, mines, agriculture, explosives, nuclear installations, radiochemical and alkali). A veritable tangle of jurisdictions, commented the Report, and it then in the most unequivocal language went on to record: "The jurisdictional boundary lines do not coincide to provide a clear and comprehensive system of official provision for safety and health at work. On the one hand, the separately-administered statutes, taken together, cover nothing like the whole of the working population. On the other hand, some of them overlap in ways that can create uncertainty and confusion. Worse, the fragmentation of the legislation and its administration makes the task of harmonising, servicing and up-dating the various statutory provisions extremely difficult; and it diffuses and compartmentalises the expertise and facilities that are available to deal with occupational hazards."

At the level of any individual workplace, the Report pointed out, one establishment may be subject to a multiplicity of safety and health provisions under a number of Acts and various regulations. At the other end of the stratum, other workplaces fall outside the scope of any protective industrial legislation at all—road haulage depots, airports, schools, hospitals, etc. In a chemical works, on the other hand, the various parts of the works might well be subject to the Factories Act 1961, the Offices, Shops and Railway Premises Act 1963, the Petroleum Consolidation Act 1928 and the Alkali Act 1906.

At the level of inspectorates, the Report said, it was often difficult for the inspector to know in the case of a factory whether a particular building or operation came within the definition of a factory, as set out in s. 175(1), Factories Act 1961. This is understandable in view of the comprehensive definition of "factory" in that sub-section; it would take a lawyer well versed in factory law with a pile of case histories to pore over quite a few hours to make up his mind on a point about which the factory inspector might have had to make a snap decision. The Report cites railway running sheds and workshops alongside building operations on railway land as subject to the Factories Act 1961, whereas works of engineering construction on existing railways are not so subject.

At the national level of policy-formulation and lawmaking, it proved impossible for any government department with responsibilities in this field to take any major initiative without close and extensive consultation with several other departments. This, the Report urged, was a retrogressive state of affairs:

"It is widely accepted that the controls over certain flammable and explosive substances provided by the Explosives and Petroleum (Consolidation) Acts ought to be extended to a number of extensively used substances with equal or greater danger potential, such as ammonium nitrate, the liquefied petroleum gases, phosgene, sulphur dioxide and liquid oxygen. To do this would require further extension, or preferably complete revision, of the Explosives and Petroleum (Consolidation) Acts. However, these Acts already overlap the field covered by the Factories Act [1961] which is the province of the Department of Employment. One major effect of revising and extending the scope of the Explosive and Petroleum (Consolidation) Acts would be to increase this area of overlap with its attendant jurisdictional problems" (para. 39).

In its appraisal of the pre-H.A.S.A.W.A. system the Report finally commented: "The most fundamental conclusion to which our investigations have led us is this. There are severe practical limits on the extent to which progressively better standards of safety and health at work can be brought about through negative regulation by external agencies. We need a more effectively self-regulating system."

The new lawmakers on health and safety

All new regulations on health and safety will be proposed by H.S.C., a body created by s. 10, H.A.S.A.W.A., with membership drawn from the C.B.I. and T.U.C. In this way any proposals made will reflect the wishes of both sides of industry. Their proposals will be submitted to the Secretary of State for Employment for presentation before Parliament in the usual way. In order to ensure that the technology of control keeps pace with the technology of industry and invention, these new laws will consist almost wholly of regulations which will generally be within the general duties of s. 2–9, H.A.S.A.W.A., but there is power vested in the Commission by subs. 15(3)(*b*) to exclude or modify in relation to any specified class of case any of the provisions of s. 2–9 or any of the existing statutory provisions (e.g. the Factories Act 1961). Moreover by virtue of subs. 15(3)(*a*) the Commission may repeal or modify any of the existing statutory provisions. The consequence of this could be that, since many of the general duties require persons to do what is reasonably practicable, the Commission could propose regulations imposing strict liability upon persons within ss. 2–9. This could happen in the case of manufacturers and suppliers of products for use at work which cause injury in the light of the E.E.C. Draft Directive on Product Liability and the fact that the Pearson Commission has recommended strict liability in relation to product defects causing injury (*see* §**46,** p. 246).

"Under one roof"

Regulations will appear under H.A.S.A.W.A. relating to particular industries, particular processes and particular trades, so that a time will arrive when all the law relating to health and safety will consist of regulations proposed by H.S.C. These regulations will be implemented by one centralised administrative agency, H.S.E., so that all the law on the matter and responsibility for its administration will be brought under one roof. Previous protective industrial legislation will gradually disappear and be replaced by regulations.

The new enforcing agency

The new enforcing agency established by s. 10, H.A.S.A.W.A., is H.S.E. This body consists of an amalgamation of the various separate inspectorates which existed before the Act. It is the function of H.S.E. to administer the laws and regulations which are to be proposed by H.S.C. In order to do this and in an effort to minimise the incidence of workplace accidents rather than stimulate interest in compensation for them after they have occurred, the Report felt that the inspectorate should be given greater executive powers to prevent accidents arising on the shopfloor. To this end the Report recommended more executive powers for inspectors, a fact reflected in the provisions of H.A.S.A.W.A. Sections 20–26 give greater powers to inspectors to stop accidents, particularly ss. 21 and 22 giving them power to serve improvement and prohibition notices (*see* §**29,** p. 188).

(*a*) *Specific powers of the Commission.* The Robens Report argued that the Commission should have the following functions:

(i) provision of advice to ministers, government departments, local authorities, employers, trades unions and others, on health and safety at work;

(ii) management of the statutory inspection and advisory services;

(iii) administering and keeping under review statutory and other provisions for health and safety at work. This would include formulating health and safety standards for promulgation either as codes of practice or in the form of statutory provisions, and for co-operating with industry in monitoring the observance of those standards;

(iv) the acquisition and provision of information and the promotion and co-ordination of research, education and training for health and safety at work; and

(v) collaboration with the C.B.I. and other employers' organisations, the T.U.C. and the trade unions and the industry-level safety bodies as well as the voluntary safety organisations, and the work of international bodies. These recommendations have now been enshrined in s. 11, H.A.S.A.W.A., setting out the general functions of the Commission.

(*b*) *Control of the Commission by the Secretary of State for Employment.* Section 12, H.A.S.A.W.A., makes it clear that the Secretary of State for Employment may approve with or without modification any proposals submitted to him in pursuance of s. 11(3)(*a*). Section 11(3)(*a*) obliges the Commission to submit to the Secretary of State for Employment particulars of what it proposes to do for the purpose of performing its functions and in addition to give effect to any directions given to it by the Secretary of State for Employment.

(*c*) *Specific lawmaking powers.* The Report having observed that the general duties of the enabling Act were designed to "write down" a duty already existing at common law, added that the general duties of the Act were to be supported by more detailed provisions in the form of (i) regulations and (ii) voluntary standards and codes of practice. Cognisance was taken of the fact that the intermediate stage of enactment into regulation would be dispensed with, with the result that in some cases processes and industries would be governed by voluntary codes of practice. This accords with the general desire expressed in the Report for more voluntary self-regulating effort on the part of both sides of industry.

Consistently with these recommendations in the Report H.S.C. has been empowered in s. 15 to propose regulations relating to health and safety except for matters relating exclusively to agriculture. These proposals will be submitted to the Secretary of State, who will lay them before Parliament, this being made clear in s. 50. Section 16 is concerned with the approval of voluntary codes of practice, recognised as very important by the Robens Report: "... many of the defects ... can be remedied by a switch in emphasis away from the extensive use of statutory regulations towards greater reliance on standards and codes of non-statutory origin. In future there should be more discrimination and selectivity in making statutory regulations. Thus the system would comprise a main Act, plus statutory regulations, plus codes of practice; but the intermediate stage of statutory regulations would often be dispensed with" (para. 134).

Section 16(1) provides that, for the purpose of providing practical guidance with respect to the requirements of any provision of ss. 2–7 or of health and safety regulations or of any of the existing statutory provisions, the Commission may,

(i) approve and issue such codes of practice as it thinks suitable; and

(ii) approve such codes of practice issued or proposed to be issued otherwise than by the Commission as it may think suitable.

(*d*) *Consent of the Secretary of State.* Sub-section 16(2), however, states that a condition precedent to the approval of a code of practice by the Commission shall be the obtaining of the consent of the Secretary of State, and, if necessary before his consent is obtained, the Commission shall consult,

(i) any government department or other appropriate body; and

(ii) such government departments as the Commission is required to consult by the Secretary of State.

(*e*) *Revision of codes of practice.* It is clear from subs. 16(4) that the Commission may revise the whole or part of any part of any code of practice or approve any such revision of the whole or any part of such code.

New statutory regulations

(*a*) *Types of regulation.* There are to be three species of regulations—of a general nature applicable to most kinds of employment, e.g. general environmental standards and notification of accidents; those dealing with particular types of hazards, e.g. electricity and toxic substances; and those dealing with particular industries, e.g. engineering, mining and agriculture.

(*b*) *Relationship of regulations and voluntary codes of practice.* Regulations and voluntary codes of practice are to be interwoven to form a permanent marriage. What is envisaged is a regulation supplemented by detailed guidance from the inspectorate or by specifications contained in approved codes of practice or in British or international standards. In this way the need to avoid building obsolescence into regulations is reconciled with the desirability of providing precise guidance for those upon whom obligations are laid. The exact legal effect of this relationship is outlined in s. 17(1), H.A.S.A.W.A., which states that a failure on the part of any person to observe any provision of an approved code of practice shall not of itself render him liable to any civil or criminal proceedings; but where in any *criminal* proceedings a party is alleged to have committed an offence by reason of a contravention of any requirement or prohibition imposed by or under a code of practice, by virtue of s. 17(2) any provision of the code of practice which appears to the court to be relevant to the requirement or the prohibition alleged to have been contravened shall be admissible in evidence in the proceedings. In other words evidence of the breach of a provision of a code of practice may be regarded as evidence of guilt which by virtue of s. 40 the accused must discharge.

27. Main provisions of the enabling Act

The general duties of H.A.S.A.W.A. are to be found in Part 1, ss. 2–9. Their purpose is to achieve the following objectives, set out in s. 1:

(*a*) securing the health, safety and welfare of persons at work;

(*b*) protecting persons other than persons at work against risks to health or safety at work or arising in connection with activities at work;

(*c*) controlling the keeping and use of explosive or highly flammable or otherwise dangerous substances, and generally preventing the unlawful acquisition, possession and use of such substances; and

(*d*) controlling the emission into the atmosphere of noxious or offensive substances from premises of any class prescribed for the purposes of this paragraph.

The scope of the general duties

The scope of the general duties ranges from the imposition of duties upon employers (s. 2) and employees (s. 7), including the self-employed (s. 3), to occupiers of workplaces or owners (s. 4) and manufacturers and suppliers (s. 6). The duties imposed upon employers and manufacturers are wide and doubtless will be amplified by regulation. Use will be made of criminal sanctions, if improvement and prohibition notices fail to achieve their objective. In its chapter on administrative sanctions the Robens Report urged "in future much greater reliance should be placed on non-judicial administrative techniques for ensuring compliance with minimum standards of safety and health at work".

The effect of this is that the chief sanction for compliance with the general duties, any subsequent regulations and any already existing statutory provisions (e.g. the Factories Act 1961), together with voluntary codes of practice, will be *administrative rather than judicial*, i.e. improvement and prohibition notices which an inspector is empowered to serve. Should this fail to achieve its objective criminal proceedings may be initiated by the inspector.

Provision is also made by the Act for civil action by s. 47. The basis of this action is negligence since subs. (4) makes it clear that the Act contemplates retention of liability in tort. Moreover by virtue of subs. 2 new statutory torts may be created, though unless the regulations stated otherwise, the liability would be based on negligence. This last fact is of considerable importance in that it is not unlikely that the regulations may state otherwise: the regulations may make provision for the introduction of strict liability in the case of certain classes of persons, e.g. producers. In the light of the Strasbourg Convention and the E.E.C. Draft Directive on Product Liability, the imposition of strict liability in the case of injuries caused by product defect, following a breach of regulation passed pursuant to s. 6, H.A.S.A.W.A., is a certainty if the United Kingdom decides to adopt the approach of these two documents.

"Reasonably practicable"

All the general duties, with the exception of ss. 7 and 8, are modified by the words "As far as is reasonably practicable". This expression was defined by Asquith, L.J., in the leading case of *Edwards* v. *N.C.B.* in 1949:

"'Reasonably practicable' is a narrower term than 'physically possible', and seems to me to imply that a computation must be made by the owner in which the quantum or risk is placed on one scale and the sacrifice involved in the measures necessary for averting the risk (whether in money, time or trouble) is placed in the other, and that, if it be shown that there is a gross disproportion between them—the risk being insignificant in relation to the sacrifice—the defendants discharge the onus on them. Moreover this computation falls to be made by the owner at a point of time anterior to the accident."

In 1954, in *Marshall* v. *Gotham Co.*, Lord Reid again compared what is reasonably practicable with what is practicable. "If a precaution is practicable it must be taken unless in the whole circumstances that would be unreasonable. And as men's lives may be at stake it should not lightly be held that to take a practicable precaution is unreasonable. . . ." That this interpretation governs the meaning of the expression in H.A.S.A.W.A. is made clear in the Scottish case of *Belhaven Brewery Co. Ltd.* v. *McLean* (*H.M.F.I.*) in 1975. B.B.C. appealed against an improvement notice, served by M, requiring secure fencing to be provided to prevent access to the transmission machinery and other dangerous parts of B.B.C.'s automatic kegging plant. The schedule attached to the notice required that the fencing be screened with doors or gates for access fitted with an interlocking device, the effect of which would be to switch off the electric power and compressed air automatically when the door was open. B.B.C. argued that construction and installation of screens with an interlocking device would prove very expensive and that they would be complying with the relevant statutory requirements (s. 2(1), H.A.S.A.W.A., and ss. 13(1) and 14(1) Factories Act 1961) if they erected safety screens that did not have such a device. Moreover they contended that their workforce was sufficiently intelligent to see that the gates were kept in position while the plant was running. Also the plant was situated in a place where there would be a high level of supervision. The company also proposed to put a notice beside the screening warning employees of the danger. The inspector, however, argued that irrespective of the level of intelligence of the employees and the degree of management supervision, only screening which included an interlocking device (which meant that an operative could not gain access to the machine while it was in motion) could establish a secure system. Failure to install this sort of device would mean the company would continue to be in breach of section s. 2, H.A.S.A.W.A., and ss. 13 and 14, Factories Act 1961. It was held by a Scottish Tribunal dismissing the appeal by the company, that the improvement notice would be affirmed without modification. The only method by which the contravention could be remedied was by fitting an interlocking device of the type specified in the schedule to the notice. On the other hand, if screening of the kind proposed by the company were to be fitted, this would not result in the transmission machinery and the dangerous parts of the machinery being securely fenced. The company would still, therefore, continue to be in contravention of the Factories Act 1961.

In addition the company would not be carrying out its duty under s. 2(1), H.A.S.A.W.A., to ensure the health, safety and welfare at work of their employees. The tribunal stated that the employer's duty under s. 2(1) to ensure the health, safety and welfare of employees was not an absolute duty but rather was limited to doing what was "reasonably practicable". On this basis, applying the test in *Edwards* v. *N.C.B.*, the risk associated with the omission to fit an interlocking device was not so insignificant in relation to the sacrifice involved in fitting such a device that the fitting of an interlocking device fell outside the parameters of what was reasonably practicable. It being reasonably practicable to fit or install an interlocking device specified in the schedule to the improvement notice, if they failed to do so, the employer would be in breach of s. 2(1), H.A.S.A.W.A.

28. Employers, employees and the self-employed

The duties of employers

The duties of employers are to be found in s. 2 of the Act, which is one of the most important sections. Sub-section 2(1) states that it shall be the duty of every employer to ensure, so far as is reasonably practicable, the health, safety and welfare at work of all his employees, whilst subs. 2 enumerates the employer's duties as follows:

(*a*) the provision and maintenance of plant and systems of work that are, so far as is reasonably practicable, safe and without risks to health;

(*b*) the making or arrangements for ensuring, so far as is reasonably practicable, safety and absence of risks to health in connection with the use, handling, storage and transport of articles and substances;

(*c*) the provision of such information, instruction, training and supervision as is necessary to ensure, so far as is reasonably practicable, the health and safety at work of his employees;

(*d*) the maintenance, so far as is reasonably practicable, of any place of work under the employer's control in a condition that is safe and without risks to health, and the provision and maintenance of means of access to and egress from it that are safe and without risks;

(*e*) the provision and maintenance of a working environment for his employees that is, so far as is reasonably practicable, safe, without risks to health, and adequate as regards facilities and arrangements for their welfare at work.

Failure to implement these duties may give rise to criminal liability under s. 33 and to a claim for damages at common law.

The nature of the duty

The duties of this Act are laid upon employers as a whole rather than upon sections of employers, as in the past when duties were laid by separate statutes upon factory occupiers, mine owners, and occupiers of offices and shops. The place where a person works, such as a factory, office or shop, is now of minimal significance. What matters is that such a person is employed. In this way more of the national workforce will be covered by this latest in the long series of protective industrial injury statutes.

Health, safety and welfare

Sub-section 2(1) enjoins the employer to ensure, as far as is reasonably practicable, the health, safety and welfare of his employees. "Health" includes not only physical but also mental health; "safety" refers to freedom from foreseeable injury; and "welfare" refers to matters such as places to hang coats, facilities for washing of face and hands, etc.

(*a*) *The wide ambit of subs. 2(2).* Sub-section 2 deals with (i) provision and maintenance of plant; (ii) arrangements for safe transport of articles; (iii) information, training and supervision; (iv) safe place of work; and (v) safe working environment.

(*b*) *Provision and maintenance of plant and equipment. Norris* v. *Syndic Manufacturing Co. Ltd.* in 1952 establishes that plant is provided when it is "ready to hand in a proper place". The sub-section also requires that plant be maintained. Here remarks made in § **21,** p. 90 in connection with the employer's duty to maintain safe plant at common law are relevant, except that breach of subs. 2(2) (*a*) attracts criminal sanction under s. 33. It is thought that cases like *Webb* v. *Rennie* and *Clarke* v. *Holmes* mentioned in § **21,** p. 91 as well as *Barkway* v. *South Wales Transport Co. Ltd.* in connection with the implementation of a defects' monitoring system may become relevant in the interpretation of subs. 2(2) (*a*) and form part of the criminal law, especially as it was Robens' wish to "write down" a duty existing at common law in H.A.S.A.W.A.

(*c*) *Safe handling and transport of goods and substances.* Sub-section 2(2)(*b*) requires every employer to see that the proper arrangements have been made for the handling, transport, storage and use of articles and substances. This includes making sure that materials made by him are packaged safely, that the vehicles used for transporting them from one depot to another or from factory to outlet point, whether dock, depot or airport or retailing point, are kept and maintained in a safe condition. Places where such articles and substances are to be stored should also be kept safe. The sub-section also contemplates the supply by the employer to his workforce of protective clothing for the purposes of handling and using the articles and substances.

(*d*) *Provision of information, instruction, training and supervision.* Sub-section 2(2)(*c*) requires the employer to train his employees in job safety. Such training may be available through the network of technical colleges, polytechnics, employers' associations and W.E.A. centres. This statutory obligation accords with the requirements of the Employment and Training Act 1973 and the obligations of the U.K. under Art. 10, European Social Charter, whereby the U.K. agrees to promote safe and healthy conditions at work, including the provision of technical and vocational training of all persons in consultation with both employers' and workers' organisations. Meanwhile the Employment and Training Act 1973 set up the Manpower Services Commission to oversee training needs and requirements. Since the passing of the Industrial Training Act in 1964 employers have to pay a levy to industrial training boards for the training of their employees and it has been customary for I.T.B.'s to issue exemption certificates if they are satisfied that the employer's training arrangements are satisfactory. This question is determined by the

Manpower Services Commission and the Secretary of State. Levies collected by I.T.B.'s are subsequently disbursed as grants to employers who carry out training to the satisfaction of I.T.B. requirements. Most industries have their own industrial training board responsible for the training needs of that particular industry and safety training is a part of this training requirement.

The importance of training

The Robens Report in para. 382 stressed the overall need for and importance of safety training. The Committee were of the opinion that safety training constituted an integral part of normal job training and that in certain cases specialised safety training was essential: "it . . . is equally clear that specialised training in safety matters is needed for particular categories such as managers, safety officers, instructors, supervisors and workers' safety representatives. The needs here are of the type that will often be best met by the provision of special off-the-job courses, organised to fit in with the general training of the grades concerned." As to whether this training should form part of normal job training or take the form of specialised instruction and separate safety training, the current view indicates that with the exception of safety officers and workers' representatives, safety training should be built into normal job training. In the case of safety officers and workers' safety representatives, the safety training would have to be specialised and geared to instruction in safety problems of the particular industry concerned.

(*a*) *Failure to train.* Failure to train or have employees trained in job safety can be a risky affair. It can lead to prosecution by an inspector under s. 33, and could have consequences in a civil action. *Shatwell's* case established that a trained and certificated skilled operative who neglected to implement job safety could end up by losing all his compensation in the event of his being injured. The corollary here seems to be that the untrained uncertificated operative could never be regarded as the sole cause of an injury that occurred to him.

(*b*) *Industry's view of training.* Industry's view of safety training is that management will be trained by employers' organisations and workers will be trained through T.U.C. facilities and W.E.A. centres, both sides making use of the network of technical colleges and polytechnics. As regards safety officers and safety representatives, the training is much more specialised and both employers and trades unions seem now to have recognised the need for safety officers and safety representatives to be trained to associate and full membership level of the Institution of Industrial Safety Officers.

(*c*) *The training of safety representatives.* Sub-section 2(4) provides for the creation of safety representatives appointed by a trade union to look after the safety and welfare of the workforce and subs. 2(6) requires an employer to consult on safety matters with safety representatives. They are also given further powers in the Safety Representatives and Safety Committees Regulations which became operational in October 1978. The T.U.C. view is that trade unions should have sole responsibility for the education and training of safety representatives as far as performance of their union activities is concerned. It rejects the idea that such training should be a responsibility jointly shared as between trade union and employer. The argument of the T.U.C. is based essentially on the fact that the Employment Protection Act 1975 makes specific provision for paid release for trade union training (see Part 3, Employment Protection Act 1975, s. 57). Courses for union-appointed safety representatives are currently being run at local technical colleges, polytechnics, W.E.A. centres and extra-mural departments of universities. There are also postal correspondence courses. The T.U.C. Course for the training of Safety Representatives covers the following aspects:

(i) *Accident prevention*. This includes knowledge of the widespread presence of hazards, the range of hazards, the effects of hazards on workers' health and safety; the need to identify hazards, the need to eliminate them; recognising a hazard, using a simple checklist of hazards; introduction to workplace inspection, introduction to handling statistics, identifying hidden hazards and their symptoms; a simple checklist for specific industries and workplaces.

(ii) *Law*. This includes knowledge of the employer's legal duty under H.A.S.A.W.A. to promote health and safety and welfare; the employee's duty of co-operation; the trade union approach to making the employer accountable, employee's duties, safety policies, safety representatives and safety committees; how to read an Act of Parliament. Knowledge of H.A.S.A.W.A., its relation to other Acts (e.g. the Factories Act 1961) and to common law; the Factories Act 1961, standards set down in the Act, standards made since the Act, regulations passed pursuant to the Act; an understanding of the general duties of H.A.S.A.W.A., the way in which it is enforced, the powers of H.S.C. to propose future regulations and other standards; the need to improve on workplace conditions beyond basic requirements, the legal status of regulations, codes of practice and other standards. A knowlege of the Offices, Shops and Railway Premises Act 1963, the scope of its coverage, standards set down in the Act, standards made since the Act and regulations, plus other Acts and regulations for specific industries and workplaces. Knowledge of legal and other standards on noise, dust, toxic chemicals, guarding machinery, mechanical handling and fire; how hazards are measured, how standards are set, the dangers of hidden hazards; checking off standards against actual conditions, measuring and monitoring. Knowledge of the safety representatives and safety committee, regulations of the union safety representative's role, role of the safety committee, need for an agreed company safety policy, plus ability to inspect accidents, dangerous occurrences and the circumstances of diseases; the role of the inspectorate, the role of the Employment Medical Advisory Service (E.M.A.S.), the role of H.S.C., the ability to use the inspectorate, E.M.A.S. and the Industrial Advisory Committees.

The 10-day T.U.C. specimen syllabus is as follows:

Legislation	14 hours
Claims and benefits	$2\frac{1}{2}$ hours
Hazard identification	$8\frac{1}{2}$ hours
Responsibilities and election of safety representatives	$2\frac{1}{2}$ hours
Responsibilities and organisation of safety committees	$3\frac{1}{4}$ hours
Consideration and preparation of checklists	$12\frac{1}{4}$ hours
Trade union attitudes	$3\frac{1}{4}$ hours
Role of the factory inspectorate	$3\frac{1}{4}$ hours
Accident reports	$3\frac{1}{4}$ hours
Forum and review	$2\frac{1}{2}$ hours
Preparation of safety committee constitutions	$2\frac{1}{2}$ hours
TOTAL	$57\frac{3}{4}$ hours

Audio-visual methods are envisaged. This accords with E.E.C. thinking on training in health and safety at work, which recommends in its guidelines the "utilisation of audio-visual information methods to improve the provision of information for accident prevention purposes for those sectors of the population especially affected. ..." Training is now a statutory duty upon all employers. In the case of smaller employers lacking a training department this might well be carried out by a group training service.

Provision of safe place of work and healthy working environment

Section 2(2)(*d*) requires an employer, as far as is reasonably practicable, to maintain any place of work under his control in a safe condition and without risks to health, including means of access and egress. Section 2(2)(*e*) requires that an employer provide and maintain a working environment for his employees that is, so far as is reasonably practicable, safe and without risks to health and adequate as regards facilities and arrangements for their welfare at work (it is thought that because the general duties "write down" a duty existing at common law, case law on employer's liability at common law mentioned in §**21** may be looked at as a guide to the meaning and ambit of the sub-sections of s. 2, subject to the proviso that liability for breach of statutory duties is criminal).

(*a*) *Preparation of company safety policy*. Sub-section 2(3) is new and requires that, except in such cases as may be prescribed it shall be the duty of every employer to prepare and as often as may be appropriate revise a written statement of his general policy with respect to the health and safety at work of his employees and the organisation and arrangements for the time being in force for carrying out that policy, and to bring the statement and any revision of it to the notice of all his employees. The duty to prepare and revise a written safety policy only applies in the case of employers employing five or more employees. In other words the smallest firms are exempt.

It is difficult to know at first what might be regarded as an adequate and suitable statement of safety policy. No detailed guidance is given on this matter by subs. 2(3) itself. It is clear, however, that the following ingredients should appear in any company safety policy:

(i) a statement of the employer's general policy with respect to the health and safety at work of his employees;

(ii) the organisation employed or appointed to implement that policy so as to be identifiable;

(iii) the arrangements for making that policy effective so as to be identifiable.

When it comes to preparing company safety policies, employers will be able to draw on codes of practice relating to the processes and operations which they carry out. Moreover, the terms of collective agreements and individual contracts of employment will identify measures that are to be taken by the workforce. Also job descriptions of safety advisers and supervisory staff will be necessary. Detailed assessment and identification of hazards in all areas affecting workforce safety will be necessary and will form a vital part of the company policy statement.

The company safety policy will serve as the starting-point for an investigation of company safety by an inspector. If there is no safety policy in existence s. 33 will apply and fines and/or imprisonment against the offenders are not unlikely. Already there are examples of this having happened, as in a recent case against Wiltshire County Council where two breaches of H.A.S.A.W.A. had occurred. An employee at the local refuse depot was trapped and crushed between a container operated by a hydraulic lorry and a hopper in which refuse was compacted, and died from his injuries. Another employee had backed his lorry carrying an empty rubbish container on to the Transpack machine. He and the deceased employee had cleared some rubbish away from the back of the lorry and he got back into the van. He was asked to pull forward about 9–12 in. for a couple of minutes, as a piece of metal was jammed, after which he looked round and reversed the container back on to the Transpack. Whilst this was happening the deceased was crushed between the container and the Transpack. On these facts the Magistrates found the County Council guilty. The Chairman of the County Council and his fellow councillors were fined £200 on each count involving two breaches of H.A.S.A.W.A.: (i) breach of s. 2(1) in that they had failed to set up a safe system of work, and (ii) breach of s. 2(3) in that they had failed to bring to the notice of their employees a statement of the general policy in respect of the health and safety at work of their employees. The company safety policy will constitute the starting-point of the inspector's inquiries in determining whether to issue improvement or prohibition notices or, failing that, to prosecute.

(*b*) *Consultation with union-appointed safety representatives.* Sub-section 2(6) provides that every employer is under a duty to consult with safety representatives with a view to the making and maintenance of arrangements which will enable him and his employees to co-operate effectively in promoting and developing measures to ensure the health and safety at work of employees, and in checking the effectiveness of such measures.

When the Act first appeared in 1974 provision was made in it for the appointment of safety representatives from recognised trade unions under subs. 2(4) and from non-unionised workforces under subs. 2(5). This latter provision has since been repealed by s. 116, Employment Protection Act 1975. This means that there is only a duty on employers of unionised labour to consult with safety representatives from amongst the workforce. In the case of non-unionised workforces there is no duty of consultation upon the employer. Alternatively there would appear to be nothing to stop an employer of a non-unionised labour force appointing safety representatives from the workforce and establishing safety committees, although such representatives would not have statutory powers nor the statutory immunity accorded to union-appointed safety representatives by the regulations.

(*c*) *The Safety Representatives and Safety Committees Regulations 1977.* These regulations were laid before Parliament on 28 March 1977 and became operational on 1st October 1978. The trade union from whose members they are appointed must be recognised by the employer for collective bargaining purposes. Recognition depends on the union being independent as defined in s. 30(1), Trade Union and Labour Relations Act 1974, which refers to the trade union being free from interference by the employer. If a trade union can satisfy the two conditions enumerated in ss. 29 and 30, Trade Union and Labour Relations Act 1974, it is entitled to apply to A.C.A.S. for recognition. This recognition can take place ultimately whether the employer consents or not, compulsory unilateral arbitration being provided for specifically in s. 16, Employment Protection Act 1975. Safety representatives may be shop stewards or convenors. The A.U.E.W. is on record to the effect that their safety representatives will be stewards. The nature of the relationship between trade union and safety representative is probably that of principal and agent. At present this fact is of no great practical significance since trade unions enjoy immunity in tort under the Trade Union and Labour Relations Act 1974, as a result of which they can incur no vicarious liability for any industrial action initiated by shop stewards. If there was a revival of the Industrial Relations Act 1971, however, the relationship would become important again.

(*d*) *Appointment of safety representatives*. Section 3(1) of the 1977 regulations states that for the purpose of s. 2(4), H.A.S.A.W.A., a recognised trade union may appoint safety representatives from amongst the employees in all cases where one or more employees are employed by an employer by whom it is recognised.

Sub-section (2) states that where the employer has been notified in writing by or on behalf of a trade union of the names of the persons appointed as safety representatives under this regulation and the group or groups of employees they represent, each such safety representative shall have the functions set out in reg. 4.

Sub-section (3) states that a person shall cease to be a safety representative for the purposes of the regulations when:

(i) the trade union which appointed him notifies the employer in writing that his appointment has been terminated; or

(ii) he ceases to be employed at the workplace but if he was appointed to represent employees at more than one workplace, he shall not cease, by virtue of this paragraph, to be a safety representative so long as he continues to be employed at any one of them; or

(iii) he resigns.

Sub-section (4) states that a person appointed under subs. (1) as a safety representative shall so far as is reasonably practicable either have been employed by his employer throughout the preceding two years or have had at least two years' experience in similar employment.

H.S.C. envisages that unions would normally appoint a group or groups of workers of a class for which the union has the negotiating rights. This does not preclude the possibility of a safety representative representing, by mutual agreement between the appropriate unions, more than one group or groups of employees (e.g. in a small workplace where the number of unions is high relative to the total number of employees). Similarly when consideration is being given to the numbers of safety representatives to be appointed in particular cases, the criteria which should be borne in mind are as follows:

(i) the total numbers employed;

(ii) the variety of different occupations;

(iii) the size of the workplace and the variety of workplace locations;

(iv) the operation of shift systems;

(v) the type of work activity and the degree and character of the inherent dangers.

(*e*) *Functions of safety representatives.* Regulation 4(1) states that in addition to his function under s. 2(4) to represent the employees in consultation with the employer under s. 2(6), H.A.S.A.W.A., each safety representative shall have the following functions:

(i) to investigate potential hazards and dangerous occurrences at the workplace (whether or not they are drawn to his attention by the employees he represents) and to examine the causes of accidents at the workplace;

(ii) to investigate complaints by any employee he represents relating to that employee's health, safety or welfare at work;

(iii) to make representations to the employer on matters arising out of the above two sub-paragraphs;

(iv) to make representations to the employer on general matters affecting the health, safety or welfare at work of the employees at the workplace;

(v) to carry out inspections in accordance with regs. 5, 6 and 7 (*see* below);

(vi) to represent the employees he was appointed to represent in consultations at the workplace with inspectors of H.S.E. and of any other enforcing authority (e.g. the public health inspector);

(vii) to receive information from inspectors in accordance with s. 28(8), H.A.S.A.W.A.; and

(viii) to attend meetings of safety committees where he attends in his capacity as a safety representative in connection with any of the above functions; but without prejudice to ss. 7 and 8 of these regulations, no function given to a safety representative shall be construed as imposing a duty on him.

Two expressions here—"welfare at work" and "workplace"—deserve closer scrutiny.

"Welfare at work" means those aspects of welfare at work which are the subject of health and safety regulations (not necessarily confined to H.A.S.A.W.A. itself). This would include existing statutory provisions relating to welfare and passed pursuant to or as part of the Factories Act 1961 or the Offices, Shops and Railway Premises

Act 1963. "Workplace" in relation to the safety representative means any place or places where the group or groups of employees he is appointed to represent are likely to work or which they are likely to frequent in the course of their employment or incidentally to it, e.g. cloakrooms.

It is important that safety representatives should be able to take up matters with management without delay. They must, therefore, have ready access to the employer or his safety adviser. Safety representatives should record when they have made an inspection, a copy of each such record should then be sent to the employer. In addition, for the purposes of s. 28(8), H.A.S.A.W.A., safety representatives are the correct persons for receiving information from inspectors. Section 28(8), H.A.S.A.W.A., provides as follows:

"Notwithstanding anything in the preceding sub-section an inspector shall, in circumstances in which it is necessary to do so for the purpose of assisting in keeping persons (or the representatives of persons) employed at any premises adequately informed about matters affecting their health, safety and welfare, give to such persons or their representatives the following descriptions of information, that is to say—

(i) factual information obtained by him as mentioned in that subsection which relates to those premises or anything which was or is therein or was or is being done therein; and

(ii) information with respect to any action which he has taken or proposes to take in or in connection with those premises in the performance of his functions; and where an inspector does as aforesaid, he shall give the like information to the employer of the first-mentioned person."

(*f*) *Training*. Sub-section 4(2) of the regulations states that an employer shall permit a safety representative to take such time off with pay during the employee's working hours as shall be necessary for the purposes of:

(i) performing his functions under s. 2(4), H.A.S.A.W.A., and the eight functions ... outlined in reg. 4(1) (*see* (*e*) above); and

(ii) undergoing such training in aspects of those functions as may be reasonable in all the circumstances having regard to any relevant provisions of a code of practice relating to time off for training approved for the time being by H.S.C. under s. 16, H.A.S.A.W.A.

(*g*) *Inspections of the workplace*. Regulation 5(1) states that safety representatives shall be entitled to inspect the workplace or a part of it if they have given the employer or his representative reasonable notice in writing of their intention to do so and have not inspected it, or that part of it, in the previous three months; and may carry out more frequent inspections by agreement with the employer.

Sub-section (2) prescribes that where there has been a substantial change in the conditions of work (whether because of the introduction of new machinery or otherwise) or new information has been published by H.S.C. or H.S.E. relevant to the hazards of the workplace since the last inspection under this regulation, the safety representatives, after consultation with the employer, shall be entitled to carry out a further inspection of the part of the workplace concerned notwithstanding that three months have not elapsed since the last inspection.

Sub-section (3) states that the employer shall provide such facilities and assistance as the safety representatives may reasonably require (including facilities for independent investigation by them and private discussion with the employees) for the purpose of carrying out an inspection under this regulation, but nothing in this paragraph shall preclude the employer or his representative from being present in the workplace during the inspection.

Inspection will normally be carried out every three months, but where there is a high-risk workplace activity or there are rapidly changing conditions at the workplace (e.g. a building site), more frequent inspections may be appropriate. Moreover H.S.C. feels that employers and safety representatives should plan a programme of formal inspection in advance, though this does not rule out independent inspections by safety representatives. The safety officer or adviser may well be called upon to give specialist advice during the inspection. The Commission considers the following types of inspection, or a combination of them, necessary for the adequate fulfilment of the functions of safety representatives:

(i) safety tours or general inspections of the workplace;

(ii) safety sampling or the systematic sampling of particular dangerous activities, processes or areas;

(iii) safety surveys or general inspections of the particular dangerous activities, processes or areas in question.

The number of representatives taking part in an inspection should be a matter of agreement with the employer. In large workplaces the inspection would probably be better carried out by breaking it up into manageable units, as the code of practice accompanying the regulations urges, e.g. on a departmental basis. Where safety representatives have made a written report to the employer in accordance with para. 5(c) of the code of practice, appropriate remedial action will normally be taken by the employer. But where remedial action is not considered appropriate or cannot be taken within a reasonable period of time or the form of remedial action is not acceptable to the safety representatives, the Commission suggests that the prescribed form should be filled in. Where remedial action is taken, the safety representative should be allowed to re-inspect the workplace and such action as has been taken should be published throughout the workplace and brought to the attention of safety committees.

(*h*) *Inspections following notifiable accidents, occurrences and diseases.* Regulation 6(1) states that where there has been a notifiable accident or dangerous occurrence in a workplace or a notifiable disease has been contracted there, and

(i) it is safe for an inspection to be carried out, and

(ii) the interests of employees in the group or groups which safety representatives are appointed to represent might be involved,

those safety representatives may carry out an inspection of the part of the workplace concerned and, so far as is necessary for the purpose of determining the cause, they may inspect any other part of the workplace; where it is reasonably practicable to do so, they shall notify the employer or his representative of their intention to carry out the inspection.

Sub-section (2) prescribes that the employer shall provide such facilities and assistance as the safety representatives may reasonably require (including facilities for independent investigation by them and private discussion with the employees) for the purpose of carrying out an inspection under this regulation; but nothing in this paragraph shall preclude the employer or his representative from being present at the workplace during the inspection.

Sub-section (3) defines "notifiable accident" or "dangerous occurrence" as "any accident, dangerous occurrence or disease, as the case may be, notice of which is required to be given by virtue of any of the relevant statutory provisions within the meaning of s. 53(1), H.A.S.A.W.A." (as for what constitute notifiable accidents and diseases, *see* § **23,** p. 143).

For the purpose of ascertaining the circumstances of a notifiable accident, dangerous occurrence or notifiable disease, it will be necessary for the safety representative to examine any relevant machinery, plant, equipment or substance in the workplace. The main purpose of this is to ascertain the causes of an accident so as to prevent a recurrence. If the employer, in the light of an inspection, has to take steps to safeguard against further hazards, he should notify the representative of this. Such examinations of machinery and plant may include visual inspection and discussions with persons who are likely to be in possession of relevant information regarding the circumstances of the accident. But the examination must not include interference with any evidence or the testing of any machinery, plant or equipment or substance which could destroy or disturb the factual evidence before the inspector has had an opportunity to investigate.

(*i*) *Inspection of documents and provision of information.* Regulation 7(1) states that safety representatives shall, for the performance of their functions under s. 2(4), of H.A.S.A.W.A., and under these regulations, if they have given the employer reasonable notice, be entitled to inspect and take copies of any document, relevant to the workplace or to the employees the safety representatives represent, which the employer is required to keep by virtue of any relevant statutory provisions within the meaning of s. 53(1), H.A.S.A.W.A., except a document relating to any health record of an identifiable individual.

Sub-section (2) prescribes that an employer shall make available to safety representatives the information, within the employer's knowledge, necessary to enable them to fulfil their functions except:

(i) any information the disclosure of which would be against the interests of national security, or

(ii) any information which he could not disclose without contravening a prohibition imposed by or under an enactment, or

(iii) any information relating specifically to an individual, unless he has consented to its being disclosed, or

(iv) any information the disclosure of which would, for reasons other than its effect on health, safety or welfare at work, cause substantial injury to the employer's undertaking or, where the information was supplied to him by some other person, to the undertaking of that other person, or

(v) any information obtained by the employer for the purpose of bringing, prosecuting or defending any legal proceedings.

Sub-section (3) states that subs. (2) does not require an employer to produce or allow inspection of any document which is not related to health, safety or welfare.

The sort of information which should be made available here by employers to safety representatives is as follows:

(i) information about the plans and performance of their factories and undertakings, in so far as they relate to health and safety matters, and any changes proposed, in so far as they affect health and safety; and

(ii) information of a technical nature about hazards to health and safety.

In this latter case where technical matters are involved, either the employer or safety representative or both are entitled, should they so wish, to call in expert advice from outside, such as from neighbouring polytechnics, technical colleges or universities. In some cases it may be necessary to call in the manufacturer of the machinery or some outside consultant with expert knowledge of such a machine, e.g. a power press, abrasive wheel or fork-lift truck.

(*j*) *Where safety representatives need not be employees.* Regulation 8(1) states that in the cases mentioned in subs. (2), safety representatives appointed under reg. 3(1) need not be employees of the employer concerned.

Regulation 8(2) states that this applies where the employees in the group or groups the safety representatives are appointed to represent are members of Equity or the Musicians' Union.

Safety committees

Sub-section 2(7), H.A.S.A.W.A., states that in such cases as may be prescribed it shall be the duty of every employer, if requested to do so by the safety representatives, to establish a safety committee having the function of keeping under review the measures taken to ensure the health and safety at work of his employees and such other functions as may be prescribed.

Regulation 9(1) states that this shall apply where at least two safety representatives request the employer in writing to establish a safety committee.

Regulation 9(2) states that where an employer is requested to establish a safety committee, it shall be done as follows:

(*a*) the employer shall consult with the safety representatives who made the request and with the representatives of recognised trade unions whose members work in any workplace in respect of which he proposes that the committee should function;

(*b*) the employer shall post a notice stating the composition of the committee and the workplace or workplaces to be covered by it in a place where it may be easily read by the employees;

(*c*) the committee shall be established not later than three months after the request for it.

The functions of safety committees, says the Commission, will include:

(i) the study of accident and notifiable diseases statistics and trends, so that reports can be made to management on unsafe and unhealthy conditions and practices, together with recommendations for corrective action;

(ii) examination of safety and audit reports on a similar basis;

(iii) consideration of reports and factual information provided by inspectors of the enforcing authority appointed under H.S.E.;

(iv) consideration of reports which safety representatives may submit;

(v) assistance in the development of works safety rules and safe systems of work;

(vi) keeping a watch on the effectiveness of the safety content of employee training;

(vii) keeping a watch on the adequacy of safety and health communication and publicity in the workplace;

(viii) the provision of a link with the appropriate inspectorates of the enforcing authority.

In addition safety committees may themselves carry out inspections, though it will be management's role to take executive action and to have adequate arrangements for regular and effective monitoring of health and safety precautions and for ensuring that the company health and safety policy is being implemented. The safety committees will supplement this role of management, they will not be a substitute.

(*a*) *Membership*. Membership is a matter to be settled as between management and trade unions' representatives and should reflect the interests of management and employees as well as safety representatives. There should be an equal number of representatives from both sides. Representation from management should include line management, works engineers and personnel managers as well as supervision. H.S.C. stresses that management representation should be aimed at ensuring:

(i) adequate authority to give proper consideration to views and recommendations; and

(ii) the necessary knowledge and expertise to provide accurate information to the committee on company policy, production needs and technical matters in relation to premises, processes, plant, machinery and equipment.

It is recommended that the company doctor (if any), an industrial hygienist (if any) and the company safety officer/adviser should be ex-officio members of the safety committee. Other company specialists, such as project engineers, chemists, organisation and methods staff and training officers, might be co-opted for particular meetings when their expertise might prove valuable. Since membership of a safety committee is to be treated as part and parcel of a person's job, no loss of pay should be incurred through this.

H.S.C. stresses that it is above all the job of the safety committee to study accidents so that they should not recur, rather than to apportion blame, which is not within their terms of reference. Safety committees should examine the facts surrounding an accident in an impartial way, consider what sort of precautions might be taken and make appropriate recommendations. This examination should include not just legally notifiable accidents (for the purposes of entitlement to industrial injury benefit under the Social Security Act 1975) but also minor injuries as well, as these can yield useful factual information.

The safety committee should thus be in a position to advise on the appropriateness and adequacy of the rules for health and safety proposed by management and/or draw attention to the need to establish rules for a particular hazardous job or operation or class of operations. Written reports made earlier by safety representatives should be studied by safety committees with a view to suitable publicity. In an effort to give teeth to safety committees, there should take place:

(i) regular meetings with effective publicity given to the discussions and other recommendations;

(ii) speedy decisions by management on the committee's recommendations, where necessary translated into prompt action backed up with effective publicity;

(iii) participation by the safety committee in periodical joint inspections between employer and safety representatives; and

(iv) development of ways of involving more employees in safety programmes.

(*b*) *Immunity from legal proceedings*. Although in certain industries the idea of safety committees is by no means a recent one (many, in fact, are traceable back to the end of the Second World War), the existence of safety representatives equipped with statutory powers is a departure from tradition. It is worth noting that by virtue of reg. 4(1) of these regulations there is no question of a safety representative being either prosecuted by an inspector or being sued by an injured workman. Thus if as a result of suggestions made by a safety representative, certain measures were introduced into the workplace which proved to be the cause of a workman's injury or disease, there cannot be any action against the safety representative for compensation, either exclusively or in conjunction with his employer. Even less can there be an action against his trade union, which enjoys immunity in a similar way (*see* §**39**, p. 233). This is not to say, however, that a safety representative could not be prosecuted as an employee under s. 7 (in other words when not performing a safety function), for breach of H.A.S.A.W.A.

This immunity of safety representatives from subsequent legal action is consistent with the immunity accorded to trade unions under a succession of enactments culminating in the Trade Union and Labour Relations Act 1974.

The duties of the self-employed

The Robens Report recognised the difficulty of applying similar rules to the self-employed as apply to employers and employees, owing to the problem of enforcement. Nevertheless it felt that the effort had to be made because the behaviour and activities of the self-employed at work not infrequently influence the behaviour of employees. The Robens Committee felt that employees might particularly be endangered by the careless conduct at work of self-employed persons, commenting that "it is undesirable that large groups of self-employed people—such as are to be found in the construction industry—should fall outside the scope of protective legislation so far as their own safety and health are concerned, as appears to be the case at present". In view of this observation there is little wonder that the court in *Ferguson* v. *Dawson & Partners Ltd.* decided that labour-only sub-contractors were, for the purposes of the Construction Regulations, employees engaged under a contract of service.

The law on health and safety at work for the self-employed is to be found in s. 3, H.A.S.A.W.A., which provides as follows:

subs. (1): it shall be the duty of every employer to conduct his undertaking in such a way as to ensure, so far as is reasonably practicable, that persons not in his employment who may be affected thereby are not exposed thereby to risks to their health or safety;

subs. (2): it shall be the duty of every self-employed person to conduct his undertaking in such a way as to ensure, so far as is reasonably practicable, that he and other persons (not being his employees) who may be affected thereby are not thereby exposed to risks to their health or safety;

subs. (3): in such cases as may be prescribed, it shall be the duty of every employer and every self-employed person, in the prescribed circumstances and in the prescribed manner, to give to persons (not being his employees) who may be affected by the way in which he conducts his undertaking the prescribed information about such aspects of the way in which he conducts his undertaking as might affect their health or safety.

Sub-section (1) refers to the situation where two separate employers with separate workforces are engaged in doing work on an identical project, there being no contractual relationship between them. Sub-section (2) refers to the self-employed operator and puts into legislative form the recommendations of Robens, whereas subs. (3) is likely to be amplified by regulations made by the Secretary of State for Employment accompanied, if necessary, by a code of practice.

The duties of landlords and persons in control of premises

There is an additional category of person upon whom this legislation lays duties, namely those persons who own or control and own premises in which persons, not their employees, work (other than domestic premises). Section 4 provides as follows:

subs. (1): this section has effect for imposing on persons duties in relation to those who—

(*a*) are not their employees; but

(*b*) use non-domestic premises made available to them as a place of work or as a place where they may use plant or substances provided for their use there, and applies to premises so made available and other non-domestic premises used in connection with them;

subs. (2): it shall be the duty of each person who has, to any extent, control of premises to which this section applies or of the means of access thereto or egress therefrom or of any plant or substance in such premises to take such measures as it is reasonable for a person in his position to take to ensure, so far as is reasonably practicable, that the premises, all means of access thereto and egress therefrom available for use by persons using the premises, and any plant or substance in the premises or, as the case may be, provided for use there, is or are safe and without risks to health;

subs. (3) where a person has, by virtue of any contract or tenancy, an obligation of any extent in relation to—

(*a*) the maintenance or repair of any premises to which this section applies or any means of access thereto or egress therefrom; or

(*b*) the safety of or the absence of risks to health arising from plant or substances in any such premises;

that person shall be treated, for the purposes of subs. (2), as being a person who has control of the matters to which his obligation extends;

subs. (4): any reference in this section to a person having control of any premises or matter is a reference to a person having control of the premises or matter in connection with the carrying on by him of a trade, business or other undertaking (whether for profit or not).

Section 4 refers to the obligations of landlords of premises or persons who *control* premises in which other persons (who are not their employees) work. Cases such as *Wheat* v. *Lacon*, *Moloney* v. *Lambeth B.C.* and *A.M.F. International Ltd.* v. *Magnet Bowling Ltd.* and *Trentham* show that occupation may be joint. The consequences of that for the purposes of s. 4, H.A.S.A.W.A., are that those who both own and manage a particular set of premises or a building are liable to be prosecuted by the inspector and sued by a person injured on those premises whilst carrying out his employment (but not for them). A typical example of the operation of s. 4 would occur where a large development company who owned a block of shops leased the shops to various concerns. Both may be liable in criminal and civil law by virtue of s. 4(1) and the common law. A similar example would occur where a development corporation let out or leased factory units to different industrial concerns.

Sub-section (3) allows lessors to contract out of obligations which they would otherwise incur under subs. 4(2). Thus it is clear that the lessee, who will be responsible for repairs, is the person who is going to be prosecuted. Thus if a development corporation leased factory units to employers, it becomes incumbent on the lessor or the development corporation to make provision for subs. (1) and (2) in the lease.

Control of noxious emissions into the atmosphere

This matter is dealt with in s. 5, the scope of which is wide and involves public safety. There is little doubt that the inspiration behind this section was the following observation of the Robens Committee: "Our conclusion is that the present orientation of the occupational safety legislation and its administration needs some modification to ensure that the public safety aspects of industrial hazards can be fully catered for" (para. 290). Paragraph 291 concludes: "Inspectors should, in the course of their normal activities, have both the power and the duty to pay due regard to the safety and health of the public."

Robens had in mind the need to protect the public as well as workers from the very large-scale hazards which have tended to characterise modern industrial operations, not infrequently involving the chemical industry.

> "The storage and use in industry of a number of intrinsically dangerous substances with highly explosive, flammable or toxic qualities, such as petroleum, liquefied petroleum gases, liquid oxygen, chlorine, phosgene and sulphur dioxide, constitutes a particular area in which significant risks can arise for the public as well as workpeople. A marked feature of industrial development over the last two decades has been the vastly increased scale on which such substances are used and stored. Our attention has been drawn to a number of locations in this country where highly explosive or flammable substances are kept in such quantities that any failure of control—however remote the possibility—could create situations of disaster-potential."

This last provision, made with public safety in mind as much as anything else, coupled with the near disasters and tragedies which occur as a result of tankers and lorries colliding and spilling their load on the highways, accounts for the coming into being of the draft regulations on Transport of Dangerous Substances by Road in June 1977.

In an effort to ensure more public safety, the Report recommended that "in the exercise of their development control powers, local authorities should have an explicit duty to take account of the public safety implications of all applications for planning permissions, and to consult the central Authority responsible for industrial safety in any case where they are in doubt". This recommendation envisages more effective liaison between H.S.E. and local authorities' planning departments. Particularly relevant here is the change of use situation not involving planning permission, indeed para. 308 of the Robens Report concludes by saying "There is a particular need for a comprehensive regime of control from the centre over the manufacture, storage and use of intrinsically dangerous substances....'. Section 5, consistently with s. 1(1)(*d*), therefore provides for this eventuality:

subs. (1): it shall be the duty of the person having control of any premises of a class prescribed for the purposes of s. 1(1)(*d*) to use the best practicable means for preventing the emission into the atmosphere from the premises of noxious or offensive substances and for rendering harmless and inoffensive such substances as may be so emitted.

subs. (2): the reference in subs. (1) to the means to be so used for the purposes there mentioned includes a reference to the manner in which the plant provided for those purposes is used and to the supervision of any operation involving the emission of the substances to which that sub-section applies.

subs. (3): any substance or a substance of any description prescribed for the purposes of subs. (1) as noxious or offensive shall be a noxious or, as the case may be, an offensive substance for those purposes whether or not it would be so apart from this sub-section.

subs. (4): reference in this section to a person having control of any premises is a reference to a person having control of the premises in connection with the carrying on by him of a trade, business or other undertaking (whether for profit or not) and any duty imposed on any such person by this section shall extend only to matters within his control.

Regulations may appear pursuant to s. 5 whereby the Secretary of State prescribes classes of premises for the purposes of such control. In addition, Schedule 4, para. 3, would seem relevant. This paragraph empowers the passing of regulations by:

sub-para. (1): prohibiting the carrying on of any specified activity or the doing of any specified thing except under the authority and in accordance with the terms and conditions of a *licence*, or except with the consent or approval of a specified authority;

sub-para. (2): providing for the grant, renewal, variation, transfer and revocation of licences (including the variation and revocation of conditions attached to licences).

Moreover by virtue of Schedule 3, para. 5, such regulations as may be passed pursuant to H.A.S.A.W.A. may require any person, premises or thing to be registered in any specified circumstances or as a condition of the carrying on of any specified activity or the doing of any specified thing.

The power to issue licences looks likely to become a very important one and operate like a planning control especially upon chemical concerns and the like in a concerted endeavour to prevent the recurrence of a Flixborough in England or a Seveso in Italy. This will mean that unless such concerns (not confined to the chemical industry) can satisfy H.S.E. of the efficacy of their safety arrangements, a licence to operate may not be forthcoming (as for the duties of manufacturers and suppliers under H.A.S.A.W.A., *see* § **51,** p. 251).

The duties of employees

(*a*) *Provision of s. 7*. Section 7 states that it shall be the duty of every employee while at work—

(i) to take reasonable care for the health and safety of himself and of other persons who may be affected by his acts or omissions at work; and

(ii) as regards any duty or requirement imposed on his employer or any other person by or under any of the relevant statutory provisions, to co-operate with him so far as is necessary to enable that duty or requirement to be performed or complied with.

It has been said that H.A.S.A.W.A. reflects the concept of total involvement. Section 7 lays duties of taking precautions and co-operating with employers upon the workforce, in this respect following the pattern set by certain previous regulations, such as the Construction Regulations, of imposing duties for their own safety, as much as for the safety of others, upon employees. Breach of such a duty may attract criminal sanctions under s. 33 and possibly involve the loss of compensation or a reduction of it in a civil action against the employer. Before an employee can be prosecuted for breach of H.A.S.A.W.A. or any other relevant statutory provisions or before he can be regarded as forfeiting his right to compensation or part of it for failing to take proper care in respect of his own health and safety, however, it will need to be shown that his employer has carried out his statutory obligations under s. 2. Before an inspector decides to prosecute a workman for doing his job in a dangerous manner, the inspector will want to satisfy himself that the workman had been instructed in the hazards of his job, had received the necessary training through the appropriate industrial training board and was effectively supervised. If s. 2 has been observed, s. 7 may apply. Assuming that the employer has done what is expected of him by subs. 2(2) and the employee disregards instructions and training in performing his job he may be prosecuted. Similarly if an employee brings an action for compensation against his employer for injury suffered at work, following cases like *Ginty* v. *Belmont Building Supplies Ltd.*, *Ross* v. *Associated Portland Cement Manufacturers Ltd.* and *Boyle* v. *Kodak Ltd.*, the court will want to know whether the employer has done what statute or regulation require of him by way of instruction, training and supervision; if he had not made provision for these things, before training became a general statutory requirement, the tendency in cases where an employee had been careless in respect of his own safety, was to treat the conduct of the employer and employee as contributing to the employee's injuries. This approach typified cases like *Ross* and *Boyle*. And in *I.C.I. Ltd.* v. *Shatwell* it was decided that where there was a statutory duty upon an employer to train and have an employee certificated as competent, as required by the shot-firing regulations (which the employer had done), the employee's conduct was the sole cause of his injuries, with the consequence that he lost all his compensation. The case was one of *volenti non fit injuria* (*see* § **22,** p. 127). Since claims in respect of injury arising from breach of the general duties of H.A.S.A.W.A. are, by virtue of s. 47, H.A.S.A.W.A., to be decided consistently with common law, it is arguable that *Shatwell's* case may apply in situations where employees have been instructed, trained and supervised and then break the law causing themselves injury. Conversely it could be argued that the decision in *Potts* v. *Reid* established that although a workman is in breach of a section [what was s. 161, Factories Act 1961], he is still entitled to compensation "criminal and civil liability [being] two separate things".

Two notable areas covered for the first time are hospitals and educational establishments, including laboratories in educational establishments and research establishments. The fact that enforcement by the inspectorate might be more difficult in the case of certain classes of employment does not mean that there will not be enforcement. Obvious examples cited in the Robens Report where enforcement of health and safety standards might prove more difficult are those of the employee postman, salesman, milkman and furniture remover who spends most of his working day away from his base of operations. To date, fines against employees have been characteristically smaller than against employers, ranging from £10 to £25.

(*b*) *Employee must take care for others and for himself.* Failure to take care in respect of his own safety and the safety of others can mean prosecution for an employee. This applies as much to the employee who, having been instructed, refuses to wear his eye protection or the supervisor who walks into an area where eye protection should be worn not wearing suitable eye protection. The same applies to the managing director who is also an employee.

Miscellaneous duties

Section 8 of the general duties prescribes that no person shall intentionally or recklessly interfere with or misuse anything provided in the interests of health, safety or welfare in pursuance of any of the relevant statutory provisions.

Section 9 further prescribes that no employer shall levy or permit to be levied on an employee of his any charge in respect of anything done or provided in pursuance of any specific requirement of the relevant statutory provisions.

29. The new powers of inspectors

In order to understand the new role of the inspectorate it is necessary to consider the reasons for reform. Having emphasised that the law should aim to stimulate and encourage more self-regulation by industry, Robens said that the basic function of the state inspection services should be the provision of advice and assistance towards progressively better standards in industrial health and safety. At the same time, the Report recognises that a certain level of indifference to safety and the safety of others both on the part of employers and employees is inevitable. There is always, therefore, a residual role for law, i.e. the criminal law. Here, however, the Report observes in some detail:

"... the criminal courts are inevitably concerned more with events that have happened than with curing the underlying weaknesses that caused them.... It is fair to say that inspectors value the threat of possible prosecution as a potent sanction, and that they attach importance to the deterrent effect of the adverse local publicity which prosecutions frequently attract. Nevertheless, the weight of the evidence points to the conclusion that the lengthy process of investigation, warning, institution of criminal proceedings, conviction and ultimate fine is not a very effective way of producing an early remedy for known unsatisfactory conditions. In sum, we do not believe that the traditional sanction commands any very widespread degree of respect or confidence in this field. Why is this? The fact is—and we believe this to be widely recognised—that the traditional concepts of the criminal law are not readily applicable to the majority of infringements which arise under this type of legislation. Relatively few offences are clear-cut, few arise from reckless indifference to the possibility of causing injury, few can be laid without qualification at the door of a particular individual. The typical infringement or combination of infringements arises rather through carelessness, oversight, lack of knowledge or means, inadequate supervision or sheer inefficiency. In such circumstances the process of prosecution and punishment by the criminal courts is largely an irrelevancy. The real need is for a constructive means of ensuring that practical improvements are made and preventive measures adopted. Whatever the value of the threat of prosecution, the actual process of prosecution makes little direct contribution towards this end. On the contrary, the laborious work of preparing prosecutions, and in the case of the Factory Inspectorate actually conducting them, consumes much valuable time which the inspectorates are naturally reluctant to devote to such little purpose. On the other side of the coin—and this is equally important—in those relatively rare cases where deterrent punishment is clearly called for, the penalties available fall far short of what might be expected to make any real impact, particularly on the larger firms. . . . What is needed is an approach which recognises that different types of situation call for different types of remedy. The sanctions available should provide scope for distinguishing between situations where the accent should be on punishment, and the more frequent situations where the accent should be on constructive remedial action. The sort of thing we have in mind is to be found in the system followed in Ontario, where substantial financial penalties imposed on employers by the Workmen's Compensation Board are frequently suspended on condition that a specified programme of remedial action is undertaken."

As regards the future use of criminal proceedings in court, the Report made two proposals that have since been implemented, (*a*) that provision should be made for the imposition of higher penalties in the case of repeated offences, and (*b*) that not only corporate bodies but also individuals such as directors, managers and operatives should be liable to prosecution.

The upshot of the Report's recommendations regarding sanctions was that, prevention being preferable to punishment, remedial action should be preferred to prosecution. Prosecution and conviction in the Magistrates' Courts and the Crown Courts was to be subordinated to the exercise of executive powers by inspectors to prevent accidents: ". . . we believe that in future much greater reliance should be placed on non-judicial administrative techniques for ensuring compliance with minimum standards of health and safety at work." The Committee was therefore of the opinion that inspectors should have the power to issue a formal improvement notice to an employer requiring him to remedy particular faults or to institute a specified programme of work within a stated time limit. At the same time the Report commented "When criminal proceedings are brought against an employer under the present Acts, the employer more often than not pleads guilty. Our proposals should result in fewer prosecutions and more 'Notice cases', and with this one might expect an increase in the number of cases where an inspector's judgment is challenged." The Report went on to advocate that appeals against improvement notices should not be heard in criminal courts but rather in industrial tribunals: "In hearings before tribunals not only relevant statutory provisions but also any relevant approved codes of practice should be admissible in evidence. In this way approved non-statutory standards and codes of practice . . . would be given considerable force in particular instances." These suggestions have largely been implemented.

After specifying that the improvement notice would serve as the inspector's main weapon in the fight against accidents at work, the Report adds that he should be given further power where the case for remedial action is particularly serious. In such cases he should be able to issue a prohibition notice. Prohibition notices are of two kinds: one which allows a reasonable period of time for remedial action, the prohibition notice only becoming effective if the remedial action is not taken within the specified time; and the other, where the danger is grave, which has the effect of bringing a particular activity or machine or part of a machine to a stop immediately. The first type of prohibition notice is called a deferred prohibition notice and the second an immediate prohibition notice. In the case of the immediate prohibition notice any appeal lodged against it does not have the effect of undermining the notice. Employers can expect that rather than be prosecuted by the magistrates, they will be given the option of either putting a defect right or of appealing against a notice to a local industrial tribunal.

Statutory powers

The new statutory powers are to be found in ss. 21–26, H.A.S.A.W.A. Section 21 states that if an inspector is of the opinion that a person—

(*a*) is contravening one or more of the relevant statutory provisions; or

(*b*) has contravened one or more of those provisions in circumstances that make it likely that the contravention will continue or will be repeated;

he may serve on him a notice stating that he is of that opinion, specifying the provision or provisions as to which he is of that opinion, giving particulars of the reasons why he is of that opinion, and requiring that person to remedy the contravention or the matters occasioning it within such period (ending not earlier than the period within which an appeal against the notice can be brought under s. 24) as may be specified in the notice. This is known as an *improvement notice*.

Section 22(1) states that this section applies to any activities which are being or are about to be carried on by or under the control of any person, being activities to or in relation to which any of the relevant statutory provisions apply or will, if the activities are so carried on, apply.

Sub-section (2) states that if as regards any activities to which this section applies an inspector is of the opinion that, as carried on or about to be carried on by or under the control of the person in question, the activities involve or, as the case may be, will involve a risk of serious personal injury, the inspector may serve on that person a notice. This is known as a *prohibition notice*.

Sub-section (3) states that a prohibition notice shall:

(*a*) state that inspector is of the said opinion;

(*b*) specify the matters which in his opinion give or, as the case may be, will give rise to the said risk;

(*c*) where in his opinion any of those matters involves or, as the case may be, will involve a contravention of any of the relevant statutory provisions, state that he is of that opinion, specify the provision or provisions as to which he is of that opinion and give particulars of the reasons why he is of that opinion; and

(*d*) direct that the activities to which the notice relates shall not be carried on by or under the control of the person on whom the notice is served unless the matters specified in the notice in pursuance of para. (*b*) and associated contraventions of provisions so specified in pursuance of para. (*c*) have been remedied.

Sub-section 4 states that a direction given in pursuance of subs. 3(*d*) shall take immediate effect if the inspector is of the opinion, and states it, that the risk of serious personal injury is or, as the case may be, will be imminent, and shall have effect at the end of a period specified in the notice in any other case.

Section 24 deals with appeals against improvement or prohibition notices, and states in subs. (2) that a person on whom a notice is served may, within such period from the date of its service as may be prescribed, appeal to an industrial tribunal; on such an appeal the tribunal may either cancel or affirm the notice and, if it affirms it, may do so either in its original form or with such modifications as the tribunal may, in the circumstances, think fit.

Sub-section (3) provides that where an appeal under this section is brought against a notice within the period allowed under the preceding sub-section, then—

(*a*) in the case of an improvement notice, the bringing of the appeal shall have the effect of suspending the operation of the notice until the appeal is finally disposed of or, if the appeal is withdrawn, until the withdrawal of the appeal;

(*b*) in the case of a prohibition notice, the bringing of the appeal shall have the like effect if, but only if, on the application of the appellant the tribunal so directs (and then only from the giving of the direction).

Sub-section (4) then states that one or more assessors may be appointed for the purposes of any proceedings brought before an industrial tribunal under this section.

Section 25 deals with the inspector's power of eliminating causes of imminent danger. Sub-section (1) provides that where, in the case of any article or substance found by him in any premises which he has power to enter, an inspector has reasonable cause to believe that, in the circumstances in which he finds it, the article or substance is a cause of imminent danger of serious personal injury, he may seize it and cause it to be rendered harmless (whether by destruction or otherwise).

Sub-section (2) states that before there is rendered harmless under this section—

(*a*) any article that forms part of a batch of similar articles; or

(*b*) any substance;

the inspector shall, if it is practicable for him to do so, take a sample thereof and give to a responsible person at the premises where the article or substance was found by him a portion of the sample marked in a manner sufficient to identify it.

Sub-section (3) states that as soon as possible after any article or substance has been seized and rendered harmless under this section, the inspector shall prepare and sign a written report giving particulars of the circumstances in which the article or substance was seized and so dealt with him by him, and shall—

(*a*) give a signed copy of the report to a responsible person at the premises where the article or substance was found by him; and

(*b*) unless that person is the owner of the article or substance also serve a signed copy of the report on the owner;

and if, where para. (*b*) applies, the inspector cannot after reasonable inquiry ascertain the name or address of the owner, the copy may be served on him by giving it to the person to whom a copy was given under the preceding paragraph.

Section 26 gives inspectors indemnity. It states that where an action has been brought against an inspector in respect of an act done in the execution or purported execution of any of the relevant statutory provisions and the circumstances are such that he is not legally entitled to require the enforcing authority which appointed him to indemnify him, that authority may, nevertheless, indemnify him against the whole or part of any damages and costs or expenses which he may have been ordered to pay or may have incurred, if the authority is satisfied that he honestly believed that the act complained of was within his powers and that his duty as an inspector required or entitled him to do it.

Improvement and prohibition notices

(*a*) *Improvement notice.* This is a notice served upon a person, often an employer, requiring him to put right a defect which the inspector considers to be a contravention of H.A.S.A.W.A. or regulation or relevant statutory provision; the new powers of serving notices can be used by inspectors to enforce the previous law. An improvement notice requires that the inspector is of the opinion that there is a legal contravention, and doubt can sometimes arise here on a point of law. Appeals can be lodged against improvement notices and the Industrial Tribunals (Improvement and Prohibition Notices Appeals) Regulations 1974 state that an appeal must be lodged with a local industrial tribunal within 21 days from the service of the notice. Appeals can be either against the time limit imposed by the inspector for putting right the defect or against the substance of the notice itself. These appeals have already attracted some case law, the result of which would appear to be that, although industrial tribunals will entertain appeals against notices and will modify the time limits imposed, they will not modify or relax the substance of the notice. There is, however, one major difference between lodging an appeal against an improvement notice and lodging one against a prohibition notice, which is, that an improvement notice does not begin to take effect until the result of the appeal is known and has been decided in favour of the inspector with or without modification.

(*b*) *Prohibition notice.* This notice may take effect immediately or upon the expiration of a period of time specified in the notice; this is known as a *deferred prohibition notice.* Appeal to an industrial tribunal does not have the effect of suspending the operation of a prohibition notice. Section 22, H.A.S.A.W.A., makes it clear that such a notice can be served when there is, in the opinion of the inspector, a risk of serious personal injury from a workplace activity.

(*c*) *Case law on appeals against improvement and prohibition notices.* There have been three important cases—*Otterburn Mill Ltd.* v. *Bulman* (*H.M.F.I.*), *Campion* v. *Hughes* (*H.M.F.I.*), and *Belhaven Brewery Co. Ltd.* v. *McLean* (*H.M.F.I.*).

Otterburn Mill Ltd. v. *Bulman* concerned a deferred prohibition notice which required the company to guard four carding machines. The company accepted that if the machines remained unguarded there would be a risk of serious personal injury to the persons operating the machines. The notice specified that the company had three months in which to take the necessary steps involved in guarding the four machines, but the company appealed against the time limit on the grounds (*a*) that the company was not in a position to afford to pay for the guarding of all the machines at once and (*b*) that the prohibition on using the machines which had not begun to be guarded would mean that the mill effectively had to close down. These arguments were not acceptable to the tribunal, which issued the following warning: "We do not think that in all these cases the difficulty which a company may experience in finding the money to carry out a necessary safety programme expeditiously should affect our minds when deciding what time should be allowed to that company to do what must be done. It will not do . . . to tell large and prosperous company

A that they must set matters right within a month because they can afford to do so, and to tell small struggling company B that they can take two years over the same job, or wait till trade looks up...."

The company's argument in favour of relaxing the time limit endeared itself more to the tribunal, and the original time limit was extended by 12 weeks so that the minimal disruption would be caused to the factory and two machines could be in operation continuously. Tribunals, it seems, do not wish to be indicted for stopping or reducing a company's output by earmarking machinery as being out of use when to do this without modification might well involve the company in loss of orders abroad and the laying off of labour at home. Modifications of time limits are likely to get a more favourable reception from tribunals than appeals against the *substance* of the notice.

In *Campion* v. *Hughes* a company appealed against the validity and the substance of an improvement notice and the time limit imposed by it. An improvement notice required changes to be carried out to the existing means of escape from fire. The notice stipulated that the fire escape be constructed on land at that time belonging to the borough council whose consent had not been obtained, in which respect the company owner or employer had been to some extent dilatory. It was thus not possible for the notice to be complied with in the stipulated time. The tribunal ruled that the changes in the fire escape would have to be implemented but the time limit would be extended by four months.

The most important of the cases is *Belhaven Brewery Co. Ltd.* v. *McLean.* A brewery company appealed against an improvement notice requiring them to fence securely transmission machinery and other dangerous parts of the company's automatic kegging plant. The schedule to the notice stipulated that this fencing was to take the form of screens with doors or gates for access fitted with an interlocking device. The effect of this device would be to switch off the electric power and compressed air automatically when the door was open. The company contended that the construction and installation of screens with an interlocking device would be very expensive and that they would be complying with the relevant statutory requirements (s. 2(1), H.A.S.A.W.A., and ss. 13(1) and 14(1), Factories Act 1961) if they erected safety screens that did not have such a device. Moreover they contended that their operatives were of sufficient intelligence to see that the gates kept in position when the plant was running. Also the plant was situated in such a place as there would be a high level of supervision. They also proposed to put a notice beside the screening warning operatives of the danger. The inspector, however, argued that irrespective of the degree of intelligence of the operatives and the level of management supervision, only screening which contained an interlocking device, so that an operative could not gain access to the machine whilst it was in motion, could establish a secure system. The tribunal, finding for the inspector, ruled that the improvement notice would be affirmed. Weighing up the extent to which the risk of injury would be eliminated, or reduced to a level at which it could be said that the statutory provisions were being complied with, the only method by which the contravention could be remedied was by fitting an interlocking device of the type stipulated in the schedule to the notice. If screening of the kind proposed by the company were to be fitted, i.e. without an interlocking device, this would not result in the transmission machinery and the dangerous parts of the machinery being securely fenced for the purposes of ss. 13(1) and 14(1), Factories Act 1961. Moreover the company would not be carrying out its duty under s. 2(1), H.A.S.A.W.A., to secure the health, safety and welfare at work of its employees. Though the employer's duty under s. 2(1) was limited to what was "reasonably practicable", applying the test formulated in *Edwards* v. *N.C.B.* and balancing the quantum of risk on the one hand, and the sacrifice involved in the measures necessary for averting the risk (whether in money, time or trouble) on the other, the risk attendant upon an omission to fit an interlocking device was not so insignificant in relation to the sacrifice involved in fitting such device that it could be said that the fitting of an interlocking device fell outside the parameters of what was reasonably practicable. It was, therefore, reasonably practicable to fit an interlocking device and if the company were to fit a device of the type proposed by them rather than of the type stipulated in the schedule to the improvement notice, there would be a continuous contravention of s. 2(1), H.A.S.A.W.A.

(*d*) *Appeals against tribunal decisions.* A person, whether corporate or not, can appeal beyond the decision of a tribunal to the High Court but only on a point of law, as where a tribunal had failed to follow a previous decision of the High Court or even the Court of Appeal or House of Lords on an identical point, or where a tribunal took a wrong view of the expression "reasonably practicable" or some other technical expression. The tribunal can also be compelled to review its decision in specified circumstances (*see* § **12,** p. 27).

(*e*) *Prosecutions and criminal sanctions.* Proceedings in criminal cases play a subordinate role to improvement and prohibition notices. If, however, a notice is not complied with or an appeal against it is unsuccessful, criminal proceedings may well follow the breach. The court before which prosecution proceedings will be brought by an inspector is normally the Magistrates' Court, though a more serious or repeated breach of health and safety legislation would go to the Crown Court, this latter court being empowered by the Courts Act 1971 to award greater penalties than the magistrates. The main provisions relating to criminal offences and criminal proceedings are to be found in ss. 33–42, H.A.S.A.W.A. These sections set up a framework of fines and/or imprisonment for breaches of H.A.S.A.W.A., any regulations passed pursuant to it and any other relevant statutory provisions such as the Factories Act 1961. Penalties vary from a maximum of £400 to £1,000 in view of the amendment of H.A.S.A.W.A. by the Criminal Law Act 1977. In certain circumstances imprisonment (and/or a fine) for a period not exceeding two years is possible, but only a Crown Court could make such an order and it would be in cases where the breach of the law was grave, as where life or limb was endangered or damage to property considerable or where the offence was repeated. The offences which carry a maximum penalty of imprisonment and/or an *indefinite* fine are as follows:

(i) an offence consisting of contravening any of the relevant statutory provisions by doing otherwise than under the authority of a licence issued by the Executive or the appropriate Agriculture Minister something for the doing of which a licence is necessary under the relevant statutory provisions;

(ii) an offence consisting of a contravention of a term of or a condition or restriction attached to any such licence as is mentioned in the preceding paragraph;

(iii) an offence consisting of acquiring or attempting to acquire, possessing or using an explosive article or substance (within the meaning of any of the relevant statutory provisions) in contravention of any of the relevant statutory provisions;

(iv) an offence under subs. 1 (*g*) above consisting of contravening a requirement or condition or prohibition imposed by a prohibition notice;

(v) an offence under subs. 1 (*j*) above, namely wrongful use of information in contravention of ss. 27(4) and 28 (*see* § **28,** p. 235).

(*f*) *Continuing offences.* Where a person is convicted of an offence under subs. (1)(*g*), i.e. contravening a requirement or prohibition imposed by a prohibition notice, if the contravention in respect of which he was convicted is continued after the conviction, he shall, subject to s. 42(3) be guilty of a further offence and liable in respect thereof to a fine not exceeding £100 for each day on which the contravention is so continued. The Criminal Law Act 1977 (Commencement No. 5) Order 1978, which came into force on July 17, 1978, increases the maximum fine which magistrates can impose under s. 33(5). The proviso machinery set up by s. 42(3) is part of the more general power accorded to the magistrates or the Crown Court to order that, either in addition to or in lieu of punishment, a person may remedy the cause of the offence. If the court chooses to exercise its right under s. 42, subs. (3) provides that a person upon whom such an order is imposed shall not be liable under any health and safety legislation creating offences, in so far as the offence continues during the time fixed by the order for the cause of the offence to be remedied (or by any extension of time affixed to the order to remedy, where an application for such extension has been made and granted by the court).

(i) *Types of offence.* There are two types of offence. Those which can result in imprisonment and/or an indefinite fine; these are outlined in subs. 33(4). Such penalties can only be ordered by the Crown Court. Alternatively there are summary as distinct from indictable offences in respect of which the magistrates have jurisdiction, and these form by far the greater part of the spectrum of criminal offences under H.A.S.A.W.A. Even in respect of the imprisonment offences, subs. 33(3) provides that s. 15(6)(*d*) and/or para. 2(2), Sched. 3, may modify the effect of subs. 33(3). Sub-section 15(6)(*d*) provides that health and safety regulations may restrict the punishment which can be imposed in respect of any offence mentioned in para. (*c*) of subs. 15(6). Sub-section 15(6) (*c*) states that health and safety regulations may exclude proceedings on indictment in relation to offences consisting of a contravention of a requirement or prohibition imposed by or under any of the existing statutory provisions or ss. 2–9 or health and safety regulations. The most important type of imprisonable offence that can be committed under H.A.S.A.W.A., subject to anything to the contrary in health and safety regulations, is therefore likely to be conduct or activities inconsistent with a prohibition notice or inconsistent with a licence or the terms or conditions attached to a licence.

(ii) *Summary offences.* Unless the offence falls within subs. 33(4) it is triable summarily and a maximum penalty of £400 may be imposed by virtue of subs. 33(2), except in the case of subs. 33(1)(*d*),(*e*), (*f*), (*h*) and (*n*), where the maximum penalty has been increased to £1,000. These offences are enumerated in subs. 33(1) as follows:

1. failing to discharge a duty to which a person is subject by virtue of ss. 2–7;
2. contravening ss. 8 or 9;
3. contravening any health and safety regulations or agricultural health and safety regulations;
4. contravening any requirement imposed by or under regulations under s. 14 or intentionally obstructing any person in the exercise of his powers under that section (power of the Commission to direct investigations and inquiries into accidents and occurrences with a view to making regulations etc);

5. contravening any requirement imposed by an inspector under ss. 20–25 (though not under s. 22, as it will be caught by subs. 33(3) (*b*) (i));

6. preventing or attempting to prevent any other person from appearing before an inspector or from answering any question to which an inspector may by virtue of s. 20(2) require an answer (for the content of subs. 20(2) *see* Appendix III);

7. contravening any requirement imposed by an improvement notice;

8. intentionally obstructing an inspector in the exercise or performance of his power or duties;

9. contravening any requirement imposed by a notice under s. 27(1) (commission or enforcing authority may, with the consent of the Secretary of State, serve a notice on a person requiring him to furnish the Commission or the enforcing authority with such information as it needs to discharge its functions);

10. making a statement which he knows to be false or recklessly making a statement which is false where the statement is made—(i) in purported compliance with a requirement to furnish any information imposed by or under any of the relevant statutory provisions; or (ii) for the purpose of obtaining the issue of a document under any of the relevant statutory provisions to himself or another person;

11. intentionally making a false entry in any register, book, notice or other document required by or under any of the relevant statutory provisions to be kept, served or given or, with intent to deceive, making use of any such entry which he knows to be false;

12. with intent to deceive, forging or using a document issued or authorised to be issued under any of the relevant statutory provisions or required for any purpose thereunder or making or having in his possession a document so closely resembling any such document as to be calculated to deceive;

13. falsely pretending to be an inspector;

14. failing to comply with an order made by a court under s. 42 (power of court to order cause of offence to be remedied either in addition to or in lieu of punishment.)

30. Criminal proceedings under H.A.S.A.W.A.

Burden of proof

Normally in criminal cases the onus of proving guilt is on the prosecution. This is not the position under H.A.S.A.W.A.; the reverse obtains. This is provided by s. 40, which states as follows: "In any proceedings for an offence under any of the relevant statutory provisions consisting of a failure to comply with a duty or requirement to do something so far as is practicable or so far as is reasonably practicable, or to use the best practicable means to do something, it shall be for the *accused* to prove ... that it was not practicable or not reasonably practicable to do more than was done to satisfy the duty or requirement, or that there was no better practicable means than was in fact used to satisfy the duty or requirement." Here, it is thought, adherence to an approved code of practice might well prove sufficient to discharge the burden, though, clearly, owing to the definition given to "practicable" as distinct from "reasonably practicable" ("practicable" meaning that the accused must show that he did everything which was possible in the light of current invention and knowledge, irrespective of cost, time and inconvenience), it will be more difficult to discharge this burden of proof than in cases where the duty orders something to be done as far as is reasonably practicable only (which is what H.A.S.A.W.A. does), although there is nothing to prevent future health and safety regulations laying down duties upon persons to be carried out as far as practicable.

Offences requiring guilty intent or strict liability

The legal mind will inquire whether a criminal offence falls into one or both of the two categories of criminal liability, (*a*) offences requiring guilty intent and (*b*) offences of strict liability. The former, like murder, manslaughter and theft, are largely of common law origin; the latter, like motoring offences and offences involving selling and using articles, are largely of statutory origin. Since offences against H.A.S.A.W.A. are of statutory origin, it might be easy to suppose that they did not require guilty intent and that they were offences of strict liability, i.e. a person could be found guilty even if he did not possess a guilty mind so long as he did the act prohibited by the statute or the regulation. This, however, is not likely to be the case.

As Glanville Williams once observed, there is a halfway house of liability between offences of guilty intent and those of strict liability, and that is negligence. It is reasonably clear from the terminology of H.A.S.A.W.A. and the recurrent use of "reasonably practicable" to qualify the nature of the general duties that the criminal liability for breach lies in negligence. In this respect the 1970 decision of the House of Lords in *Sweet* v. *Parsley* is instructive. This case (schoolteacher prosecuted for allowing students to smoke cannabis in her rented premises, unbeknown to her) is generally regarded as having established that as a matter of statutory interpretation *fault is an essential ingredient in a criminal offence* unless it can be demonstrated that Parliament intended to enact otherwise. It is doubtful therefore whether an offence can be committed under H.A.S.A.W.A. without either intent, or more often fault or negligence.

Offences by bodies corporate

Although the emphasis is on individual responsibility and hence individual liability, there is still provision in s. 37 for prosecutions to be brought by an inspector against a body corporate, whether public or private company or public corporation or local authority. Prosecutions against corporate bodies have been possible since the decision in *I.C.R. Haulage Ltd.* v. *R.* in 1944. The offence must, however, be punishable only by a fine or by a fine or imprisonment but not punishable only by imprisonment, owing to the obvious physical difficulty of imprisoning a company or corporation. What also emerges from this case, as well as from *D.P.P.* v. *Kent & Sussex Contractors Ltd.*, is that the offence must have been committed by the alter ego (the other self) of the corporate body, i.e. by the brains as distinct from the hands, or the controlling mind of the company. This means the board of directors. Hence a prosecution against a company is a prosecution against the board of directors of the company. This is provided for in s. 37(1) which states "Where an offence under any of the relevant statutory provisions committed by a body corporate is proved to have been committed with the consent or connivance of, or to have been attributable to any neglect on the part of, any director, manager, secretary or other similar officer of the body corporate or a person who was purporting to act in any such capacity, he as well as the body corporate shall be guilty of that offence and shall be liable to be proceeded against and punished accordingly".

Since responsibility for health and safety within a company, as evinced by the company safety policy, will ultimately rest with the controlling mind of the company or the board of directors or even the director of health and safety personally, s. 37(1) envisages prosecutions both against the company (the board of directors) and/or any individual director. Since they are the link at the end of the managerial chain, they are the ones who must eventually bear the responsibility and the liability, as a recent case (*Herron* (*Procurator Fiscal*) v. *Armour 1976*) involving Strathclyde Regional Council and its Director of Roads shows clearly (it must be emphasised, however, that this was a decision of a lower court on a prosecution brought by the Procurator Fiscal (the Factory Inspector cannot prosecute in Scotland for breaches of H.A.S.A.W.A., *see also* Appendix I). While a bridge over the River Clyde was being repainted under a direct labour contract with the Roads Department of the Strathclyde Regional Council, two painters fell from the scaffolding which had been erected for the work. One was killed. The prosecution alleged that both the Regional Council and their Director of Roads had failed to comply with H.A.S.A.W.A. and the Construction (Working Places) Regulations 1966 and the Factories Act 1961. The case against the Director of Roads was taken under s. 37(1), H.A.S.A.W.A., to which he pleaded not guilty. It was held by the court that the Director of Roads (along with the Regional Council) was guilty of breaches of the Factories Act 1961, ss. 127 and 155 and of the Construction (Working Places) Regulations 1966, both of which were relevant statutory provisions within the meaning of s. 53, H.A.S.A.W.A., as mentioned in s. 37(1). Although the defence argued that, since subs. 2(3) placed the duty to prepare a written statement on the employer, the accused director therefore had no duty to do so, the Sheriff disagreed, saying that where the employer is a corporation it cannot act except through its officials and that a high official like the accused was identifiable with the corporation, its alter ego. Moreover the prosecution could prove that the director had been negligent in failing to perform a duty of which he knew or ought to have known by showing that the Council had issued a statement of safety policy consistently with subs. 2(3). This statement had been circulated to the Region's various departments and part of it instructed directors and heads of departments, etc., to ensure safe conditions of work and a supplementary "Advisory Bulletin" told them that each department should issue a safety statement specifically related to the work being done and that this should include training and instruction in safe practices. The Council was fined £500 and the Director of Roads £125 (*see* Appendix I, the case of *Armour* v. *Skeen*, for the appeal in this case).

A case which throws a good deal of light on the meaning of s. 37(1), H.A.S.A.W.A., is *Tesco Ltd.* v. *Nattrass*, occurring under the 1968 Trades Descriptions Act. A store manager employed by Tesco failed to carry out his duties by omitting to check the wording of a "special offer" poster. Tesco was charged with giving an indication that goods were offered for sale at a price less than that at which they were offered, contrary to subs. 11(2) of the Act. The company was convicted by the magistrates and Tesco appealed to the Divisional Court who affirmed the conviction. On an appeal to the House of Lords Tesco was found not guilty. During the course of giving judgement Lord Reid said:

"I must start by considering the nature of the personality which by a fiction the law attributes to a corporation. A living person has a mind which can have knowledge or intention or be negligent and he has hands to carry out his intentions. A corporation has none of these; it must act through living persons, though not always one or the same person. Then the person who acts is not speaking or acting *for the company*. He is acting *as the company*, and his mind which directs his acts is the mind of the company. There is no question of the company being vicariously liable. He is not acting as a servant, representative, agent or delegate. He is an embodiment of the company or, one could say, he hears and speaks through the persona of the company, within his appropriate sphere, and his mind is the mind of the company. If it is a guilty mind, then, that guilt is the guilt of the company. It must be a question of law whether, once the facts have been ascertained, a person in doing particular things is to be regarded as the company or merely as the company's servant or agent. In that case any liability of the company can only be a statutory or vicarious liability. . . ."

Delegation

Many statutes including H.A.S.A.W.A. impose statutory duties upon companies and other bodies corporate. In effect these statutory duties are imposed upon the board of directors. There is, however, nothing to prevent a company or board of directors from delegating the *performance* of a statutory duty, as distinct from the duty itself, to other individuals lower down the company tree. As Lord Reid observed in *Tesco Ltd.* v. *Nattrass* "A board of directors can delegate part of their functions of management so as to make their delegate an embodiment of the company within the sphere of delegation. But here the board never delegated any part of their function. They set up a chain of command through regional and district supervisors but they remained in control. The shop managers had to obey their general directions and also to take their orders from the supervisors. The acts or omissions of shop managers were not acts of the company itself."

In situations such as those referred to by Lord Reid where the person who is committing the offence is not sufficiently senior, the company may be vicariously liable for the commission of the statutory offence. The decision in *Allen* v. *Whitehead* in 1930 is instructive, establishing that where a person has delegated to another duties imposed upon him by statute, he may be held liable for the acts of that other. Though it should be remembered that whether a company or employer will be found vicariously liable, will depend on the terminology of the statute or regulation. Particularly where a statute creating an offence does not require guilty intent, the company or employer is more likely to be found vicariously liable. It would appear following the dicta of Sheriff Macphail in the Strathclyde case that where company departments fail to issue safety statements in line with company safety policy, including provision for training and instruction of employees in safe practices (there being neglect here on the part of the departmental head or the director), the company would be vicariously liable for offences committed by less senior employees.

Commission of offences by "another person"

Section 36(1), H.A.S.A.W.A., provides "Where the commission by any person of an offence under any of the relevant statutory provisions is due to the act or default of some other person, that other person shall be guilty of the offence, and a person may be charged with and convicted of the offence by virtue of this sub-section whether or not proceedings are taken against the first-mentioned person."

This sub-section mirrors exactly the problem that occurred in *Tesco Ltd.* v. *Nattrass* where it was held by the House of Lords that the shop manager was "another person" for the purposes of subs. 11(2), Trades Descriptions Act 1968. As explained by Lord Reid, the performance of statutory duties placed upon Tesco had not been delegated to him at all, he was not in a position to exercise discretion in the performance of his duties. Fortunately for Tesco under the provisions of the Trades Descriptions Act 1968 there was a built-in statutory defence in s. 24(1) to the effect that "in any proceedings for an offence under this Act, it shall be a defence for the person charged to prove (*a*) that the commission of the offence was due to a mistake or to reliance on information supplied to him or to the act or default of another person, and (*b*) that he took all reasonable precautions and exercised all due diligence to avoid the commission of such offence by himself or any person under his control".

Relating *Tesco Ltd.* v. *Nattrass* to ss. 36(1) and 37(1), H.A.S.A.W.A., it would appear that where offences have been committed by a person not sufficiently senior in the corporate structure for it to be said that performance of a statutory duty has been delegated to him, that person, e.g. a training officer or safety officer or supervisor or manager will, for the purposes of subs. 36(1), qualify as "another person". This is not to intimate, however (as was the position in the *Tesco* case), that the company itself and individual directors will be absolved from liability. There is no built-in defence in H.A.S.A.W.A. (though there was in the original Bill) such as exists in the Trades Descriptions Act 1968, with the practical consequence that the company and individual directors can be prosecuted as well, in circumstances where an offence is committed by "another person". This is provided for in subs. 36(1) to the effect that proceedings may be taken against "another person" whether or not proceedings are taken against the company and/or individual directors under subs. 37(1). This leaves the door open to the exercise of fairly wide discretion on the part of inspectors in deciding where and at what level to prosecute. Certainly in cases like the Strathclyde case where there was neglect on the part of the Director of Roads, an inspector would be justified in prosecuting the director as well as another person under s. 36. Where, however, there is no neglect on the part of the director, a prosecution against the company under s. 37 on a straight vicarious liability basis would seem unfair, especially in the light of the observations of Lord Reid in *Tesco Ltd.* v. *Nattrass*: "Making an employer criminally responsible even when he has done all he could do to prevent an offence, affords some additional protection to the public because this will induce him to do more. But if he has done all he can, how can he do more?"

Court's power to order the offence to be remedied either in addition to or in lieu of the punishment

This extra jurisdiction is given to magistrates and Crown Courts by s. 42 and it is suspected that it is one that will be exercised frequently, as it is in keeping with the Robens Report.

Sub-section (1) states that where a person is convicted of an offence under any of the relevant statutory provisions in respect of any matters which appear to the court to be matters which it is in his power to remedy, the court may, in addition to or instead of imposing any punishment, order him, within such time as may be fixed by the order, to take such steps as may be specified in the order for remedying the said matters.

Sub-section (2) states that the time fixed by an order under subs. (1) may be extended or further extended by order of the court on an application made before the end of that time as originally fixed or as extended under this sub-section, as the case may be.

Sub-section (3) states that where a person is ordered under subs. (1) to remedy any matters, that person shall not be liable under any of the relevant statutory provisions in respect of those matters in so far as they continue during the time fixed by the order or any further time allowed under subs. (2).

Sub-section (4) states that subject to the following sub-section, the court by or before which a person is convicted of an offence such as is mentioned in s. 33(4)(*c*) in respect of any such explosive article or substance as is there mentioned may order the article or substance in question to be forfeited and either destroyed or dealt with in such other manner as the court may order.

Sub-section (5) states that the court shall not order anything to be forfeited under the preceding sub-section where a person claiming to be the owner of or otherwise interested in it applies to be heard by the court, unless an opportunity has been given to him to show cause why the order should not be made.

Prosecution statistics

The total number of prosecutions initiated by the inspectorate during 1976 decreased over previous years, 1,157 compared with 1,433 in

1975 and 1,826 in 1974. This may have had something to do with the *Wallsend Slipway & Engineering Co.* case, where the inspectorate appealed against the ruling of the Wallsend magistrates to the effect that H.S.E. was not constitutionally empowered to prosecute. The total amount of fines imposed during 1976 was £189,652.

Breakdown of prosecutions per section

Section	*Information*	*Conviction*	*Withdrawn*	*Dismissed*
2(1)	113	101	7	5
2(2)(*a*)	51	50	–	1
2(2)(*b*)	3	3	–	–
2(2)(*c*)	21	16	3	2
2(2)(*d*)	11	11	–	–
2(3)	2	2	–	–
3(1)	29	25	1	3
3(2)	9	8	1	–
4(1)	3	3	–	–
4(2)	9	5	–	4
6(1)(*a*)	9	9	–	–
6(1)(*c*)	1	1	–	–
6(4)(*a*)	1	1	–	–
6(4)(*c*)	3	3	–	–
7(*a*)	16	13	2	1
8	2	2	–	–
Total	283	253	14	16

31. Civil liability under H.A.S.A.W.A.

Section 47 deals with civil liability or actions for compensation by injured employees in respect of breaches of the general duties of the Act. Particularly interesting are subs. (1) (*a*) and (*b*), subs. (2) and subs. (4). In addition, subs. (5) may have a significant effect in the future on business contracts.

Sub-section (1) provides as follows:

Nothing in this Part shall be construed—

(*a*) as conferring a right of action in any civil proceedings in respect of any failure to comply with any duty imposed by ss. 2–7 or any contravention of s. 8;

(*b*) as affecting the extent (if any) to which breach of a duty imposed by any of the existing statutory provisions is actionable;

(*c*) as affecting the operation of s. 12 of the Nuclear Installations Act 1965.

Sub-section (2) provides that a breach of a duty imposed by health and safety regulations or agricultural health and safety regulations shall, so far as it causes damage, be actionable except in so far as the regulations provide otherwise; and subs. (4) states that subs. (1)(*a*) (2) of s. 47 are without prejudice to any right of action which exists apart from the provisions of this Act.

Finally subs. (5) states that any term of an agreement which purports to exclude or restrict the operation of subs. (2), or any liability arising by virtue of that sub-section, shall be void, except in so far as health and safety regulations or, as the case may be, agricultural health and safety regulations, provide otherwise.

Negligence as basis of civil liability

When outlining the scope of the general duties we acknowledged the obvious resemblance to the common law, and it would be surprising if common law had no role to play in actions for injury at work involving breach of the general duties. Section 47, H.A.S.A.W.A., makes common law the fulcrum of liability for actions for personal injury. The two relevant sub-sections are subs. (1) and (4).

Sub-section (1), as modified by subs. (4), to the effect that civil actions are to be decided by reference to "any right of action existing apart from the provisions of this Act", means that actions for injury at work involving breach of the general duties of H.A.S.A.W.A. are to be determined by negligence. In this respect, H.A.S.A.W.A. follows the Factories Act 1961, where actions for injury involving breach required proof of negligence for their success. Unlike the Factories Act 1961, however, there is to be no action for breach of statutory duty under H.A.S.A.W.A. Retention of liability in tort means, however, that the employer when sued for injury by an employee involving a breach of the general duties will be able to raise the defences he could have raised at common law, i.e. contributory negligence and possibly *volenti*. In this connection the Royal Commission on Civil Liability and Compensation for Personal Injury has recommended that tort liability be retained in addition to the provision made by an improved industrial injuries scheme operated by the state coupled with pensions calculated in the same way as the new state pensions.

H.A.S.A.W.A. binds the Crown

Sub-section 48(1) makes it clear that the provisions of H.A.S.A.W.A. bind the Crown, though not ss. 21–25 and 33–42 (notices and criminal penalties) although these sections do bind persons in the public service of the Crown.

Insurance against offences committed under H.A.S.A.W.A.
Both individual managers and the company itself can be liable under ss. 36 and 37, H.A.S.A.W.A. Although it would be inconsistent with public policy for an insurance company to offer indemnity in respect of a fine imposed under H.A.S.A.W.A., nevertheless it is now possible to insure against the legal expenses involved in any court action arising out of a prosecution under the Act. This protection plan means that the employer would take out policies, one for employers and one for employees, and as regards the latter the policy would cover all employees, not just those most at risk.

32. Regulations pursuant to H.A.S.A.W.A.
Apart from the regulations relating to safety representatives and safety committees, other regulations have appeared. These include the regulations on Packaging and Labelling of Dangerous Substances 1978, the Fire Certificates (Special Premises) Regulations 1976 and the Draft Regulations relating to the Transport of Dangerous Substances by Road 1977.

Content of Health and Safety Regulations (*see also* Appendix III)

Sub-section 15(3) states that health and safety regulations—

(*a*) may repeal or modify any of the existing statutory provisions;
(*b*) may exclude or modify in relation to any specified class of case any of the provisions of ss. 2–9 or any of the existing statutory provisions;
(*c*) may make a specified authority or class of authorities responsible, to such extent as may be specified, for the enforcement of any of the relevant statutory provisions.

Sub-section (5) provides that health and safety regulations—

(*a*) may provide (either unconditionally or subject to conditions, and with or without limit of time) for exemption from any requirement or prohibition imposed by or under any of the relevant statutory provisions;
(*b*) may enable exemptions from any requirement or prohibition imposed by or under any of the relevant statutory provisions to be granted (either unconditionally or subject to conditions, and with or without limit of time) by any specified person or by any person authorised in that behalf by a specified authority.

Sub-section (6) states that health and safety regulations—

(*a*) may specify the persons or classes of persons who, in the event of a contravention of a requirement or prohibition imposed by or under the regulations, are to be guilty of an offence, whether in addition to or to the exclusion of other persons or classes of persons;

(*b*) may provide for any specified defence to be available in proceedings for any offence under the relevant statutory provisions either generally or in specified circumstances;

(*c*) may exclude proceedings on indictment in relation to offences consisting of a contravention of a requirement or prohibition imposed by or under any of the existing statutory provisions, ss. 2–9 or health and safety regulations;

(*d*) may restrict the punishments which can be imposed in respect of any such offence as is mentioned in para. (*c*) above.

THE FACTORIES ACT 1961

33. Main provisions

Although the repeal of the Factories Act 1961 is an ongoing process, with certain sections having already been repealed (including s. 76 under which many of the regulations under the Factories Act 1961 were made by the Minister), some sections are still in force and constitute some of the most important statutory provisions relating to industrial health and safety. The fencing requirements of the Act, for example, especially s. 14(1), are still being enforced assiduously by H.S.E. It is these sections, and s. 14 in particular, around which a body of stateable doctrine has grown, and an inspector can enforce these provisions of the Factories Act 1961 by his new powers of improvement and prohibition notice under H.A.S.A.W.A. Before the provisions of the Factories Act 1961 can apply, the premises where work is being done must qualify as a factory within the meaning of that expression in s. 175 of the Act. This provision attracted much case law, but in view of the much wider ambit of H.A.S.A.W.A. as applicable to all places of work, it is of no great practical value to ascertain whether premises are a factory or not. As long as work is being done in the premises, an inspector can serve notices or prosecute for breach of the general duties of H.A.S.A.W.A.

The Factories Act 1961 is a penal measure designed to punish factory occupiers whose standards fall below those of the Act. Most of the authoritative case law on this statute, however, has arisen in civil cases and forms part of the civil law. The reason for this is that prosecutions under the Factories Act 1961 tend to be brought before magistrates where conviction rather than legal argument prevails, although it would be of interest to know how far the interpretation of sections in civil cases were authoritative as regards criminal actions under the Factories Act 1961. On this basis there could not be any successful prosecutions under s. 14(1) of the Act unless the machinery was dangerous as defined in *Walker* v. *Bletchley-Flettons Ltd.*; there would have to be risk of foreseeable injury. Some light was shed on this dark corner by the House of Lords in *Potts* v. *Reid* in 1943 when it was said "Criminal and civil liability are two separate things.... The legislation might well be unwilling to convict an owner who failed to carry out an absolute statutory duty of a crime with which he was not himself directly concerned, but still be ready to leave the civil liability untouched." In this case a workman who was found guilty of a breach of s. 161, Factories Act 1937, and fined accordingly was nevertheless entitled to compensation in respect of an injury suffered in consequence of the breach. This suggests that civil cases could not be regarded as authoritative when the same sections have to be interpreted in prosecutions.

The provisions of the Factories Act 1961 fall into three parts, health, safety and welfare, but we shall concentrate on the safety provisions, ss. 12–17.

Fencing provisions

(*a*) *Prime movers*. Section 12(1) states that every flywheel directly connected to any prime mover and every moving part of any prime mover, except such prime movers as are mentioned in subs. (3), shall be securely fenced, whether the flywheel or prime mover is situated in an engine-house or not.

Section 12(2) states that the head and tail race of every water wheel and of every water turbine shall be securely fenced.

Section 12(3) states that every part of electric generators, motors and rotary converters, and every flywheel directly connected thereto, shall be securely fenced unless it is in such a position or of such construction as to be safe to every person employed or working on the premises as it would be if securely fenced.

A prime mover is defined by s. 176(1), Factories Act 1961, as "Every engine, motor or other appliance which provides mechanical energy derived from steam, water, wind, electricity, the combustion of fuel or other source".

It will be noted that the provision in subs. 12(1) does not require the prime mover to be completely fenced, but rather the flywheel and moving parts. Safety by position or by construction, permissible in the case of electrical gadgetry by subs. 12(3), is not possible with respect to subs. 12(1); in the case of electrical gadgetry mentioned in subs. 12(3) the fencing or safety by position or construction applies to every part, not just to the moving parts as in subs. 12(1).

(*b*) *Transmission machinery*. Sub-section 13(1) states that every part of the transmission machinery shall be securely fenced unless it is in such a position or of such construction as to be safe to every person employed or working on the premises as it would be if securely fenced.

Sub-section 13(2) states that efficient devices or appliances shall be provided and maintained in every room or place where work is carried on by which the power can promptly be cut off from the transmission machinery in that room or place.

Sub-section 13(3) states that no driving belt when not in use shall be allowed to rest or ride upon a revolving shaft which forms part of the transmission machinery.

Sub-section 13(4) states that suitable striking gear or other efficient mechanical appliances shall be provided and maintained and used to move driving belts to and from fast and loose pulleys which form part of the transmission machinery, and any such gear or appliances shall be so constructed, placed and maintained as to prevent the driving belt from creeping back on to the fast pulley.

Transmission machinery is defined by s. 176(1) of the Factories Act 1961 as "Every shaft, wheel, drum, pulley, system of fast and loose pulleys, coupling, clutch, driving-belt or other device by which the motion of a prime mover is transmitted to or received by any machine or appliance".

Prime movers and transmission machinery are regarded by the law as dangerous *per se* and hence the obligations laid down by ss. 12 and 13 are absolute and must be carried out regardless of time, cost or money and irrespective of whether the danger is foreseeable or not. Of course, safety by position or safety by construction would constitute a valid defence to the factory occupier when charged with breach, and the benefit of the obligations of ss. 12 and 13 extend not only to employees but also to other persons working in the premises, e.g. outside contractors or sub-contractors or demonstration personnel. But persons doing their own work inside the factory, such as firemen, policemen and factory inspectors and what might be called "visitors", following the decision in *Flannigan* v. *British Dyewood Co. Ltd.* in 1970, would be excluded from protection.

A leading case on s. 13 is *Carroll* v. *Barclay & Sons Ltd.* (actually occurring under the 1937 Factories Act but nevertheless regarded as a leading authority on transmission machinery). A driving belt broke in the course of working and hit a worker. It was held by the House of Lords that the factory occupier was not liable for a breach of s. 13, as the purpose of the fencing requirement was "to keep the worker out, not the machine or its product in". This rule, although resulting in a limited interpretation of s. 13, seems consistent with the principles of statutory interpretation, particularly penal statutes. There is an action at common law for negligence in these circumstances following the decision in *Hindle* v. *Birtwhistle*, where it was held that a shuttle which flew out and hit an operative was a dangerous part of machinery, since such occurrences had happened previously and were reasonably foreseeable. This explains the existence of the double-barrelled action in such cases (*see Kilgollan* v. *Cooke & Co. Ltd.* and *Bux* v. *Slough Metals Ltd.* in §**3**, p. 39.

(*c*) *Duty to fence dangerous parts of all other machinery.* This duty is to be found in s. 14, Factories Act 1961, which is probably the most important section and the one most used by the inspectorate. The justification behind the section is that, while prime movers and transmission machinery are dangerous *per se* and so should be securely guarded, this is not necessarily so with respect to other machinery in the factory, which, not being dangerous *per se*, does not have to be securely fenced unless dangerous (as defined by case histories).

Sub-section 14(1) states that every dangerous part of any machinery, other than prime movers and transmission machinery, shall be securely fenced unless it is in such a position or of such construction as to be safe to every person employed or working on the premises as it would be if securely fenced.

(*d*) *Automatic safety device.* Sub-section 14(2) states that in so far as the safety of a dangerous part of any machinery cannot by reason of the nature of the operation be secured by means of a fixed guard, the requirements of s. 14(1) shall be deemed to have been complied with if a device is provided which automatically prevents the operator from coming into contact with that part.

Sub-section 14(3) states that where the Minister is satisfied that there is available and suitable for use in connection with machinery of any class any type or description of safety device which—

(i) prevents the exposure of a dangerous part of machinery while in motion; or

(ii) stops a machine forthwith in case of danger;

he may make regulations directing that the type of description of device shall be provided for use in connection with such class of machinery as may be specified in the regulations.

Sub-section 14(4) establishes that it is a defence to show that "a device at least equally effective was being used".

Sub-section 14(5) states that any part of a stock-bar which projects beyond the head-stock of a lathe shall be securely fenced.

Section 14 has attracted a wealth of case law, some of it irreconcilable but for the most part showing that the section is not intended nor is to be construed as protecting the employees or others working in the premises from things and pieces flying out of the machinery. Not so where a person comes into contact with the machinery. A succession of cases has established this principle, ranging from *Nicholls*

v. *Austin Ltd.*, *Eaves* v. *Morris Motors Ltd.*, *Sparrow* v. *Fairey Aviation Co. Ltd.*, *Walker* v. *Bletchley-Flettons Ltd.* to *Close* v. *Steel Company of Wales* and *Callow (Engineers) Ltd.* v. *Johnson.* Indeed it is largely because of cases such as these that subs. 14(6) existed. This sub-section stated "The Minister may, as respects any machine or any process in which a machine is used, make regulations requiring the fencing of materials or articles which are dangerous while in motion in the machine". No regulations were ever made. The sub-section was repealed by the Factories Act 1961 (Repeals and Modifications) Regulations 1974.

(*e*) *No duty to fence material in the machine?* The 1961 decision in *Eaves* v. *Morris Motors Ltd.* established that there is no duty to fence material in a machine when the danger is due to the material. E was working a milling machine and saw that it was behaving irregularly. He pulled back his hand, grazing his finger on a bolt being milled in the machine. It was held that the bolt was not a part of the machine and need not be fenced. *Sparrow* v. *Fairey Aviation Co. Ltd.* decided shortly afterwards established that s. 14 does not give protection, nor is it so intended, to prevent injury arising as a result of an employee coming into contact with a dangerous part of the machine through contact being made by a tool which he is using. S was engaged on operating a lathe during his employment with F.A.C. One day whilst machining a metal disc and whilst he was removing the burrs on the disc with a hand tool as the disc was revolving in the lathe, the tool caught against the jaws of the chuck and he was injured as a result of his hand being thrown against the disc although his hand did not come into contact with the jaws of the chuck themselves. Claiming damages, he alleged breach of s. 14(1), Factories Act 1937 (now s. 14(1), Factories Act 1961). It was held that the factory occupiers had complied with s. 14(1), although the fencing was not such as to prevent the hand tool coming into contact with the jaws of the chuck, which had caused the plaintiff's injury.

(*f*) *Machine and material in the machines.* Where the injury is due to both the machine and a material in the machine, the decision in *Midland & Low Moor Iron & Steel Co. Ltd.* v. *Cross* in 1965 established that machinery had to be fenced where, in the course of its operation, material was brought into contact with a moving part and created a danger. The 1970 decision of *Callow (Engineers) Ltd.* v. *Johnson* by the House of Lords established that wherever there is interaction between two parts with material or some other object in between, there is a duty to fence. Here a rotating workpiece and an almost imperceptibly moving bar of a centre lathe on which J was machining a stainless steel casting caught his fingers, mangling them. The 1977 decision of the House of Lords, however, in *Wearing* v. *Pirelli Ltd.* has given a more liberal interpretation to the terminology of subs. 14(1), and although, as Lord Hailsham observed in *Callow (Engineers) Ltd.* v. *Johnson*, the gap created by *Eaves* v. *Morris Motors Ltd.*, *Walker* v. *Bletchley-Flettons Ltd.* and *Close* v. *Steel Company of Wales* can only be closed by legislation as regards injuries caused by bits and pieces flying *out* of machines, nevertheless until this gap is plugged either by subs. 2(2), H.A.S.A.W.A. or by regulations proposed by H.S.C., it is not going to get any bigger. In *Wearing* v. *Pirelli Ltd.* W worked on a machine which moulded rubber for car tyres. His job involved folding over a drum a sheet of rubber $\frac{1}{16}$ in. thick, which extended about 6 in. beyond each end of the drum. He then put a beading of rubber and metal over the rubber "skirt" extending over the end of the drum. The drum was then revolved at high speed and W, using a hand tool, folded the extended rubber skirt back on itself, over the beading. Once when applying the hand tool to the rubber, he felt a jerk and his hand was pulled down, rubbing against the beading which caused a friction burn and broke a bone in his wrist. He sued his employer, alleging breach of the statutory duty to fence dangerous machinery coupled with negligence at common law in failing to set up a safe system of work. The House of Lords took the view that the fact that the revolving drum brought W into contact with the material covering it was sufficient to amount to a breach of duty on the part of the employers to fence. The point about a safe system of work at common law was not pursued in the House of Lords, though Viscount Dilhorne thought that there was a breach of common law duty on this ground. Moreover this case, like *Kilgollan* v. *Cooke & Co.* and *Bux* v. *Slough Metals Ltd.* before it, serves to underline the attractiveness to injured workers of the double-barrelled action for breach of statute coupled with common law negligence.

(*g*) *Foreseeability*. The celebrated quartet of cases—*Close* v. *Steel Company of Wales Ltd.*, *Walker* v. *Bletchley-Fletton Ltd.*, *Eaves* v. *Morris Motors Ltd.* and *Burns* v. *Terry*—have introduced notions of common law into s. 14(1) by defining "dangerous" in terms of foreseeability. This body of case law has regarded the duty imposed by s. 14(1) not as an absolute duty but as imposing liability only where the injury caused by the machinery was reasonably foreseeable. Put another way, a "dangerous" part of machinery is a part of machinery which gives rise to an injury that is reasonably foreseeable. In *Close* v. *Steel Company of Wales Ltd.* in 1961 an employee of S.C.W. was working an electric drilling machine when the bit of the drill shattered and a piece flew into his eye. He claimed damages for breach of s. 14(1), Factories Act 1937 (now s. 14(1), Factories Act 1961). Although bits had shattered before there was no evidence that such an accident had happened before. It was held by the House of Lords that the employer was not liable under s. 14(1) because the danger from the bit was not "reasonably foreseeable". Nevertheless Lord Goddard felt, although the point did not arise to be decided, that the workman would have had a good action at common law. This sort of approach had been earlier indorsed in impeccable language by du Parcq, J., in 1937 in *Walker* v. *Bletchley-Flettons Ltd.*, when he said "A part of machinery is dangerous if it is a possible cause of injury to anybody acting in a way in which a human being may be reasonably expected to act in circumstances which may reasonably be expected to occur". Now that foreseeability has emerged as the basis of liability under s. 14(1), do the courts mean that foreseeability of accident or foreseeability of the operative's conduct is the basis of liability? In *Burns* v. *Terry & Sons*, decided in 1950, the former view was taken. A 17-year-old boy, who was an employee of T, was engaged in keeping the floor clean round a chocolate-making machine, and in collecting any chocolate beans that spilled on to the floor. On Friday afternoon when the machine was stopped he helped to clean it, during the course of which he used to climb a ladder and collect any beans which had spilled on to a shelf. At the back of the machine was a cogged pulley wheel 26 in. in diameter. This wheel engaged through a cogged pinion and horizontal shaft 5 ft. 10 in. from the ground. The shelf was about 9 in. above the pulley wheel. A circular wire mesh guarded the pulley shaft on the pinion, but only from the front, leaving the sides exposed. When no-one was looking, the boy took a ladder, and placed it against a rotating shaft above the top of the machine so as to be able to collect any beans left on the shelf. The ladder slipped sideways and whilst trying to grasp some means of support, the boy's hand was crushed between the pulley wheel and the pinion. Claiming damages from T, he alleged breach of s. 13(1), Factories Act 1937 (now s. 13(1) Factories Act 1961). It was held by the Court of Appeal that there was no liability in the employer. The test of liability was whether the transmission machinery was not so securely fenced as to give security from such danger as may be "reasonably foreseeable". The court ruled that the machinery was securely fenced according to that test, emphasising that it is the *foreseeability of accident* which is important for liability purposes. In *Uddin* v. *Associated Portland Cement Manufacturers Ltd.* in 1965, on the other hand, the same court took a different view, that it was *forseeability of the behaviour of the operative* that was important. The plaintiff, whilst attempting to retrieve a pigeon in a place where he was not authorised to be and doing something which he was not authorised to do, was injured. The plaintiff, a Pakistani, was a machine minder over a cement packing plant and he suffered injury whilst in the dust extracting plant in the factory. It was held by the Court of Appeal that, although his conduct was unauthorised, yet it was not unforeseeable and as a result he was entitled to recover partial damages.

Although the case histories surrounding *Close* v. *Steel Company of Wales Ltd.* have generally been regarded as better law, there are a cluster of cases which establish that foreseeability is irrelevant to the question of liability under s. 14(1), notable among which are *Summers & Sons Ltd.* v. *Frost*, *Millard* v. *Serck Tubes Ltd.* and *Allen* v. *Aeroplane & Motor Aluminium Castings Ltd.*, the first two of which are cited below. In *Summers & Sons Ltd.* v. *Frost* in 1955 the respondent was a maintenance fitter employed by the appellant in a steelworks. One day he was grinding a piece of metal on a power-driven grinding machine and was injured when his thumb came into contact with the revolving grindstone. The grindstone consisted of a horizontal shaft at each end of which there was a circular grinding wheel and the shaft was powered by an electric motor. The machine was fitted with an efficient hood and the only part of the revolving grindstone that was exposed was an arc about 7 in. long; there was a gap of $\frac{3}{8}$ in. between the hood and the revolving arc of the stone. It was held that the grinding stone was a dangerous part of machinery for the purposes of s. 14(1) and there was an absolute obligation to fence irrespective of whether the operative was careless or inattentive. This obligation had been breached here and so there was liability, even though the result of the decision meant that grinding machines became commercially unusable (the retrograde effect of this decision was later rectified by the Abrasive Wheels Regulations 1970, *see* § 5, p. 10). In the second case of *Millard* v. *Serck Tubes Ltd.* decided in 1969, the appellant was employed by the respondent on a machine used for drilling holes in metal billets. His hand, which was gloved, was caught by some swarf and was dragged in and caught by the moving drills of the machine. The machine was fenced, but there was a gap through which his hand had been drawn. The employers argued that although they were in breach of s. 14(1), in that the machine was not securely fenced, yet they were not liable for the man's injury because his injury occurred in a completely unforeseeable way. It was held by the Court of Appeal that the employer was liable, the court observing "It is argued that because the defendants could not reasonably have foreseen that an accident would occur in this particular way, there is no liability.... It seems to me that the fact that the accident occurred in an entirely unforeseeable way is wholly irrelevant in this case." The court went on to say that foreseeability was only of importance when the question before them was: was a particular part of a machine dangerous? In other words, when the question was: ought a particular part of the machinery to have been fenced? the court felt that that was not the dominant issue and that it had been conceded that the machine had not been properly and securely fenced and there was liability.

(*h*) "*Machinery.*" The expression "machinery" is obviously of key importance to the operation of ss. 12, 13 and 14, Factories Act 1961, as well as other provisions relating to fencing. This expression got judicial scrutiny in the leading case of *Parvin* v. *Morton Machine Co. Ltd.* in 1952. The employee, who was just under 18 years old, was an apprentice fitter employed by M.M.C. He was injured when he was cleaning a doughbrake machine which had been manufactured and assembled in the factory belonging to M.M.C. This machine was not guarded and the employee claimed damages for breach *inter alia* of s. 14(1). On an appeal to the House of Lords, it was held that the expression "machinery" did not extend to machinery manufactured in the factory but only applied to machinery used in the product-making process, in other words to machinery from which the dough-brake machine would have been made. Moreover when new machinery is being installed, the fencing requirements apply as soon as the installation is complete, this point having been established by *Irwin* v. *White Tomkins & Courage Ltd.* in 1964, where a hoist was regarded as machinery even though it was to be used in connection with a conveyor belt that was under construction.

(i) "*In use or motion.*" Section 16 deals with the question of machinery in "use or in motion". It is felt s. 16 is a chronological misplacement, in that if machinery is not in "use or in motion" ss. 12, 13 and 14 are irrelevant anyway. Section 16 has attracted some case law, though nowhere near as much as ss. 13 and 14. An early authority was *Mitchell* v. *Westin Ltd.* occurring under the Factories Act 1937. The employee was a fitter and turner in the employment of W. The machine which he was working on was not running properly and in order to see what was wrong he took off a guard fence and in an effort to reach some screws in a collar securing a gear he rotated the machine by switching it on and off virtually simultaneously. His finger was trapped in the moving gear wheels and he suffered injury. The employee claimed damages for breach of ss. 14 and 16, Factories Act 1937 (now ss. 14 and 16, Factories Act 1961). It was held by the Court of Appeal that the lathe was not in motion, since the machine was only partly in motion when being switched on and off. This was in no way comparable with the normal working of the lathe and so there was no liability in the employer. This point had been made by the earlier decision of the House of Lords in *R.T.B.* v. *Cummings* in 1955. The employee was injured when he was adjusting a machine which was normally driven by belts passing over pulleys. The machine was undergoing repairs and so the power had been shut off, but the cover guarding the pulleys and belts had been removed.

In order to carry out the adjustment, he rotated the machine by pulling the belt with his hand and was injured. He claimed damages for breach of s. 13, Factories Act 1937 (now s. 13, Factories Act 1961), and the employers argued that the machine was not in use or in motion. They contended that s. 16 did not apply to movement of machinery by hand. It was held that the machine was not in motion and so there was no liability in the employer for the man's injury. The mere fact, however, that machinery is being prepared for use does not mean that it is not "in motion". Thus in *Stanbrook* v. *Waterlow & Sons Ltd.* in 1964 a printing machine was being got ready for printing purposes. In order to do this, it had to revolve quickly over short distances by being switched on and off. It was held that the machine was "in motion" within the meaning of s. 16 and the employee was entitled to recover for his injury. But it has been established that machinery is not "in use" where it is being cleaned or examined, hence it seems that before the machine can be regarded as being "in use" it must also be "in motion" and vice-versa. When a machine is being installed the fencing requirements begin to apply as soon as installation is complete, this point having been established by *Irwins* v. *White Tomkins & Courage Ltd.*

After prescribing that fencing is absolutely essential whilst the machinery or parts thereof are in use or in motion, s. 16 goes on to say "except when such parts are necessarily exposed for examination and for any lubrication or adjustment shown by such examination to be immediately necessary, and all such conditions as may be specified in regulations made by the Minister are complied with". Such regulations have been made in the form of the Operations at Unfenced Machinery Regulations 1938–1946, the effect of which is that such examination, lubrication or adjustment can only be carried out by certificated machine attendants. When this work is being done overalls must be worn, another person must stand by in case of emergency and the person must be over 18 and male. This last requirement may well soon change and the prohibition on the cleaning of machinery by women imposed by s. 20, Factories Act 1961, will probably be removed as a result of a review of protective legislation for women by H.S.E. and the Equal Opportunities Commission. Section 20 will probably be replaced by regulations made under s. 15, H.A.S.A.W.A.

(ii) "*Machinery*" *constructed after 29th July 1937*. Sub-section 17(1) provides "In the case of any machine in a factory being a machine intended to be driven by mechanical power—

(*a*) every set-screw, bolt or key on any revolving shaft, spindle wheel or pinion shall be so sunk, encased or otherwise effectively guarded as to prevent danger; and (*b*) all spur and other toothed or friction gearing, which does not require frequent adjustment while in motion, shall be completely encased unless it is so situated as to be as safe as it would be if carefully encased".

Sub-section 17(2) also makes liable sellers or hirers if they supply for use in a factory a machine which fails to comply with subs. (1). They may incur a maximum fine of £200. The decision in *Biddle* v. *Truvox Engineering Ltd.* in 1952 has established that breach of this section by the manufacturer does not render the manufacturer liable to damages for injury caused to the workman by machinery not complying with subs. 17(1) (although the manufacturer as co-defendant was held not liable along with the employer in *Biddle's* case it is felt that this decision might not be followed today in the light of the reform of the law on product liability, *see* §**47**, p. 246).

Health provisions

These are contained in ss. 1–11, Factories Act 1961. Some of the more important are mentioned below.

Section 1 deals with cleanliness and states that every factory shall be kept in a clean state.

Section 2 deals with overcrowding and states that whilst work is being carried on, a factory shall not be so overcrowded as to cause risk or injury to the health of persons employed therein.

Section 3 deals with temperature and provides that a reasonable temperature must be secured and maintained in every workroom. If work or a large part of it is done sitting and does not involve serious physical effort, after the first hour a temperature of not less than 60°F shall be deemed reasonable.

Section 4 deals with ventilation and states that every workroom shall be provided with effective and suitable means for the circulation of fresh air and a method adopted to render harmless any dust, fumes or other impurities generated as a result of the work. The method envisaged as suitable by s. 4 for rendering harmless impurities is ventilation only; this was made clear in *Ebbs* v. *Whitson Ltd.* in 1952. The plaintiff was employed by the defendant as a coach builder. During the course of his work he had to scrape and sandpaper monsonia wood (West African walnut). He caught dermatitis from the wood dust in respect of which he claimed damages for breaches of ss. 4(1) and 47(1), Factories Act 1937 (now ss. 4(1) and 63(1), Factories Act 1961). As far as the breach of s. 4(1) was concerned, it was held that there was no liability in the employer, as subs. 4(1) related only to ventilation (and to nothing else) as a method of rendering dust harmless. The object of s. 4 is to get rid of impurities and dust and fumes which get into the air when methods envisaged in s. 63 have failed to operate successfully.

Section 5 deals with lighting and states that sufficient and suitable lighting, either natural or artificial, must be provided and all windows and skylights used for the lighting of workrooms must, as far as practicable, be kept in a clean condition and free from obstructions. The Minister may by regulation prescribe the standards of lighting.

Section 6 deals with drainage and states that an efficient method of drainage must be provided and maintained where any process is carried on which makes the floors wet so as to be capable of being drained.

Section 7 deals with sanitary conveniences and provides that a sufficient and suitable number and type of sanitary conveniences must be provided and kept in a clean condition and there must be separate accommodation for both sexes.

(*a*) *Dust and fumes*. Apart from s. 4 dealing with the circulation of fresh air to get rid of impurities in the air, s. 63, Factories Act 1961, states that in every factory in which there is a process which gives off dust, fumes or other impurities likely to be injurious or offensive to the persons employed, measures must be taken to prevent the dust, fumes or other impurities from being inhaled or accumulating, and where practicable, exhaust facilities shall be provided near the point of origin of these impurities. Section 63 may be excluded by special regulations, but if its operation is not excluded, these measures must be carried out as far as is *practicable*. This last point was well illustrated in the case of *Richards* v. *Highway Ironfounders Ltd.* in 1955. The plaintiff was employed as a moulder by the defendant and during the course of his employment he contracted silicosis through inhaling dust created by the "knocking out" process. This related to the removal of grains of sand adhering to the castings. He claimed damages for breach of what is now s. 63(1), Factories Act 1961. It was held by the court that the employer was not liable because when the employee contracted his illness it was not generally known that silicosis could arise from this process. The employer would have to keep abreast of current knowledge and invention to ascertain whether there was a link between the process or not. If such information was not available, it was hardly realistic to expect that the employer would know about it.

The expression "practicable" and the implementation of duties as far as "practicable" was considered by the court in the leading case of *Schwalb* v. *Fass Ltd.* in 1946, where it said that where the duty that requires to be done is what is practicable, if it is *feasible*, even though difficult or inconvenient, it must be done in the light of current knowledge and invention.

(*b*) *Floors, steps, stairs and passages.* Section 28, Factories Act 1961, deals with the condition and structure of the factory as regards its floors, steps, gangways and passages. The duty created by subs. 28(1) is partly strict and partly based on negligence. Section 29, on the other hand, is concerned with means of access to work and the place of work itself. The duty created by subs. 29(1) is based on negligence and requires to be done what is reasonably practicable.

Sub-section 28(1) provides "All floors, steps, stairs, passages and gangways shall be of sound construction and properly maintained and shall, so far as is reasonably practicable, be kept free from any obstruction and from any substance likely to cause persons to slip". The duty to have floors, steps, stairs, passages and gangways soundly constructed and properly maintained is a strict one and a factory occupier could not be absolved from liability by arguing that the defect in construction or maintenance was not discoverable on reasonable inspection. The duty does not depend on negligence or the exercise of reasonable care. The second part of the duty, however, is based on negligence and was written in to accommodate the decision in *Latimer* v. *A.E.C. Ltd.* in 1953. Factory occupiers could not be expected to make provision for an act of God, i.e. a transient condition like a thunderstorm of exceptional severity. If the defect causing the obstruction or causing persons to slip would not be detectable by the exercise of reasonable care, there will be no liability.

(*c*) *Access to place of work.* Sub-section 29(1) provides "There shall, so far as is reasonably practicable, be provided and maintained safe means of access to every place at which any person has at any time to work, and every such place shall, so far as is reasonably practicable, be made and kept safe for any person working there".

The scope of s. 29 is wide and applies to everything that could constitute a means of access to work. Thus everything from roofs to ditches, chimneys to planks and stepping stones is within the ambit of this sub-section. Liability depends upon failure to exercise reasonable care, i.e. inadvertent negligence. If the defect could not have been discovered on the exercise of reasonable care and skill, the factory occupier is not liable for injury suffered. An interesting case is *Hosking* v. *de Havilland Ltd.* in 1949 where an oncoming independent building contractor, engaged to do work at the defendant's factory, dug a ditch and bridged its gap with an unsound and defective plank. The plaintiff was injured whilst using the plank to gain access to the factory. It was held that the factory occupier was liable (as a matter of fact, if work was being done under the R.I.B.A. Contract in cases such as this, the factory occupier as building owner could claim indemnity automatically from the contractor, which in this case he did.) The sub-section gives statutory protection to "any person" who has at any time to work in the factory premises, on them, over them or under them. Thus in the case of *Lavender* v. *Diamints Ltd.* in 1949 factory occupiers were held liable for injuries to a visiting windowcleaner who, in order to carry out his work and clean windows in the factory, had to climb across an asbestos roof and, as a result of its defective condition, fell through. Under this sub-section factory occupiers would also find themselves liable, as well as the employer, for injuries suffered by employees of main and sub-contractors working on the factory premises. Again contractual indemnity would normally be forthcoming from the contractor or sub-contractor via the main contractor if there was any negligence on the part of the latter. If, however, the negligence is solely that of the factory occupier, the liability will be his under s. 29(1) to all contract labour employed on the factory premises. The scope of s. 28(1) is equally wide in that the duty is owed to "persons" (likely to slip). But neither would give protection to persons doing their own job on the factory premises, e.g. factory inspectors and firemen, following *Flannigan* v. *British Dyewood Co. Ltd.* (*see* p. 208).

34. Regulations pursuant to Factories Act 1961

Regulations used to be made under s. 76, Factories Act 1961. This section was repealed by the Factories Act 1961 (Repeals and Modifications) Regulations 1974 which came into force on 1st January 1975. Regulations made under s. 76, however, are continued in force by reg. 7(3) of the 1974 Regulations. In future regulations affecting health and safety at work will be made under s. 15, H.A.S.A.W.A., a provision which empowers H.S.C. to propose regulations to the Secretary of State for Employment. Important regulations have been made under s. 76 including the Power Presses Regulations 1965 (*see* Appendix IV), the Asbestos Regulations 1969 (*see* Appendix V), the Abrasive Wheels Regulations 1970 (*see* Appendix VI) and the Protection of Eyes Regulations 1974 (*see* Appendix VII).

35. The law and injuries to eyes

Of all the types of industrial injuries the one which has attracted most case law is injury to eyes. Much legislation relates to eyes either directly or indirectly. This takes the form of s. 65, Factories Act 1961, and a series of regulations dating back to 1938. More generally protection of eyes would be covered by ss. 2–7, H.A.S.A.W.A., as supplemented by s. 33 in that all employers should take steps to ensure the health, safety and welfare of all their employees, and this includes measures for protecting the eyes of their employees wherever they may work, irrespective of the specific requirements of the Protection of Eyes Regulations 1974 (*see* Appendix VII). Also the general safety provisions of ss. 12–17, Factories Act 1961, by requiring machinery to be securely fenced, are designed to prevent injury to the eyes of operatives working the machinery. Much of the case law on s. 13 and 14, Factories Act 1961, including *Close* v. *Steel Company of Wales Ltd.*, concerned injury to eyes. *Bux* v. *Slough Metals Ltd.* and *Kilgollan* v. *Cooke & Co. Ltd.* were concerned with injury to eyes arising from faulty supervision.

Specific legislation

Section 65, Factories Act 1961, states that in the case of any process specified in regulations "which involves a special risk of injury to eyes from particles or fragments thrown off in the course of the process suitable goggles or effective screens shall in accordance with ... the regulations be provided to protect the eyes of persons employed in the process". This section (or its predecessor in the 1937 Act, s. 49) attracted some curious case law.

In 1954 in *Whalley* v. *Briggs Motor Bodies Ltd.* An employee of a contractor was breaking up concrete in a factory occupied by B.M.B. While he was using a pneumatic pick, the property of his employer, he was struck in the eye by a piece of concrete. No goggles had been provided either by his employer or by the factory occupier, as was required by s. 49, Factories Act 1937 (now s. 65). It was held by the court that:

(*a*) the process of breaking up the concrete was not the process of B.M.B. nor was W their employee. B.M.B. was not, therefore, liable for an injury to an employee of an independent contractor; and

(*b*) since W was an employee of the contractor and the contractor was the owner of the pneumatic pick it was the contractor who was the factory occupier; hence B.M.B. was not liable.

Another case where the decision was similar was *Walsh* v. *Allweather Mechanical Grouting Ltd.* An employee of a firm of contractors who were nominated to break up and relay concrete on an apron near to an aircraft hangar (which was a factory) was held not to be entitled to compensation for injury to his eye when a piece of concrete dislodged by the pneumatic drill struck him in the eye. The reason for the decision was that, although the apron was in the curtilage of the factory, as required by s. 175, Factories Act 1937 (definition of a factory), it was not being used for purposes carried on in the factory.

Had both these cases come up for decision today the result would have been different. *Whalley* v. *Briggs Motor Bodies Ltd.* and analogous situations are now covered by reg. 6, Protection of Eyes Regulations 1974, which extends protection to persons who, although not employed in the process, are nevertheless at risk. Today *Whalley* would have fallen into that category. Moreover s. 2, H.A.S.A.W.A., would have applied to his situation today, with a possible fine for the employer. There could also have been a prosecution by an inspector under the Protection of Eyes Regulations 1974. Virtually the same position would have obtained in *Walsh*'s case. The fact that the place where the injury occurred was not a factory is irrelevant and prosecutions could have been brought for breach of s. 2, H.A.S.A.W.A., and reg. 6, Protection of Eyes Regulations. In addition a civil action in respect of the injury would probably have been successful at common law. Furthermore the eye injuries might not have taken place today in the light of s. 7, H.A.S.A.W.A., and reg. 11, Protection of Eyes Regulations.

The expression "suitable" has been defined in a number of civil cases as meaning suitable in that the goggles protect the eyes. However, *Daniels* v. *Ford Motor Co. Ltd.* in 1955 establishes that this section does not impose an absolute duty that the goggles shall protect the eyes. The section seems to contemplate that the goggles be well adapted to process and to wearer in the sense that they must fit him. As to how far the section and the Protection of Eyes Regulations require that individual requirements are met, it is difficult to be sure. If there is an employee with a particular disability or idiosyncrasy common law requires that the employer cater for his individual needs, as will be seen from the decision in *Paris* v. *Stepney B.C.*

The Protection of Eyes Regulations 1974

The history of eye protection regulations goes back to 1938 when the first regulations were passed. The requirements were general because appropriate specifications for eye protectors did not exist. Various industries where eye hazard was especially great created more stringent regulations; this applied particularly to the chemical, iron and steel, foundries, shipyards and construction industries. All these regulations have now been unified and superseded by the single Protection of Eyes Regulations 1974 which specify processes for which eye protection is compulsory and require persons engaged in such processes and persons at risk from such processes to wear approved eye protection. Under these regulations which came into force on 10th April 1975 all premises covered by the Factories Act 1961 are designated eye hazard areas. Moreover since s. 2, H.A.S.A.W.A., requires that an employer set up a safe system of work and a safe place of work, it is arguable that the Protection of Eyes Regulations 1974 apply to all places of work where there is a risk of foreseeable injury to eyes from the work being done. Depending on the type of hazard the regulations stipulate appropriate protection, including spectacles, goggles, face shields and helmets to be worn by workers as well as fixed shields for mounting on dangerous plant, such as welding and metal-spraying equipment. The regulations do not distinguish between plano spectacles (i.e. with plain lenses) and prescription spectacles (with ophthalmic lenses for correction of eyesight), but they do require that eye protectors are suitable for the user. The requirement of suitability is of crucial significance, and it is thought that the wearing of a face mask over normal prescription spectacles where a face mask would not otherwise be obligatory is not "suitable". The need for protective prescription spectacles seems to follow from these regulations.

The regulations also say that eye protectors must be of an approved type. Certified approval by the Chief Inspector of Factories is issued only if the eye protector complies with the relevant British Standards, BS 679: 1959; BS 1542: 1960; BS 2092: 1967.

Duties on employers and employees

Under the Protection of Eyes Regulations 1974 the employer of persons employed in processes mentioned in Sched. 1 must provide *eye protectors* if those persons are employed in processes mentioned in Part 1, Sched. 1, e.g. blasting or erosion of concrete or the cleaning of buildings. In the case of persons employed in processes mentioned in Part 2, e.g. use of an exposed electric arc, the employer must provide a *shield* or a sufficient number of *fixed shields*. And as regards persons mentioned in Part 3, Sched. 1 processes, e.g. hot fettling of steel castings, the employer must provide *eye protectors* or a *shield* or a sufficient number of fixed shields. This is mandatory by virtue of s. 5. By virtue of s. 6 an employer must provide either eye protectors or shields or a sufficient number of fixed shields for persons, e.g. outside contractors, who, *although not employed* in the processes, are *at risk* from processes mentioned in Sched. 2 of the regulations. Under reg. 11 of the 1974 Regulations there is an obligation upon every person provided with eye protectors or a shield to use or wear the eye protector at all times while employed in the specified process or whenever there is a risk of foreseeable injury to his eyes from processes mentioned in Sched. 2, para. 3 (work with drop hammers, power hammers, etc.) except when fixed shields have been provided.

Compensation for injury

There are three sources of compensation in respect of eye injury suffered at work—common law; breach of statutory duty, either the Protection of Eyes Regulations 1974 or some other regulations, e.g. the Non-Ferrous Metals (Melting and Founding) Regulations 1962; and industrial injury benefit or industrial disablement benefit under the Social Security Act 1975. The first two heads of civil liability will be examined.

(*a*) *Common law.* Where an employee suffers eye injury at work he can sue his employer for negligence at common law, as part and parcel of the latter having failed to set up and effectively maintain a safe system of work. That this course of action in the circumstances is preferable to action for breach of statutory duty was true before H.A.S.A.W.A. was passed, as shown by the cases of *Whalley* v. *Briggs Motor Bodies Ltd.* and *Walsh* v. *Allweather Mechanical Grouting Co. Ltd.* The position now is that a civil action is brought in respect of eye injury on the basis of common law, at least until such time, if any, as a "no fault" system of liability is introduced. An instructive case is *Paris* v. *Stepney B.C.* in 1951. An employee of S.B.C. engaged on vehicle maintenance had only one good eye and his employer knew this. One day whilst he was attempting to remove a bolt from the chassis of a vehicle with a hammer, a chip of metal flew into his good eye, with the result that he became completely blind. He sued his employer for damages at common law in that he had not been supplied with goggles and that in consequence he had sustained the eye injury. The employer showed that it was not normal practice in the trade to supply goggles, at least to workmen with two eyes. The House of Lords held that there was a much greater risk of injury to a one-eyed man than to employees with two good eyes. The employer should have realised this and taken the necessary steps to avert the danger to the one-eyed man. The fact that they did not do so meant that they were liable for negligence.

(*b*) *Breach of statutory duty*. The action in respect of eye injury for breach of statute will be based on s. 65, Factories Act 1961, and/or the Protection of Eyes Regulations 1974.

THE CONTRACTOR'S LIABILITY

36. The dual role of the contractor

On a building site the main contractor is an employer and occupier as regards health and safety at work. As regards his own employees he is an employer, but as regards the employees of sub-contractors under his overall control, whether nominated or domestic, he is an occupier. With regard to the health and safety of the general public he is an occupier. The effect of this is that as an employer his liability is governed by s. 2, H.A.S.A.W.A., coupled with the common law, the Factories Act 1961 and the Construction Regulations. The duty cast upon an occupier by the law is, on the other hand, a considerably less onerous one, as his relationship is essentially one involving *visitors* rather than employees. The duty laid upon an occupier by the Occupiers' Liability Act 1957 is one of taking reasonable care as far as the structural condition of the premises is concerned, to all lawful visitors and with regard to operations on the occupier's premises.

Liability is never strict. The liability of an occupier is normally underwritten by a public liability policy regarding injuries to visitors, whether employees of other contractors working on the building site or members of the public generally.

The concept of occupation

The concept of occupation is crucial to an understanding of the liability of a building contractor. The cases seem to establish that it is not merely physical control of the premises which constitutes occupation for the purpose of liability of building contractors but rather that occupation refers to a legal and equitable interest in the premises or land. Since various persons can have legal and equitable interests in one piece of land the consequence is that occupation of land, such as a building site or premises on land, may well be shared. Exclusiveness of occupation is practically effete. This point was well illustrated by the 1965 House of Lords decision in *Wheat* v. *Lacon*, where it was held that both an owner of a public house and a manager would have been liable under the Occupiers' Liability Act 1957, in the absence of a technicality, to the widow of a man killed on the premises. The same attitude was taken a year later in 1966 in *Moloney* v. *Lambeth B.C.*, where a child was badly injured when falling through a gap in a balustrade in a flat owned by L.B.C. and rented from them by X, both therefore being construed as occupiers.

The case of *London Borough of Hounslow* v. *Twickenham Garden Developments* established that the proper relationship of main contractor and building owner was one of licensee – licensor and, since the sub-contractor as well has control over a certain part of the building operations, his relationship with the main contractor is one of licensee – licensor also. Overall occupation rests with the main contractor or contractors rather than with the sub-contractor or sub-contractors. Because of the principle in *Wheat* v. *Lacon* both building owner and main contractor or contractors share the control of the building site and hence the occupation. To that extent they share the liability. If a standard form contract has been entered into between the building owner and the main contractor and between the main contractor and the sub-contractors, the contract will make express provision for indemnity from liability of the building owner by the main contractor and of the main contractor by the sub-contractor or sub-contractors.

The practical consequences of occupation

It is reasonably clear that the safety of an operative working on a building site or supplying goods and plant to a building site rests, as far as a safe system of work is concerned, upon his immediate employer, but as regards matters concerning health and safety outside the compass of the immediate employer – employee relationship, upon the criterion of occupation, each occupier being responsible for his sphere of operations. Suppose that Z, an employee of a sub-contractor nominated to do steelwork, was injured as a result of a co-employee having erected steelwork negligently—here the liability is that of the employer/sub-contractor nominated for the steelwork. This is because the injury is referable to the employer–employee relationship. Suppose, however, that the same employee was injured as a result of bad palletisation of bricks by a fork-lift truck driver employed by T, the forks trying to carry more than their specified load, here T's employer is vicariously liable to Z. This is shown by the decision in *Smith* v. *Vange Scaffolding & Engineering Ltd.* in 1970, a case of vicarious liability. Alternatively there might be a case of an occupier being liable for operations being performed negligently within his sphere of operations on a building site. Here it is pointless, however, for the injured employee to prove fault under the Occupiers' Liability Act 1957 when he can sue an employer under the doctrine of vicarious liability. There will, however, be occasions when an injury may not be traceable to the fault of an employee with a loss-distributor employer waiting to pay. This could occur where injury takes place through the defective structural condition of premises or the generally unsafe condition of the building site. It is in these cases that the injured person must prove negligence against the occupier or the multiple occupiers. Suppose that before demolition work was due to begin on old factory premises, a steeplejack went to inspect the chimney, climbing up an erected scaffold, and was subsequently injured through the chimney collapsing owing to a latent defect; here the principle in *Smith* v. *Vange* cannot be invoked, as the injury is not due to the negligence of an employee, and the position as stated in *Indermaur* v. *Dames* will apply: the visitor taking reasonable care on his part will not be exposed to *unusual* dangers from the premises. In other words, the visitor will not be exposed to dangers which he would not normally encounter in the course of his work on the assumption that the occupier could not have known about them. In the case of the steeplejack it is doubtful if the occupier of the factory would have been liable to him at common law, since it is unlikely that the occupier would have known about a latent defect in the chimney's condition, unless the occupier was knowledgeable with respect to chimneys. On the other hand, the factory occupier might be liable under s. 29, Factories Act 1961. Employers must take positive steps towards ensuring the safety of their employees, whereas occupiers can assume that an employee coming to work on their premises will look out for himself in connection with the carrying out of his work.

The importance of Smith *v.* Vange

In *Smith* v. *Vange* an employee of a sub-contractor was injured through the negligence of an employee of a main contractor. It was held that the main contractor was liable on the basis of vicarious liability. Exactly the same result would have followed if the person injured had been a member of the public. What emerges from this is that, as far as injuries to employees of sub-contractors are concerned or to members of the public, whether passers-by or policemen or firemen or ambulance drivers coming on to a building site to give aid to an injured workman, if the fault is traceable to the act or omission of an employee, the employer is liable and injury may be compensable under the Employers' Liability (Compulsory Insurance) Act 1969, whereas if there is no employer–employee relationship to which the injury is referable, the basis of the action will be occupation, which may result in a multiciplity of defendants. A fireman or ambulance driver or policeman or member of the public injured by a gas explosion on a building site (assuming *res ipsa loquitur* is inapplicable) should bring his action against all in occupation of the building site, though effectively this means the building owner and the main contractor or contractors. Similarly an employee of a sub-contractor injured in the same way should frame his action against building owner, main contractor and sub-contractor. The injured person should always look for a loss distributor first; only in the absence of this should occupation be regarded as possible means of compensation.

Contractual relationships on a building site

It is a well-known fact that the relationship between main contractor and sub-contractor is contractual and governed by both Green and Blue Forms of Sub-Contract. Similarly the relationship between building owner and main contractor is contractual, being governed by the Form of Main Contract. This fact is important as far as liability for personal injuries to operatives and members of the public is concerned. Supposing that the building work is Standard Form work, the Form of the Main Contract between building owner and main contractor will contain an insurance instruction clause followed by an indemnity clause. The same machinery appears in the Green and Blue Forms embodying the contractual relationship between main and sub-contractors. This means that, as far as personal injuries to operatives and members of the public are concerned, the building owner can call for an indemnity from the main contractor and the latter can do so from the sub-contractors, assuming they have taken out the requisite insurance coverage. Ultimately, the one at fault pays, Insurance is in the form of an employer's liability policy (which is compulsory) and a public liability policy in respect of injuries to persons other than employees, including members of the public.

Trespassers

The expression "trespasser" applies not only to the member of the public who goes on to a building site when not supposed to do so but also to operatives on the site who do what they are not supposed to do or go where they are not supposed to go. *Uddin's* case and *Westwood* v *Post Office* are examples. An employer's duty to his own employees who trespass in the course of their employment is strict, as we saw in *Uddin* and *Westwood*; this is not surprising since the employer is a loss distributor and his liability is not dependent on fault, at least only in a technical sense. An occupier's duty to persons, whether operatives on a site or members of the public, is not strict. Under the Occupiers' Liability Act 1957 an occupier is under no duty to see that the structural state of the premises is reasonably safe from dangers in respect of trespassers. They take the state of the premises as they find them, though retributive acts against simple trespassers are forbidden. Where, however, operations are being carried on on premises, e.g. building operations, the occupier, as far as trespassing members of the public are concerned, will be governed by the rule of conscientious humanity posited in *British Railways Board* v *Herrington*. Here, it will be recalled, Lord Reid proclaimed "So the question whether an occupier is liable in respect of an accident to a trespasser on his land would depend on whether a conscientious, humane man with his knowledge, skill and resources could reasonably have been expected to have done or refrained from doing before the accident something which would have avoided it. If he knew before the accident that there was a substantial probability that trespassers would come I think that most people would regard as culpable failure to give any thought to their safety." His Lordship went on to point out that discharge of the duty will depend on the resources of the occupier. A wealthy occupier might be expected to take precautions against trespassers coming on to a site and, failing that, to adopt measures designed to prevent their being injured in the event of their trespassing. The same duty could not be expected to be so easily discharged by an impecunious occupier (see *British Railways Board* v. *Herrington*). As for the question whether an occupier can contract out of liability, in other words whether he can absolve himself from liability for injury to operatives of other employers on the site or to members of the public, the case of *Burnett* v. *British Waterways Board* suggests that, since employees cannot exercise freedom of choice, a notice or any other form of contracting out cannot absolve a contractor from liability to employees of other employers where they have suffered injury as a result of his negligence as an occupier. Moreover as regards injury to other employees and to members of the general public the efficacy of notices in law would appear to be in some doubt, at least as far as the notice seeks to exempt the contractor, sub-contractor or occupier from liability in respect of activities to be carried out but not yet carried out. *Dicta* in both *Burnett's* case and *White* v. *Blackmore* in 1972 indicate this clearly. The use of notices to exclude or limit liability of occupiers is now prohibited by the Unfair Contract Terms Act 1977.

37. The Construction Regulations

It is no part of this book to analyse the Construction Regulations. Rather it is proposed to mention their objectives along with the more important cases and provisions.

The Construction Regulations are in three parts, the Construction (General Provisions) Regulations 1961, the Construction (Lifting Operations) Regulations 1961 and the Construction (Working Places) Regulations 1966. These represent the operation of factory legislation in a modified form to building sites. In s. 127, Factories Act 1961, it is stated that its provisions apply to "building operations" and to "works of engineering construction". Moreover by subs. 127(4) a building site is a *notional* factory and the Minister has power to make regulations with respect to building sites and works of engineering construction, since the contractor or civil engineering firm is the factory occupier for these purposes.

Probably one of the most important provisions of the Construction Regulations is reg. 3, Construction (Working Places) Regulations 1966, the object of which is to divide responsibility between contractors and employers. The basis of the division of responsibility under regulations is not like that at common law, equating liability with control. Regulation 3(1)(*a*) deals with contractors as employees, whilst reg. 3(1)(*b*) is concerned with contractors in the course of carrying out operations.

Regulation 3(1)(*a*) provides that it shall be the duty of every contractor, and every employer of workmen, who is undertaking any of the operations or works to which these regulations apply to comply with such of the requirements of the following regulations as affect him, or any workman employed by him, namely regs. 6–23, 25, 30, 31, 32, 36 and 38 and, in so far as they relate to the falling or slipping of persons, regs. 24, 26, 27, 28, 29, 33, 34 and 35.

Regulation 3(1)(*b*) provides that it shall be the duty of every contractor, and every employer of workmen, who is undertaking any of the operations or works to which these regulations apply to comply with such of the requirements of the following regulations as relate to any work, act or operation performed or about to be performed by any such contractor or employer of workmen, namely regs. 37 and 39 and, in so far as they relate to the falling of materials and articles, regs. 24, 26, 27, 28, 29, 33, 34 and 35.

Regulation 3(1) then continues "and it shall be the duty of every contractor and every employer of workmen who erects or alters any scaffold to comply with such of the requirements of these regulations as relate to the erection or alteration of scaffolds having regard to the purpose or purposes for which the scaffold is designed at the time of erection or alteration; and of every contractor and every employer of workmen who erects, installs, works or uses any other plant or equipment to which any of the provisions of these regulations applies, to erect, install, work or use any such plant or equipment in a manner which complies with those provisions".

Regulation 3(2) goes on to prescribe that it shall be the duty of every person employed to comply with the requirements of such of these regulations as relate to the performance of or the refraining from an act by him and to co-operate in carrying out these regulations and if he discovers any defect in the plant or equipment to report such defect without unreasonable delay to his employer or foreman, or to a person appointed by the employer under Part 2 of the Construction (General Provisions) Regulations 1961.

Regulation 3(1)(a)

It was established in *Smith* v. *George Wimpey Ltd.* that this duty on the part of the contractor of complying with the above regulations was restricted as far as the bringing of a civil action was concerned (and, presumably, this means to prosecutions as well), to immediate employees of the contractor. Thus an employee of a sub-contractor could not sue nor could a member of the public or a fireman or a policeman coming on to the site for injury caused by breach of the regulation. This would in no way impair the right to sue as far as persons who were not immediate employees were concerned under the Occupiers' Liability Act or at common law. In view, however, of the 1976 Court of Appeal decision in *Ferguson* v. *Dawson & Partners Ltd.*, a labour-only sub-contractor will now qualify as an employee for the purposes of reg. 3(1)(*a*). This seems to indorse the recommendation of the Robens Committee that lump labour should be brought within the framework of the health and safety laws. The regulations to which reg. 3(1)(*a*) applies are listed in Appendix IX; the Construction (Working Places) Regulations 1966 are listed in Appendix VIII.

The purpose of the special regulations at the end of reg. 3(1) is, according to *Smith* v. *Vange*, to lay an extra duty on those *erecting scaffolds* to *all workmen*, whereas the duties under reg. 3(1)(*a*) and (*b*) were to the *contractor*'s workmen only.

The Construction Regulations give rise to prosecution and civil actions.

Chapter 4

Safety and Industrial Relations

INDUSTRIAL RELATIONS: THE EARLY LEGAL BACKGROUND

38. Employers' associations and trade unions

Voluntarism or autonomy for the parties to the industrial contract is the main characteristic of British industrial relations. Collective bargaining is a reserve jealousy watched by the two sentinels of industry, employers' associations and trade unions, into which the law treads with great care and with a minimum of self-advertisement. The reasons for this are historical and traceable back to a series of measures passed by Tory and Whig governments in the nineteenth century. Employers had begun to federate and workers had begun to form combinations (later in Gladstone's Trade Union Act 1871 both were to become known as trade unions), long before this *fait accompli* had been grasped by Parliament. When Parliament finally got round to legislating for the proletariat, it was apparent that benefits which would have come to the working classes through legislation had already been obtained for the workers by their own trade unions. In England, therefore, the law did not create and nurture collective bargaining. Rather it grafted itself on to it.

Two cases decided at the turn of the century by the House of Lords made it clear that the attitude of the law towards combinations of employers and workers was one of signal non-intervention.

The Rule in Lumley *v.* Gye

There could not have been better evidence of this than s. 4, Trade Union Act 1871, which, having recognised formally the existence of collective agreements, went on to enact that, although such agreements were perfectly lawful, their enforcement could not be entertained in a court. The law of tort posed what at first sight might have appeared a serious obstacle to the development of voluntary collective bargaining. The rule in *Lumley* v. *Gye* (1853), that to interfere with existing contracts was a tort, and so illegal, taken out of its original humble context, was almost fortuitously fashioned into the sharp weapon of inducement of breach of contract of employment. Employers in federation or trade unions in combination would be liable in tort if they conspired or combined to cause economic harm to other employers or other workers.

The Mogul Steamship *case*

In the 1892 House of Lords' decision in *Mogul Steamship Co. Ltd.* v. *McGregor Gow & Co.*, however, a group of shipowners, engaged in transporting goods from ports in China, decided to form a federation with the object of "cornering" the Hong Kong carrying market. The effect of this act of federating was, and was intended, to drive competitors out of business, or alternatively, to force them to join the federation. One might have thought such conduct contrary to the rule in *Lumley* v. *Gye*, but to the dismay of the competitors of the federation, the House of Lords took a more sanguine view. It was held that the defendants' conduct was in the best interests of the members of the group and was done with the lawful object of increasing their profits and extending their trade.

Allen *v.* Flood

Allen v. *Flood*, six years later in 1898, showed the other side of the coin and posed the question—what if workers combined to bring about a similar economic result? Would such conduct be actionable in tort or would it qualify for judicial clemency? The judiciary could not be seen to be partial and so what was good for the capitalist goose would have to be good for the socialist gander. The appellant and others were employed as shipwrights on a day-to-day basis by the G company which manufactured iron. The respondent and others, who were officials of a trade union, intimated to the company that if the company did not dismiss the shipwrights, the ironworkers would withdraw their labour and cause economic loss to the company. This intimation the company could not treat lightly and they dismissed the shipwrights, who then sued the respondent and others acting in combination for conspiracy to injure according to the rule in *Lumley* v. *Gye*. The House of Lords, though not unanimously, ruled that the respondent and the other officials had done no wrong and had not committed the tort of conspiracy to injure. By the turn of the century, therefore, both capital and labour were uniformly structured within the law whose chief characteristic was absence of formalism. Employers' associations and trade unions were self-governing as regards collective bargaining.

39. 1906 and after

Registered and unregistered trade unions

There was one respect, however, in which companies and trade unions differed markedly. It had been recognised for some time that limited companies were separate legal entities from the members who at any given time comprised the company, this had the consequence that companies were the subject of both legal rights and legal duties. Companies could be sued in contract and tort, but the position was nowhere near as clear as regards trade unions. To make matters more complicated two types of trade unions existed, the registered trade union and the unregistered trade union or organisation of workers. *Temperton* v. *Russell* had established in 1893 that an unregistered trade union was not a legal entity but was more in the nature of an unincorporated association, similar to a club formed for political or social objects, with the practical result that union funds could not be attacked in tort for conduct amounting to inducement of breach of contract within the rule in *Lumley* v. *Gye* or for conspiracy to injure; the law gave them and their shop stewards licence to take or threaten to take industrial action, the effect of which could be economically damaging for employers and groups of non-unionised workers.

The Taff Vale *judgment*

The 1901 House of Lords' decision in *Taff Vale Railway Co.* v. *Amalgamated Society of Railway Servants*, however, struck a substantial blow at organised labour. This case established that a registered trade union was a legal entity like a company and was subject to legal duties as well as legal rights, it treated shop stewards as the employees of the union with the result that the union was vicariously liable for torts committed by shop stewards which caused economic harm to employers and non-unionised workers.

The consequence of *Taff Vale* was that an incoming Liberal government passed the Trades Disputes Act 1906, the effect of which has been to stabilise the law relating to trade unions for 70 years and more. The Trade Union and Labour Relations Act 1974 is only a refurbishing of the Trades Disputes Acts 1906 and 1965, whilst at the same time removing from industrial law certain developments associated with American labour doctrine, such as the tort of interference with business relations and the interlocutory injunction immortalised in *Ford Motor Co. Ltd.* v. *A.E.F. and Others* in 1969. This last case was an attempt on the part of F.M.C. to get the courts to pronounce a collective agreement as legally binding, which they declined to do.

THE TRADE UNION AND LABOUR RELATIONS ACT 1974 AND THE EMPLOYMENT PROTECTION ACT 1975

This 1974 Act grants immunity from legal action, in the event of a trade dispute, to either a trade union or an employers' association or any members or officials of either organisation, and to persons acting or threatening to act unofficially. The Act, therefore, permits the official and unofficial strike and lock-out. A strike or lock-out will be official if authorised or ratified by the executive of the trade union or the T.U.C. and the executive of the employers' association or the C.B.I. The Act says, in effect, "if you do this the law will not exercise sanctions"; it does not say, "the law encourages you to do this". The freedom to strike or take some other form of industrial action is, therefore, a negative one, in the form of an immunity from sanction rather than of a formal offer of encouragement. The adoption of a positive attitude of encouragement to organise into trade unions and demand *recognition* from employers has been left to the Employment Protection Act 1975, many of whose objectives would appear to be in keeping with the Treaty of Rome and the European Social Charter.

40. The Employment Protection Act 1975

History will cite the Employment Protection Act as the foundation of a workers' legal charter in the sense that it restricts managerial freedom. Through a superstructure of quasi-judicial machinery in the form of the Advisory Conciliation and Arbitration Service (A.C.A.S.), the Act positively encourages groups of workers to get their employers to *recognise* them as belonging to a trade union. The Certification Officer can only register such union as recognised if it is independent within the meaning of s. 30(1), Trade Union and Labour Relations Act 1974; if necessary, by virtue of s. 15, Employment Protection Act, a union, through application to A.C.A.S., can force an employer into recognition if the union satisfies the requirement of s. 30(1), Trade Union and Labour Relations Act 1974. Ultimately, therefore, recognition can be by way of unilateral compulsory arbitration.

The certificate of dependence issued by the Certification Officer:

(*a*) allows a union freedom from legal action in tort for industrial action or the threat of it;

(*b*) allows its officials time off with pay for training in connection with the job;

(*c*) allows its officials time off with pay for carrying out duties concerned with industrial relations between the employer or the employers' association and the trade unions (s. 57(1));

(*d*) makes consultation with the union necessary in the case of redundancies from the workforce;

(*e*) allows union members to sue the employer for unfair dismissal under certain conditions outlined in the Trade Union and Labour Relations Act; by implication this allows the employer fairly to dismiss non-members;

(*f*) allows unions and their officials to require the disclosure of information subject to certain conditions outlined in s. 18, Employment Protection Act, relevant to collective bargaining.

41. Collective agreements and individual contracts of employment

The law accommodates these by providing for the importation of terms of a collective agreement into individual contracts of employment either in the form of express or implied terms. Because of s. 2(3), H.A.S.A.W.A., and the obligation on employers to issue a company safety policy, statements made in this policy may qualify as express and implied terms.

42. Safety representatives and the disclosure of information

Section 28(8), H.A.S.A.W.A., prescribes that "Notwithstanding anything in the preceding sub-section an inspector shall, in circumstances in which it is necessary to do so for the purpose of assisting in keeping persons (or the representatives of persons) employed at any premises adequately informed about matters affecting their health, safety and welfare, give to such persons or their representatives the following descriptions of information, that is to say—

(*a*) factual information obtained by him which relates to those premises or anything which was or is therein or was or is being done therein; and

(*b*) information with respect to any action which he has taken or proposes to take in or in connection with those premises in the performance of his functions; and where an inspector does as aforesaid, he shall give the like information to the employer of the first-mentioned person".

The safety representative is therefore entitled to disclosure of information relevant to health and safety which has been obtained by an inspector. Whether or not such information as is disclosed would qualify as a trade dispute, as defined in s. 29(1), Trade Union and Labour Relations Act 1974, is a matter best left to a court to decide. This sub-section does, however, state that a trade dispute *inter alia* relates to "terms and conditions of employment or the physical conditions in which workers are required to work".

43. Safety regulations and unfair dismissal

The law relating to unfair dismissal in the Trade Union and Labour Relations Act 1974 is set out in Sched. 1, para 6. The exact effect of H.A.S.A.W.A. and regulations on the law of unfair dismissal is uncertain, but several recent tribunal decisions shed some light. In *Wood* v. *Brita Finish Ltd.* in 1976 W was dismissed from his job as an apprentice anodiser for refusing to work unless he was provided with overalls in addition to acid-proof gloves, goggles, wellingtons and a protective apron (all factory inspector approved) which he had already been provided with by his employer. Evidence showed that the contribution made to his safety in this case by the provision of overalls was negligible. The Tribunal held that the dismissal was fair.

A case which may prove a source of comfort to employers is the 1977 tribunal decision in *Howard* v. *Overdale Engineering Co. Ltd.* H was dismissed for refusing to work in O.E.C.'s new factory where dust was being generated by engineers drilling cable lines in the floor, allegedly exposing him to a health risk. The tribunal having taken the view that there was no breach on the part of the employer of either ss. 4 or 63, Factories Act 1961 (preventing impurities getting into the air, on the one hand, and the accummulation of dust and fumes on the other; and circulating fresh air if they should do so), the employee who refused to work in those conditions was fairly dismissed. Though it is fair to say the tribunal felt that a prior warning of dismissal should be given by the management in such circumstances, for without the warning the employee might well have a good case for unfair dismissal. Moreover the drilling operation lasted only one hour after H's refusal to work, the doors at both ends of the factory were kept open for ventilation purposes and two other employees working closer than H to the dust were not affected by the conditions generated.

Employee a danger to himself or to others

In *Marsh* v. *Judge International Housewares Ltd.* it was held to be fair to dismiss an employee whose hair was a safety hazard to himself and who refused to do anything about it. M's hair was 2 ft. 6 in. long and the management had been told by the factory inspector that if M was injured as a result of his hair getting caught in one of the machines he was working with, the inspector would consider prosecution against the employer. Warnings to him about the danger from his hair having been ignored, the company provided the employee with a hat which he refused to wear. It was held that the dismissal was fair and lawful. A similar point arose in *Spalding* v. *Port of London Authority* where the employee was a mechanical fitter for a company which had been taken over by P.L.A. P.L.A offered him a job as a fitter, one of the conditions being that he took a medical examination. The employee started his job but later failed the medical examination owing to a form of deafness. On this basis the employee was dismissed and he sued for unfair dismissal. Evidence showed that P.L.A.'s medical standards were set for the benefit of and to ensure the safety of P.L.A.'s employees. These standards, the tribunal found, were not unreasonable. Medical standards were higher for operatives of mechanical equipment than for labourers, it being necessary to have good hearing because orders and advice were passed on by word of mouth.

Spalding's case is an example of an employee being dismissed because he was a danger to others. In *Yarrow* v. *Q.I.S. Ltd.* the employee was dismissed because he was a danger to himself. Here the employee suffered from psoriasis, and because he worked with radiography equipment, he was subject to the Ionising Radiations (Sealed Sources) Regulations 1969. It was held that the dismissal was fair. Another instructive case is *Parsons* v. *Fisons Ltd.*, which underlines the fact that although the employee's condition has not yet caused an accident, it may still be lawful to dismiss the employee. The employee was given early retirement because she suffered from bad sight, having only one eye and bad tunnel vision in that. On several occasions the company doctor had seen her missing vehicles in the company yard at considerable risk to herself, there was no other suitable job for her in the company and she was dismissed. It was argued that her poor sight had not yet led to any accidents and that therefore her dismissal was unfair, but the tribunal ruled that the dismissal was fair.

Refusal to wear protection

Other tribunal decisions related to safety and unfair dismissal are *Mayhew* v. *Anderson Ltd.* and *Henry* v. *Vauxhall Motors Ltd.* In the former case a sewing machinist who had suffered a number of accidents to her eyes through needle breakage was advised to wear eye protection. This was urged by the company's insurers, who threatened to refuse to insure the company if this was not done. Later the company purchased a pair of goggles for 78p but the union representative said that he did not think that the employee would wear such goggles because of the discomfort and the appearance. He pointed out to the management that custom-made eye protectors could be obtained at a price varying between £33 and £35. The employee refused to wear the goggles bought by the company and she was warned that, if she continued to do so, she would be dismissed. At the end of May 1977 she was dismissed. It was held by a Liverpool tribunal that the dismissal was unfair, because the employee had never refused to wear reasonable eye protection. The fact was that the goggles offered to her were uncomfortable and irritated her eyes. During the course of giving judgment the tribunal said "The Tribunal have come to the unanimous decision that it was unreasonable to expect the applicant to wear the goggles they provided at 78p in the circumstances when more acceptable alternatives were available, even at the risk of creating a precedent with other employees which the respondents claimed was the reason why they could not purchase the alternative glasses".

In *Henry* v. *Vauxhall Motors Ltd.* H was sent to work in the lead sanding booth. Employees on this work were required to wear the following items of safety equipment—a smock, a safety helmet and gloves to protect them from hazards associated with lead absorption. H's usual helmet was missing that particular day and he found the one provided uncomfortable. Hence he did his work without the necessary protective head gear. H was then approached by the production foreman, who thought that the helmet in question was satisfactory and ordered H to wear it, but he refused to do so. After the involvement of higher management H was eventually dismissed for gross misconduct. It was held that the dismissal was unfair because management had failed to investigate sufficiently the reasons for H's refusal to wear the helmet. The tribunal stated "Throughout his employment, the applicant had been a satisfactory employee, had never made any complaint before about working conditions and one would have thought that, in these circumstances, his employers would have regarded his complaint as having some substance to it, that it was not just a matter of being uncomfortable in itself, that it was not just a matter of being awkward but that there must have been some deeper underlying reason to it all.... One suspects that if the company had found the true reason for the applicant's claim (as they could with reasonable investigation) the whole matter could no doubt have been settled satisfactorily." Moreover H had not been properly represented at his interview with management. As he was an immigrant with a limited command of English, management, the tribunal said, should have taken more care. H's compensation was reduced by 60 per cent because he was partly to blame for his own dismissal; he could have volunteered more information as to how the helmet was uncomfortable.

Refusal to go on training course

The effect of refusal to go on training courses was spotlighted in two recent cases. In *Brodie* v. *Elmwood Sensors* B was employed as a press operator, but to increase flexibility the employers asked the press operators to undergo a two-week training programme on another operation. B persistently refused. Finally she was dismissed and the dismissal was held to be lawful. More pointedly in *Wilson* v. *Stephens & Carter Ltd.* W refused to attend a safety training course which involved a week away from home. When he sued for unfair dismissal, it was held that the dismissal was fair.

Disregard of safety duties

More recently still in *Ashworth* v. *Needham & Sons Ltd.* A was employed in a sand reclamation plant. Underneath the plant there was a cellar which was covered by either steel or concrete plates so that no-one could fall in. It was A's job to see that the plates were in position. One day A left work early because he was not feeling well. Instead of replacing the plates he put up a collapsible fence. The foundry manager discovered this and, after discussing the matter with the safety officer and union representatives, decided to dismiss A. This dismissal was in accordance with the company's disciplinary procedure which listed "flagrant disregard of safety duties" as gross misconduct. During the course of giving judgment the tribunal stated "Here was a hole in the floor down which somebody could fall 20 ft. and he had not put the proper fencing back, and if somebody had fallen down and broken their neck, then he would have been the person entirely responsible and the respondents themselves would have had to pay the necessary damages. In those circumstances, we have reached the conclusion that the respondents acted reasonably in treating it as a sufficient reason for dismissing the employee."

44. Suspension from work on medical grounds

This matter is dealt with in ss. 29–33, Employment Protection Act 1975. Section 29 provides that an employer must pay remuneration to an employee who is suspended from work on medical grounds, such remuneration being payable for not longer than 26 weeks.

Qualification for remuneration

In order to qualify for this remuneration an employee must show that the suspension is due to a provision of

(*a*) a statutory enactment; or
(*b*) a code of practice.

Schedule 2, Employment Protection Act 1975 sets out the statutory provisions which are relevant for the purposes of suspension under s. 29. This is not exhaustive and, in the light of the expanding expression "prescribed industrial disease" in the Social Security Act 1975, can be added to by the Secretary of State for Employment. At present they are as follows:

Paints and Colours Manufacture Regulations 1907;
Yarn (Dyed by Lead Compounds) Heading Regulations 1907;
Vitreous Enamelling Regulations 1908;

Tinning of Metal Hollow-ware, Iron Drums and Harness Furniture Regulations 1909;
Lead Smelting and Manufacture Regulations 1911;
Lead Compounds Manufacture Regulations 1921;
India Rubber Regulations 1922;
Chemical Works Regulations 1922;
Electric Accumulator Regulations 1925;
Lead Paint Regulations 1927;
Pottery (Health and Welfare) Special Regulations 1950;
Factories Act 1961, including the extension of that Act by s. 128;
Ionising Radiations (Unsealed Radioactive Substances) Regulations 1968;
Ionising Radiations (Sealed Sources) Regulations 1969; and
Radioactive Substances (Road Transport Workers) (Great Britain) Regulations 1970.

If an employee is suspended as a result of these provisions or any attendant code of practice but is otherwise able to work, such employee is entitled to remuneration for not longer than 26 weeks from his employer during his suspension.

Section 29 makes it clear that an employee is not entitled to remuneration if he is incapable of work by reason of "disease or bodily or mental disablement". If, therefore, an employee after being medically suspended falls sick with influenza, he forfeits remuneration until he is well again. And the remuneration will continue until the period of 26 weeks has elapsed or the suspension ceases, whichever is the earlier. Section 29 applies to the situation where an employee is still employed but is not provided with work or does not carry out the work he normally carries out in view of one of the above statutory regulations or a code of practice.

Where the employer does not have to pay remuneration

There are five situations outlined in s. 30 in respect of which an employer does not have to pay remuneration during medical suspension, *viz.*:

(*a*) where an employee has not been continuously employed for four weeks ending with the last complete week before the suspension begins;

(*b*) where an employee is employed under a fixed term contract for 12 weeks (unless the employee has been employed for more than 12 weeks);

(*c*) where an employee is or becomes incapable of work through disease or bodily or mental disablement;

(*d*) where the employee is offered "suitable alternative work", whether or not it is work which the employee was employed to do under his contract. If the employee then unreasonably refuses to carry out this work, he is not entitled to remuneration under s. 29; and

(*e*) where an employee fails to comply with reasonable requirements imposed by his employer to ensure that his services are available; in other words he must be ready and willing to come to work in addition to being well.

Chapter 5

Product Liability

INTRODUCTION TO PRODUCT LAW

Product liability refers to a system whereby a producer may incur criminal and civil liability for a defective product. Criminal law expects the producer to do what is reasonably practicable to see that his products are safe when used properly. It is not necessary for injury to occur; risk of such injury is sufficient. Civil liability varies according as the defect causes injury or not. If the defect renders the product unmerchantable or unfit for normal purpose, the user will have to sue in contract, assuming that he has a contract. If he bought or hired directly from a retailer, he can sue on the contract and the Supply of Goods (Implied Terms) Act 1973 will assist him by implying certain terms as to fitness and merchantability into the contract. On the other hand, if the defect caused the user injury, whether he has a contract with anyone or not, the liability of the producer is based on negligence.

This chapter is concerned with products which are defective in the sense that they cause injury and give rise to liability in tort as well as criminal liability. It is suggested that there may be a parallel here between liability in tort for product defects and the American doctrine of unreasonable danger, in other words a product is defective if it is dangerous. The classic definition of "dangerous" was probably that of du Parcq, J., in *Walker* v. *Bletchley-Flettons Ltd.* where he said, referring to factory machinery, "A part of machinery is dangerous if it is a possible cause of injury to anybody acting in a way in which a human being may be reasonably expected to act in circumstances which may be reasonably expected to occur." This shows that liability for defective products causing injury is based on lack of reasonable care, though it should be emphasised that this may become an area of strict liability. Certainly the E.E.C. Draft Directive on Product Liability and the Strasbourg Convention on Product Liability recommend a change to strict liability on the part of producers for defects in their products causing injury, the former including damage to property, whilst the latter is confined to personal injury. Moreover it is within the brief of the Pearson Commission to recommend strict liability for product defects and this has happened. The Report states that the Commission were "to consider to what extent, in what circumstances and by what means compensation should be payable in respect of death or personal injury (including ante-natal injury) suffered by any person ... through the manufacture, supply or use of goods or services ... having regard to the cost and other implications of the arrangements for the recovery of compensation, whether by way of compulsory insurance or otherwise". It is clear from this that the Pearson Commission might have made recommendations for the recovery of compensation for injury caused by product defect outside the present interparty or adversary system by means of an extension of the social security system. But this has not happened. The Law Commission, on the other hand, suggested that strict liability for product defect might take the form of a change in the burden of proof applicable to product defect cases, it becoming incumbent on the manufacturer to establish that he took reasonable care in making the product and that the defect did not exist at the time the product left his possession. This would amount to an introduction of *res ipsa loquitur* into product defect cases, the interparty system being retained. The E.E.C. Draft Directive, the Strasbourg Convention and the Law Commission recommend some form of "channelling" of the strict liability, though not necessarily to the head manufacturer exclusively but to sub-manufacturers and first distributors as well (*see* §**47**, p. 247).

45. Civil liability of manufacturers

Whereas a retailer or even a wholesaler would normally incur liability in contract for a defective product, a manufacturer or sub-manufacturer would normally incur liability in tort, since there is not generally a contractual relationship with the ultimate user or consumer. The same would apply to an installer or repairer or service engineer if injury was caused as a result of bad workmanship. The law was laid down in the leading case of *Donoghue* v. *Stevenson* in 1932. In this case a young lady bought a bottle of ginger beer for her friend who, after drinking the contents, became ill as a result of seeing the decomposed remnants of a snail in the base of the bottle. The case is generally regarded as authority for the rule that a manufacturer of products is liable to the ultimate user if:

(*a*) the defect in the product is one that may result in injury to the consumer's life or property;

(*b*) the defect existed at the time the manufacturer parted with possession of it;

(*c*) the defect was one that the manufacturer could not reasonably have expected the consumer or some third party to detect and correct before it could do harm;

(*d*) the existence of the defect was attributable to lack of reasonable care on the part of the manufacturer.

During the course of giving judgment Lord Atkin uttered his famous dictum regarding the position in civil law of manufacturers, saying "a manufacturer of products, which he sells in such a form as to show that he intends them to reach the ultimate consumer in the form in which they left him with no reasonable possibility of intermediate examination, and with the knowledge that the absence of reasonable care in the preparation or putting up of the products will result in an injury to the consumer's life or property, owes a duty to the consumer to take reasonable care".

This decision establishing a system of liability in manufacturers based on failure to take reasonable care was a breakthrough in manufacturers' liability extending beyond situations involving products dangerous *per se*. Until then in England, as distinct from North America, a manufacturer had only been liable to a user of his products in three limited situations—where the product was inherently dangerous (such as a gun), where it was known that the product was dangerous, and where dangerous products had been put into the possession of an incompetent person. Moreover an action in negligence was not thereafter restricted to cases where injury was caused from eating bad food, as in *Donoghue* v. *Stevenson*, but extended to all manner of consumer products from cars to dishwashers, and liability in tort had the added attraction that users of products who were not the original buyers could maintain in action, including anyone, such as a passer-by, who was the victim of a reasonably foreseeable injury. In *Stennett* v. *Hancock & Peters Ltd.* in 1939 the plaintiff whilst walking along the pavement was injured by the flange flying off a wheel on a lorry belonging to H. The lorry had earlier been repaired at the premises of the second defendant, P, who had been required to service the wheel. It was held that the repairer was liable to the plaintiff for having failed to exercise reasonable care. Passers-by and employees employed to repair and service products as well as passengers in cars which had been newly manufactured or repaired are thus able to claim redress in respect of their injuries.

46. Developments in American products' law

While the courts in England, prior to *Donoghue* v. *Stevenson*, were restricting liability for defective products to that limited category of products dangerous *per se*, American courts were imposing liability in tort for carelessly manufactured and carelessly designed products. The starting point was the decision of a New York court in *Macpherson*

v. *Buick Motor Co.* in 1916. The plaintiff had bought a Buick car from a dealer and was injured when a wheel collapsed while he was driving it at low speed. The motor car was not regarded as dangerous *per se* and the wheel had been bought in from a reputable manufacturer. There was evidence, however, that the defect could have been discovered on reasonable inspection by the assembler. It was held that the motor manufacturer was liable. Giving judgment the court said, "If the nature of a thing is such that it is reasonably certain to place life and limb in peril when negligently made, it is then a thing of danger. . . . If to the element of danger there is added knowledge that the thing will be used by persons other than the purchaser, and used without new tests, then, irrespective of contract, the manufacturer of this thing of danger is under a duty to make it carefully."

Macpherson's case concerned liability for failure to take reasonable care. Later, as manufacturers sought to promote their products through advertising, the notion of *liability under express warranty* was evolved by the courts. This development reached its peak in 1932 in the case of *Baxter* v. *Ford Motor Co.* where, as a result of having seen Ford cars advertised in brochures circulated by Ford, the plaintiff relied on a statement that the windscreens were shatterproof and bought one of their cars. He was injured whilst out driving as a result of a pebble being thrown up by another car. It was held that Ford was liable on the basis that it would be unjust to allow manufacturers to advertise their products as having certain qualities, whilst at the same time depriving users of a right of action on the ground of privity of contract after injury had occurred as a result of the product lacking the quality advertised. In more recent times liability on express warranty has extended to cases of pure economic loss, assuming that the user is a private individual and not a commercial undertaking which would be expected to insure against such commercial losses, as evidenced by *Seely* v. *White Motor Co.* Another example of liability grafting itself upon representation or advertisement rather than a contract between two persons occurred in the case of *Henningsen* v. *Bloomfield Motors (U.S.) and Chrysler Corporation* in 1960. The plaintiff bought a Chrysler car from a dealer. His wife was injured whilst driving the vehicle as a result of faulty steering on the car. It appeared that there was no principal–agent relationship between the manufacturer and the retailer. It was held that the plaintiff's wife was entitled to redress for her injuries, which seemed to show that liability on a warranty was founded on tort not contract, as there was no contract between the manufacturer or the dealer and the plaintiff's wife.

At present American product law favours the notion of strict liability in tort for product defects causing injury. Once a manufacturer has put a product on the market which is unlikely to be inspected for defects by a distributor, e.g. a tin of fish or a bottle of coca-cola, if the product has a defect which causes injury, he will be liable for that defect, even though he could not have known of its existence. The emergence of strict liability in tort for product defects is generally associated with the case of *Greenman* v. *Yuba Power Products* in 1963. A man was injured when using a power drill that his wife had bought him for a Christmas present. The manufacturer was held liable and the court said "A manufacturer is strictly liable in tort when an article he places on the market, knowing that it is to be used without inspection for defects, proves to have a defect that causes injury to a human being ... the liability is not one governed by the law of contract warranties but by the law of strict liability in tort".

Some states, notably New York and California, have gone further than this. In *Ford Motor Co.* v. *Mathis* the head manufacturer and assembler was held liable for injury caused by a defective dimmer switch assembled into one of their cars and supplied by a submanufacturer, even though an adequate system of inspection could not have discovered the defect. In California everyone involved in the design, manufacture and distribution of a product, including a credit company, is strictly liable for defects in products. This is known as *enterprise liability*. If Pearson's recommendations in favour of strict liability in tort are implemented here, the position may well be as in *Greenman* v. *Yuba Power Products*.

THE REFORM OF PRODUCT LAW IN BRITAIN

As we said earlier, liability for defective products is under review at present and changes in the law relating to product defects seem inevitable. Changes of different kinds in the law have been recommended by the Law Commission, the E.E.C. Draft Directive on Product Liability and the European Convention on Product Liability in regard to Personal Injury and Death, the so-called Strasbourg Convention. In addition the Pearson Commission was empowered to examine the present law relating to personal injury and death, including ante-natal injury, caused by defective products. All of these seem to be in favour of some kind of strict liability for producers of defective articles.

47. Strict liability

Before examining in outline the different solutions proposed by these bodies, it may be appropriate to compare the different meanings of strict liability. Strict liability can firstly refer to a system whereby the burden of proving that reasonable care was taken lies upon the manufacturer, who must also show that the defect in the product did not exist when it left his possession, in other words that the existence of a defect in a product is evidence of lack of reasonable care on the part of a manufacturer or sub-manufacturer. This argument might work well in the case of products where there is no reasonable expectation of inspection by an intermediary, as in the case of a bottle of beer or a tin of peas. It would also solve the problem posed by *Evans* v. *Triplex Safety Glass Ltd.* where a plaintiff who had bought a Vauxhall car and was injured when the windscreen shattered was held not entitled to recover because he could not establish whether the fault lay with the head manufacturer in assembling the windscreen into the car or with the maker of the windscreen who may have used defective glass. By reversing the burden of proof, the task of proving that the defect was not in the product will lie with the assembler or sub-manufacturer.

Strict liability can refer secondly to a system whereby the head manufacturer is responsible for defects in sub-manufactured products which could have been detected on reasonable inspection. This would imply the exercise of indemnities by the head manufacturer as against those at fault and would require the head producer to insure against claims from injured users and would be an example of the liability being "channelled" to him. This accords with the present position relating to assemblers of sub-manufactured products, with the exception that the burden of proving who was liable for the defect, assembler or sub-manufacturer, is on the user under the present fault system and there is no provision for channelling.

Strict liability may refer thirdly to a system where liability is channelled to the head producer, irrespective of whether he took reasonable care. This would accord with the American position in *Ford Motor Co.* v. *Mathis*. Under such a system indemnity should be exercisable against the person at fault.

Strict liability can refer finally to enterprise liability, where everyone involved in production and distribution from designer to retailer and finance company is strictly liable, as in California. This last is the most extreme form.

The Law Commission has recommended a change in the burden of proof. The E.E.C. Draft Directive and the Strasbourg Convention, however, embody proposals for a different kind of strict liability. The E.E.C. Draft Directive, Art. 1, states that the producer of an article shall be liable for damage caused by a defect in the article, whether or not he could have known of the defect. The producer shall be liable even if the article could not have been regarded as defective in the light of the scientific and technological development at the time when he put the article into circulation. This then contemplates a system of liability not referable to failure on the part of a manufacturer to take reasonable care. Liability is to be channelled to the final producer subject to a right of indemnity against those responsible for the fault. The Strasbourg Convention, on the other hand, contemplates the channelling of liability to the final producer and, if a component producer fails to disclose the identity of the final producer within a reasonable time, to the producer of the component part and even the importer or first distributor in the case of a foreign product. The channelling of liability to producers can give rise to problems where the product is multi-purpose, such as a nut or bolt, because here the producer of the component part will be liable for defects in his product even if it is put to a use which could not be reasonably expected by the producer of the component. The Strasbourg Convention makes provision for this by stating that it shall be a defence for the maker of the component to show that the defect arose from the design or specification required by the makers of the product into which the component was incorporated. A prescription period of 10 years is available for the bringing of actions against producers from the time when the article was put into circulation under the Strasbourg Convention. Whether this is better than the present three-year period from when the claimant becomes aware of his injury is debatable since under the proposed 10-year period he may not become aware of his injury until after the product has been in circulation for more than 10 years.

48. Components

The present law relating to liability in respect of defective components was well put in *Taylor* v. *Rover Ltd.*, where the plaintiff, an employee of R, was injured when using a chisel which had been supplied to R by X and put out for hardening to Y. Although the employer was found liable because of a *novus actus* on the part of his foreman in failing to detect the defective chisel through the exercise of reasonable care, the court summed up the liability of a head manufacturer in respect of defective components as follows: "A manufacturer's duty is not limited to those parts of his product which he makes himself. It extends to component parts, supplied by his sub-manufacturers or others, which he uses in the manufacture of his products. He must take reasonable care, by inspection or otherwise, to see that those parts can properly be used to put his product in a condition in which it can be safely used or consumed in the contemplated manner by the ultimate user or consumer." If in the present state of the law the user cannot show that the manufacturer failed to take reasonable care in the above sense, he must sue the maker of the component part.

49. Duty of intermediaries to inspect and test

Sometimes a product is prepared and packaged in a way that it is likely to reach the user in the form in which it left the manufacturer, e.g. a tin of peas. Here it is not likely that there would be a duty on an intermediary to inspect, and an action by a user would be against the manufacturer. The converse is usually the case with respect to plant and machinery supplied to workplaces for use at work. So much is clear from *Biddle* v. *Truvox Engineering Ltd.* Moreover since s. 2(2), H.A.S.A.W.A. (*see* § **28,** p. 164), retains the duty on the part of employers to inspect and test plant and machinery, this fact may snap the chain of causation between manufacturer and employee (it should be remembered that at the present moment liability of a manufacturer is determined in accordance with the rule in *Donoghue* v. *Stevenson* and causation plays an important part). This will also be the case, it is thought, where s. 6(8) applies and the machinery has been built and designed to the specifications of the user. Some cases, however, notably *Griffiths* v. *Arch* Engineering *Ltd.*, where an employee was injured when his employer hired a portable grinding machine, have established that liability for injury here is joint between manufacturer and employer.

50. Possible effect of strict liability

Assuming that a regime of strict product liability is introduced, there are certain key issues which deserve special attention. Firstly would the defences of contributory negligence and *volenti non fit injuria* still continue to apply? If we are to have a system of liability which channels liability to the final producer based on his failure to exercise reasonable care, the answer may be yes. Second, because the introduction of any strict liability system for product defects will force premiums up, should there be a ceiling on the amount of no fault compensation over and above which fault would have to be proved? Third, only American product liability law makes producers liable for defects in their products which the lack of technology and scientific development could not have permitted them to know of. Would this be so in English law? It seems not for the present. In *Wright* v. *Dunlop Ltd.* a drug, Nonox S, which had been produced and marketed by I.C.I. had been supplied by them to Dunlop for nearly 30 years, the latter using it as an anti-oxidant. The drug was withdrawn from circulation in 1949 on the ground that it might be a carcinogen. W started to work for Dunlop in 1946 and in 1966 he developed bladder cancer. It was held that the supplier was liable, there being knowledge available to indicate that the drug was dangerous even in 1945. Accepting that the liability of a producer in negligence arises in respect of production defects, design defects and lack of or inadequacy of warnings and directions for use, a system based on reasonable care argues that a producer would not be liable for dangerous qualities which his product might have unless he could have known of this by reference to the state of the art. But would this be the position under a system of strict liability? Would a motor car manufacturer be liable for fitting a replacement part to a whole batch of cars, once a defect in a component came to light, including cars *previously* supplied as well as those to be supplied? It is impossible to be sure about this, but there is no reason why strict liability cannot have this meaning.

Another major problem arises in connection with design of products. It is well known that design liability is one of the most hazardous areas of underwriting, since bad design may affect a whole range of products as distinct from one product being badly constructed. If liability is to be strict, would not this add to the underwriting problems unless a ceiling were fixed for no fault liability?

The main arguments against the introduction of strict liability are that inability to chart liability by reference to failure to take reasonable care will be a disincentive to further development and production, especially of new products such as drugs, and that the producer who operates a good quality control system would be penalised just as much as the producer with indifferent quality control standards, there being no distinction between the two in strict liability. Moreover such a system would lead to an increase in the number of claims, including spurious ones. On the other hand, the proponents of strict product liability claim that it would lead to higher quality control standards, a reversal of an unfair burden of proof coupled with a reduction in legal expenses, since there would be fewer transfers of indemnity. In addition this view favours strict liability because the manufacturer not the user should be liable for loss caused by his product.

51. Criminal liability

The criterion of "dangerous"

This refers to situations that could give rise to prosecution by H.S.E. under s. 6, H.A.S.A.W.A. It is not necessary that injury have occurred, but the article or substance must be dangerous and create a risk of injury. So far there is no authoritative case law on s. 6, but it will be interesting to see how far, if at all, previous authorities from civil law will play a part in the interpretation of this section and particularly whether the criterion of "dangerous" will turn out to be du Parcq, J.'s, dictum in *Walker* v. *Bletchley-Flettons Ltd.* If so, as far as criminal liability is concerned, we shall have a product liability system comparable with the American regime of unreasonable danger.

Section 6, H.A.S.A.W.A.
There are ten sub-sections to this section, providing as follows:

(1) It shall be the duty of any person who designs, manufactures, imports or supplies any article for use at work—

(*a*) to ensure, so far as is reasonably practicable, that the article is so designed and constructed as to be safe and without risks to health when properly used;

(*b*) to carry out or arrange for the carrying out of such testing and examination as may be necessary for the performance of the duty imposed on him by the preceding paragraph;

(*c*) to take such steps as are necessary to secure that there will be available in connection with the use of the article at work adequate information about the use for which it is designed and has been tested, and about any conditions necessary to ensure that, when put to that use, it will be safe and without risks to health.

(2) It shall be the duty of any person who undertakes the design or manufacture of any article for use at work to carry out or arrange for the carrying out of any necessary research with a view to the discovery and, so far as is reasonably practicable, the elimination or minimisation of any risks to health or safety to which the design or article may give rise.

(3) It shall be the duty of any person who erects or installs any article for use at work in any premises where that article is to be used by persons at work to ensure, so far as is reasonably practicable, that nothing about the way in which it is erected or installed makes it unsafe or a risk to health when properly used.

(4) It shall be the duty of any person who manufactures, imports or supplies any substance for use at work—

(*a*) to ensure, so far as is reasonably practicable, that the substance is safe and without risks to health when properly used;

(*b*) to carry out or arrange for the carrying out of such testing and examination as may be necessary for the performance of the duty imposed on him by the preceding paragraph;

(*c*) to take such steps as are necessary to secure that there will be available in connection with the use of the substance at work adequate information about the results of any relevant tests which have been carried out on or in connection with the substance and about any conditions necessary to ensure that it will be safe and without risks to health when properly used.

(5) It shall be the duty of any person who undertakes the manufacture of any substance for use at work to carry out or arrange for the carrying out of any necessary research with a view to the discovery and, so far as is reasonably practicable, the elimination or

minimisation of any risks to health or safety to which the substance may give rise.

(6) Nothing in the preceding provision of this section shall be taken to require a person to repeat any testing, examination or research which has been carried out otherwise than by him or at his instance, in so far as it is reasonable for him to rely on the results thereof for the purposes of those provisions.

(7) Any duty imposed on any person by any of the preceding provisions of this section shall extend only to things done in the course of a trade, business or other undertaking carried on by him (whether for profit or not) and to matters within his control.

(8) Where a person designs, manufactures, imports or supplies an article for or to another on the basis of a written undertaking by that other to take specified steps sufficient to ensure, so far as is reasonably practicable, that the article will be safe and without risks to health when properly used, the undertaking shall have the effect of relieving the first-mentioned person from the duty imposed by subs. 1 (*a*) above to such extent as is reasonable having regard to the terms of the undertaking.

(9) Where a person ("the ostensible supplier") supplies any article for use at work or substance for use at work to another ("the customer") under a hire-purchase agreement conditional sale agreement or credit sale agreement, and the ostensible supplier—

(*a*) carries on the business of financing the acquisition of goods by others by means of such agreements; and

(*b*) in the course of that business acquired his interest in the article or substance supplied to the customer as a means of financing its acquisition by the customer from a third person ("the effective supplier"),

the effective supplier and not the ostensible supplier shall be treated for the purposes of this section as supplying the article or substance to the customer, and any duty imposed by the preceding provisions of this section on suppliers shall accordingly fall on the effective supplier and not on the ostensible supplier.

(10) For the purposes of this section an article or substance is not to be regarded as properly used where it is used without regard to any relevant information or advice relating to its use which has been made available by a person by whom it was designed, manufactured, imported or supplied.

52. Responsibility for product safety

Section 6 is in two parts, the first part dealing with liability for *articles* for use at work and the second part concerned with *substances* for use at work. Both these expressions are defined in s. 53, the interpretation clause, as follows:

"Article for use at work" means—

(*a*) any plant designed for use or operation (whether exclusively or not) by persons at work, and

(*b*) any article designed for use as a component in any such plant.

"Substance for use at work" means "any substance intended for use (whether exclusively or not) by persons at work".

Here the key expression in the case of both articles and substances is "for use at work". By analogy with *Parvin* v. *Morton Machines Co. Ltd.* (1952) (*see* §**33**, p. 214) it could be argued that, as with the expression "any machinery" for the purposes of s. 14(1) Factories Act 1961, the expression "for use at work" in s. 6 would not extend to articles and/or substances which are the finished product of the manufacturing process but rather would apply to machinery and plant and component parts used in the manufacturing process. Thus in the case of a motor manufacturing company assembling the component parts to be distributed as a finished motor vehicle to dealers, s. 6 would have no application to the finished product, the motor vehicle, since a consumer relationship is involved at this point, which is better dealt with by consumer protection legislation. But until the finished product emerges from the assembly line, the requirements of s. 6 apply, e.g. to braking systems and car bodies and other machinery used in the production process. Similarly, with regard to s. 6(3) dealing with installation and erection of articles for use at work, by analogy with *Irwin* v. *White Tomkins & Courage Ltd.* (1964) (*see* §**33**, p. 214), since the fencing requirements of the Factories Act 1961 apply as soon as installation is complete, once installation is complete, s. 6 and its requirements cease to be applicable, the matter at that point being within the province of s. 2, H.A.S.A.W.A.

The chain of responsibility

Product safety for the purposes of s. 6 centres round a chain of responsibility. Section 6 means that designers, manufacturers and installers will normally be directly responsible for ensuring the safety of their products; whereas importers and suppliers, in order to discharge their responsibilities under s. 6, must obtain the assistance of designers and manufacturers for the purposes of ascertaining and describing the nature and properties of the products which they supply. In other words building safety into products is the responsibility of designers and manufacturers rather than importers and suppliers and installers, who are not involved in the production process. Taking subs. 6(1) in conjunction with subs. 6(6), it is clear that a major part of the testing and research necessary will take place at the production stage. And there is no further obligation (s. 6(6)) upon suppliers and importers to duplicate testing already carried out, provided that the information and results of this which they are given are reliable. Hence suppliers and importers of articles and substances for use at work will be in a position to reduce or eliminate their commitment to testing and research by the use of accurate reference material supplied by the manufacturers and designers. What the supplier and importer must do, as far as is reasonably practicable, is to satisfy themselves that the information which they have been given about the product is accurate. In other words they are under a duty to *supply* safe goods and products. Since not all designers and manufacturers will have the facilities to carry out this sort of testing and research, H.S.E. recommends that either they consider combining for such purpose or that they make use of existing commercial and government test centres.

Research and testing

The research which designers and manufacturers of articles for use at work must carry out is meant to do two things, firstly to discover any risks that might be associated with the design of the article or substance, and second to find ways of eradicating and/or reducing the incidence of known and discovered dangers. This would include research into such matters as safe working loads, the effect of stresses, and the need for built-in safeguards. Above all, as far as is reasonably practicable within the meaning of *Edwards* v. *N.C.B.*, the manufacturer has to give the supplier information about the product, both as to its composition and construction and the uses to which it can be put, coupled with the necessary warnings and directions for use on the article itself, so as to enable the supplier to assure the user-employer that his employees will be safe when using the article and/or substance properly.

Designers and manufacturers cannot regard a single isolated provision of information as a complete discharge of their responsibilities. There is an implicit continuing commitment to remain aware of research and progress in the relevant field concerning the product and to bring any advances to the notice of the user. This might in the case of certain products such as vehicles include training, e.g. in the case of fork-lift trucks. Manufacturers must as standard practice issue adequate information about machines and vehicles and drugs particularly, and users must demand it, so as to meet their obligations under s. 2. It is thought that manufacturers have a statutory responsibility to insist that machines must be operated by trained personnel.

Moreover it is not unrealistic to suppose that they must specify the training need.

This is not to argue that the designer and manufacturer must supply information with every piece of equipment which he supplies, but he should ensure for the purposes of the user's obligation under s. 2 that there will be ready access to such information. The user's management must then communicate this information obtained from the designer and manufacturer to the shop floor, and safety representatives, it is suspected, will be acting within their rights when they demand it either from an inspector or from management. Articles for use at work should be accompanied when they are supplied with data sheets or labelled in a durable fashion or the user may be referred to appropriate documentation, e.g. British Standards Specifications. H.S.E., it seems, do not regard data sheets as obviating the need for labels.

Transfer of responsibilities

An important provision is subs. 6(8), the exact scope of which it is difficult to define. It is reasonably clear, however, that it provides for the transfer of obligations in specified circumstances. When someone gives a written undertaking that he will take specified steps to ensure that an article passed to him in the supply chain will be safe, the designer or manufacturer or both or the importer or supplier passing it to a user is relieved of his duties to the extent that the terms of the undertaking are reasonable. It should be emphasised, however, that this does not in any way over-ride the duty to supply adequate information under subs. 6(1)(*c*); the view widely held is that the purpose of subs. 6(8) is to enable customers to order plant made to their own specifications or to order unfinished machinery with the intention of putting special guards upon it (these being supplied by a different company) without putting the suppliers of the original article at fault.

Chapter 6

Fire

STATUTE LAW AND FIRE

The over-riding purpose of statute law with respect to fire is to set down measures which must be carried out by employers and occupiers of workplaces in which people work, in order to prevent fire breaking out. This includes the issuing of fire certificates in respect of such premises. The object of civil law or tort is to posit rules regarding liability in the event of statutory measures having failed and fire having broken out. Liability for the escape of fire is often strict.

53. Premises affected by statute or regulation

The main statute relating to fire is the Fire Precautions Act 1971. This Act supersedes the fire precautions in the Factories Act 1961 and the Offices, Shops and Railway Premises Act 1963 by virtue of the fact that the Fire Precautions (Factories, Offices, Shops and Railway Premises) Order 1976 (*see* Appendix XI) renders the Fire Precautions Act 1971 applicable to certain factories, offices, shops and railway premises. On the other hand, there are certain premises, known as special premises, which may partake of the nature of factories, offices, shops and railway premises to which the Fire Certificates (Special Premises) Regulations 1976 (*see* Appendix X) render H.A.S.A.W.A. applicable. Since 1st January 1977, there are now two types of premises for the purposes of fire precautions—firstly premises designated in the Fire Precautions (Factories, Offices, Shops and Railway Premises) Order 1976, for which fire certificates from the fire authority are required, and second, premises for which fire certificates are required from H.S.E. as designated in the Fire Certificates (Special Premises) Regulations 1976.

Fire Certificates (Special Premises) Regulations 1976

These regulations came into force on 1st January 1977 and were made pursuant to s. 15, H.A.S.A.W.A., and other enabling provisions of that Act. The major effect of this legislation is to require that fire certificates be obtained from H.S.E. in respect of premises referred to in Sched. 1. The regulations are set out with the other fire regulations in Appendix X. The greater the hazard existing on the premises the more likely it is that a fire certificate is required from H.S.E., and the more likely it is that the Fire Certificates (Special Premises) Regulations 1976 will apply in such cases. Particularly significant in these regulations is subs. 5(2) which empowers H.S.E. to impose any conditions they consider appropriate upon the issue of fire certificates.

Fire Precautions (Factories, Offices, Shops and Railway Premises) Order 1976
This order also became operational on 1st January 1977. Its principal effect is to designate, for the purposes of s. 1, Fire Precautions Act 1971 (premises for which a fire certificate is required from the fire authority), *all* factory, office, shop and railway premises in which persons are employed to work *except*:

(*a*) those premises in which the persons employed at any one time do not exceed the numbers specified in Art. 4 of the Order (these premises are subject to the requirements of the Fire Precautions (Non-Certificated Factory, Office, Shop and Railway Premises) Regulations 1976 (*see* Appendix XII); and

(*b*) those special premises specified in the Fire Certificates (Special Premises) Regulations 1976 in respect of which a fire certificate is required from H.S.E. (*see* Appendix XI).

Fire Precautions (Non-Certificated Factory, Office, Shop and Railway Premises) Regulations 1976.
These regulations came into force on 1st January 1977 and provide for certain fire precautions in relation to escape in the case of fire to be taken in the smaller factories, offices, shops and railway premises that do not require a fire certificate from the fire authority under s. 1, Fire Precautions Act 1971. The effect of this legislation is that premises excluded from the operation of the Fire Precautions (Factories, Offices, Shops and Railway Premises) Order 1976 and specified in Art. 4 of these regulations are subject to the provisions of the Fire Precautions (Non-Certificated Factory, Office, Shop and Railway Premises) Regulations; these are the smaller factories, offices or shop and railway premises in which:

(*a*) not more than twenty persons are employed to work at any one time; or

(*b*) not more than ten persons are so employed elsewhere than on the ground floor.

The regulations relating to fire are set out in Appendices X–XII.

54. Liability for the escape of fire

Liability at common law for injury to persons at work is based in most cases on negligence. This applies also to actions for breach of statutory duty as against employers and factory occupiers. Where, however, fire escapes and does damage, the liability is often strict.

The rule in Rylands *v.* Fletcher

The escape of fire and consequent damage may be governed by the rule in *Rylands* v. *Fletcher* (1868), where the following doctrine was laid down by Blackburn, J.: "Where a person for his own purposes brings and keeps on land in his occupation anything likely to do mischief if it escapes, he must keep it in at his peril, and if he fails to do so he is liable for all damage naturally accruing from the escape."

Since liability under this rule is strict the circumstances in which the rule applies are restricted by the courts to situations where damage is caused to property of someone having an interest in land, possibly resulting in personal injury as well, as a result of the escape of fire from property nearby. If, therefore, someone suffers merely personal injury, there being no escape of fire from one set of premises to premises occupied by someone else, it appears that he would have to bring his action in negligence. This has been the view for many years, but there are some cases to the contrary. In *Perry* v. *Kendricks Transport Ltd.* in 1956 the defendant left a coach on a parking space belonging to him. The coach was disused and the petrol had been drained off it with a cap screwed over the entrance pipe. A boy of 10 was injured as he went into the parking lot by an explosion of petrol fumes from the coach. Evidence showed that some time earlier the cap had been removed by someone else who could not be traced and that a lighted match had been thrown down the pipe. The Court of Appeal ruled that the facts were within *Rylands* v. *Fletcher*, though because the explosion was caused by the act of a stranger, the occupier of the land was not liable. On the other hand in *Read* v. *Lyons Ltd.* in 1947 the House of Lords took a more circumscribed view. An employee of the Ministry of Supply was inspecting munitions in the course of her employment. Whilst doing this she was injured by the explosion of a shell which was being manufactured in the premises she was inspecting. Her claim for liability based on the rule in *Rylands* v. *Fletcher* failed, there being no escape of a dangerous thing from someone's premises to those of someone else. This has always been regarded as an authoritative decision and has meant that strict liability for the escape of fire is restricted and does not apply to personal injury cases *per se*.

The law, therefore, with respect to the escape of fire is that liability is strict where

(*a*) a thing capable of self-combustion and self-activation (e.g. petrol in storage), which

(*b*) is not naturally on the land or premises (e.g. petrol in storage),

(*c*) escapes
(*d*) to the premises of another
(*e*) causing damage to both persons and property or either.

Meaning of strict liability for the purposes of the rule.
In *Rylands* v. *Fletcher* itself F employed in 1868 an independent engineering contractor to construct a reservoir on his land. The contractor was competent and employed competent employees. Nevertheless water escaped from the reservoir, percolating through some disused mine shafts, and flooded R's mine. It was held that, although F had not been negligent, yet he was liable for the negligence of the contractors. It was then incumbent upon the employer to seek indemnity by way of contract from the contractor after the loss had been redressed.

Chapter 7

Compensation and Insurance

THE IDEA OF INSURANCE

The idea of insurance is that a person may purchase protection against certain unpleasant contingencies which may arise during the course of his employment, whilst he is driving a motor vehicle, using a product he has bought or even whilst he is on holiday. There are numerous varieties of insurance to afford protection against almost any kind of contingency which may eventuate. The policy holder is buying protection against the occurrence of future happenings which may cause him harm, in the sense that he is injured or has got to make financial reparation for the harm which he has caused, either directly or vicariously, to another.

A good example is a motor vehicle policy. Cover is provided against damage which a motorist may do to a third party whilst driving his car (this is known as third-party insurance) and against damage to his own vehicle and himself (known as comprehensive insurance). In the case of comprehensive insurance the motorist is buying insurance to protect himself and not others. In addition insurance will be taken out by employers in respect of injury to their employees and other third parties, by business men in respect of loss of profits (known as profits insurance), by householders in respect of liability which they may incur as occupiers, by employers and other persons engaging in works (often building and civil engineering) which may cause injury to persons other than their employees (known as a public liability policy) and by sellers and suppliers and manufacturers of products (known as a product liability policy).

55. Insurance is a contract

There is a contract between the person who is buying protection, known as the policyholder or the insured, and the insurer or insurance company. This contract is a special sort of contract *uberrimae fidei* or a contract of the utmost good faith. In consequence of this the proposer or the insured is required to make complete disclosure of all material information to the insurer. Failure to do this enables the insurer to invalidate the contract and not pay out the insurance moneys in the event of the contingency occurring. A fact is deemed to be material if it would serve to influence a prudent insurer in deciding whether to accept the risk and at what rate he would fix the premium.

Because insurance, whether liability or first-party, consists of a contract between insurer and insured, it follows that the relationship is governed by the rules relating to contract. Thus although an employer's liability to his employee, as far as injuries at work are concerned, is tortious, whether an injured employee will receive compensation ultimately from the insurer may depend upon the validity of the contract between the employer-policyholder and the insurer. This fact could prove to be important, in that if the contract can be invalidated by the insurer, i.e. for lack of *uberrima fides* or absence of utmost good faith on the part of the proposer, the insurer can refuse to indemnify the employer against his liability towards his employee, with the practical result that the employee will have to sue the employer at common law and/or for breach of statutory duty in order to obtain redress.

56. Grounds on which the contract can be invalidated

Failure to make complete disclosure of material information

Material information is information which would assist the insurer in deciding whether to accept the risk. For instance, if the employer was engaged in the production of articles which required for their manufacture the use of some substance which was associated with a prescribed industrial disease, e.g. asbestos, for which the prescribed industrial disease is mesothelioma, and the employer failed to disclose this fact, when this omission comes to light, the insurer may refuse to indemnify the employer in respect of liability towards his employees even if the employee has contracted another disease, about which the employer did not keep silent when filling in the proposal form, or even if the employee suffered a physical injury at work rather than a disease. Not infrequently, however, the insurer may elect to accept the risk on receipt of an additional premium.

Material mis-statement

The law relating to contracts of the utmost good faith requires that a proposer disclose accurate answers to questions posed in the proposal form of insurance, for it is on the basis of those answers that a policy is later drawn up. Thus when filling in a proposal form for life assurance, a proposer may well be asked whether he is suffering from certain medical conditions or has ever done so. If he genuinely states that he is not and this is confirmed by medical examination, the policy could still be invalidated if it was found that the assured was suffering from the first stages of a prolonged cancer. Such an application of the *uberrima fides* rule creates hardship and for that reason most proposal forms require the proposer to state such information accurately to the best of his knowledge.

The basis clause rule

This rule prescribes that if the proposal form for insurance contains a basis clause, it is irrelevant whether the information which the proposer failed to disclose is material or not. In other words any misdescription, however slight, can result in the insurer invalidating the policy and refusing to indemnify the policyholder. A typical basis clause in a proposal form might be as follows: The Proposer warrants the truth of the statements he has made and the answers that he has given and the proposal and answers given therein shall form the basis of the contract and shall be deemed incorporated into the contract. A case illustrating how disastrous the effect of this rule can be on claims for indemnity was *Dawsons Ltd.* v. *Bonnin*, where D insured a lorry used in his business with B. The proposal form for insurance asked where the lorry would be garaged and D's secretary inadvertently stated that it would be garaged at the address of the business premises in Cadogan Street, Glasgow. There were no garaging facilities at that address. Later the lorry was destroyed in a fire at the Cadogan Street address and D claimed under the policy. It was held that his claim must fail. The contract of insurance contained a basis clause to the effect that D warranted the truth of the statements made in the proposal form and that they should form the basis of the contract. During the course of giving judgment in the House of Lords Viscount Cave observed: "The meaning and effect of the basis clause, taken by itself, is that any untrue statement in the proposal, or any breach of its promissory clauses, shall avoid the policy, and if that be the contract of the parties, the question of materiality has not to be considered."

The risk increases after the premium is fixed

Where this happens the insured must inform his insurer of any changes relating to the state of the premises, which would involve an increase in the risk. Failure to inform the insurer entitles the insurer to treat the policy as invalidated. If, therefore, an employer was using certain substances in his manufacturing process, none of which at the time of contract was known to be dangerous, but at a later date and during the currency of the policy it was established that one or more of the substances were carcinogenic, it is incumbent on the employer to inform his insurer of this change; otherwise the insurer may elect to treat the policy as invalid. As the risk increases, so does premium.

It must be emphasised that in all cases where a policy is invalidated the injured employee must sue his employer for breach of common law and/or statutory duty.

57. Other consequences of the contract

The whole picture of actions for compensation in respect of industrial injuries, and in respect of road injuries, has an air of unreality about it. Although it is the employer who is liable in tort to his employees, yet generally his insurer pays the damages. Indeed it is generally quite uncommon for an injured workman to press his claim by going to court, though if he has trade union representation it is more likely than not that the matter will go to court if the settlement between the union negotiator and the insurance or claims assessor fails to produce a satisfactory solution. If, however, a claim is pressed and the matter (both liability and extent of damages) is litigated, it is against the employer not against the insurance company that the action is brought. Yet it is the latter who pays. It is all very unreal.

Plaintiff who takes legal action cannot normally obtain judgment direct against insurer

It may be that a third party who has suffered injuries at the hands of an insured person may be dissatisfied with the settlement offered and in consequence refuses to accept the settlement. If he takes the insured person to court and wins his case, he still normally cannot proceed to enforce the judgment against the insurer in cases where the insured is impecunious. There is one notable exception to this rule, in the case of road accident victims, where by virtue of subs. 148(1), Road Traffic Act 1972, an injured third party can proceed direct against the insurer to obtain redress, which includes interest and costs. In such a case, however, he must give seven days' notice to the insurer of his action against the insured person. And in these cases where road-accident victims are claiming directly against the insurer, the insurer cannot avoid liability even by proving that the insured's conduct was intentional.

This privilege of proceeding against the insurer directly is denied to the injured employee. If, therefore, his employer is bankrupt and his employer's insurer refuses to indemnify the employer, the injured employee is without redress save for social security in the form of industrial injuries benefit.

The law does not allow persons to buy first-party insurance for others

Because the relationship between insurer and insured is a contractual one and because of the doctrine of privity of contract, which precludes third parties suing on contracts for contractual benefits, the law goes to considerable lengths to stop people buying first-party insurance for the benefit of others. There are one or two minor exceptions to this, as where a passenger injured in a car is allowed to sue the insurer, even though the contract was made between the insurer and the motor-vehicle owner, and secondly in the case of endowment policies, a kind of life assurance whereby on the happening of a certain date a sum of money may become payable to a person other than the insured. But these two situations differ from the situation mentioned above, where as a result of privity of contract a person is precluded from buying first-party insurance for the benefit of his employees. First passenger claims are an example of third-party liability insurance rather than first-party insurance; second endowment policies are examples of life assurance which is not indemnity insurance. Liability insurance and first-party personal accident insurance and profits insurance, on the other hand, are indemnity insurance. Their purpose is the replacement of loss suffered.

The case which established that it was not legally possible to buy first-party insurance for the benefit of others was *Green* v. *Russell*, and similar sentiments had been expressed earlier in *Re Schebsman*. Why this judicial reluctance to prevent people buying insurance for others? The answer would seem to be that our system of injury insurance has grafted itself on to the bedrock of tort liability. In other words cover offered by a policy is designed to protect the policyholder from liability to others and the policy seeks to cover the policyholder against as many areas of legal liability as possible. It is because the cover purported to be offered by the policy is drawn up on the basis of avoiding legal liability that it is not always easy to be sure whether the policy cover is adequate. It is seriously suggested that these problems could be short-circuited by changing from cover based on third-party liability to cover based on first-party protection (for the difference in meaning between these two expressions, *see* § **61,** p. 268).

58. Insurance is a contract: tort is ancillary

Now that the practice of insuring has become universal, and certain types of insurance, e.g. third-party motor-vehicle insurance and employers' liability insurance, are compulsory, it is not difficult to see that the real source of compensation is liability insurance, with tort liability playing an ancillary role. Indeed evidence shows that only about 2 per cent of all personal injury cases are litigated; by far the greater proportion of them are settled out of court by negotiation between the plaintiff or his solicitor or his trade union and the insurer.

Where no insurance contract was made

This situation can create problems for injured third parties and for injured employees more than for injured road-accident victims. Although third-party motor-vehicle insurance is compulsory and although the same is true of most employers as regards their liability towards their employees (since the passing of the Employers' Liability (Compulsory Insurance) Act 1969), despite the existence of criminal sanctions there will always be those who will not insure. Where this happens in the case of a motorist and a third party is injured by an uninsured motorist, he can make a claim on the Motor Insurer's Bureau which, being a conglomerate of the various insurance companies, agrees to indemnify the injured person. Such payments as are made in these cases are on an *ex gratia* basis. There is no corresponding institution in the field of employers' liability; as yet an employers' liability bureau has not developed, the inference being that employers are more claims-conscious than motorists.

Where the employer goes into liquidation or becomes bankrupt

By virtue of s. 1, Third Parties (Rights against Insurers) Act 1930, an injured third party does not have to prove in the bankruptcy or liquidation of his employer (or if a road-traffic victim in the bankruptcy of the motor-vehicle user) to obtain judgment. Were he forced to do so he would have to be satisfied along with the other creditors with a dividend. As it is, however, he may proceed directly against the bankrupt insured's insurance company and obtain judgment in full from them, though if the policy were invalid against the policy-holder for reasons mentioned earlier, e.g. material misdescription, it would also be invalid against the third party.

Where the insurer goes into liquidation

Where this happened prior to 1975 the injured party had to rely on the bona fides of A.B.I.A., the Association of British Insurance Agents. Now, however, the likelihood of insurance companies going into voluntary or compulsory liquidation, as risk becomes more and more expensive to underwrite in the light of inflation, has itself become a real possibility. That this trend is likely to continue, with new liabilities being laid upon employers, is evidenced by the Congenital Disabilities (Civil Liability) Act 1976 and the proposed legislation on product liability. This perhaps argues for fixed ceilings of indemnity as now happens with nuclear fall-out where insurers refuse to insure for more than £5m in respect of any one claim under the Nuclear Installations Act 1965. In order to protect insured persons in the event of liquidation of insurance companies through being unable to meet these high claims, the government passed the Policyholders' Protection Act 1975, the effect of which has been to establish the existence of a Policyholders' Protection Board, whose role is to indemnify policyholders in the event of a collapse of an insurance company.

TYPES OF INSURANCE

59. Voluntary and compulsory insurance

Private insurance is both voluntary and compulsory. In other words certain kinds of private insurance are voluntary, whereas other kinds are compulsory and must be taken out under threat of a criminal sanction, e.g. employers' liability insurance. Similarly third-party cover for users of motor vehicles, as well as passenger cover, is compulsory. Personal accident insurance and other types of consequential loss insurance, such as loss of profits insurance, are voluntary, though advisable if operating a business. Decisions like *Weller & Co.* v. *Foot & Mouth Disease Research Institute* and *Spartan Steel & Alloys Ltd.* v. *Martin (Contractors) Ltd.* are ample evidence of this (*see* § **20,** p. 54). Householders' insurance and products' liability insurance are voluntary.

Social insurance, that is national or state insurance, is compulsory. The state takes the view that certain contingencies or misfortunes are common to everyone and, as a result, it compels us to insure against them. The contingencies against which it is necessary to insure are:

(*a*) retirement in respect of which the state provides a pension;
(*b*) sickness in respect of which the state provides benefit;
(*c*) unemployment in respect of which the state provides benefit;
(*d*) industrial injuries in respect of which the state provides benefit;

in addition there are certain other benefits of lesser importance, though of great significance to the claimant, e.g. unemployability allowance, invalidity benefit, constant attendance allowance and death benefit (*see* § **23,** p. 149).

Private insurance takes the view that the purpose of compensation is the complete replacement of loss, and hence seeks to compensate in full any loss which has been suffered. Personal accident insurance seeks to do likewise, though it is more expensive to recoup a loss in this way. The conditions for the recovery of compensation are (*a*) in the case of liability insurance that there is legal or tortious liability in the insured and (*b*) in the case of first-party personal accident insurance that the insured had the foresight to take out the insurance. (Indeed, this may well be the only way for a self-employed professional man engaged in potentially dangerous work, e.g. a safety consultant, to insure himself in the event of injury at work followed by long-term disablement, though this sort of insurance as a provision for long-term disablement is quite expensive, although if the self-employed professional is a director in his own company he is technically an employee and so entitled to industrial injuries benefit or industrial disease or death benefit.)

State insurance or social security does not take the same view of the function of insurance. It argues that injury and disease are common to everyone and the expense associated with personal suffering requires all employed persons to be insured against sickness and injury. It used to assume that benefits should be the same for everyone. The social security system does not aspire to make full compensation to a few (those who can prove fault) and used to provide flat-rate benefit (the same for all), except in the case of industrial injuries benefit, which has always been earnings-related since the Workmen's Compensation Act 1897. As from 1975 when the Social Security Act was passed, other benefits including pensions have become earnings-related.

60. Private and social insurance

Private insurance is much older than social security, the former reflecting a laissez-faire approach to misfortunes which might befall people (at least originally till certain types of liability insurance became compulsory), whilst the latter is more consistent with a socialist approach to the problem of people's misfortunes. It is this divergence in ideology that accounts for the fact that state insurance is compulsory. The state considers that certain misfortunes and contingencies of a long-term nature and requiring the expenditure of vast amounts of money are potentially common to everyone. That being so, it requires and compels everyone in his own interest to insure against the possibility of these harmful and tragic consequences. In other words, most of us have no choice but to contribute to the social security system. Employed persons make a contribution along with their employers and the taxpayer in respect of (*a*) industrial injuries benefit; (*b*) unemployment benefit; (*c*) sickness benefit; (*d*) retirement pension. Self-employed persons make contribution along with the general body of taxpayers in respect of retirement pensions and, if directors in their own company, are entitled to industrial injuries and industrial disease and death benefit, this resulting from limited liability. With the exception of industrial injuries benefit and industrial disease benefit we saw that state benefits were flat-rate not earnings-related. The criterion of social security, as seen by Beveridge, was need rather than risk. Thus even if the risk to one person or class of persons or class of workers was greater than to another person or class of persons or workers, the contribution rate was the same. Everyone contributed the same regardless of position or means and everyone took the same out except for injury or disease at work. This meant that, as with the tort system, two men injured at work could receive different amounts of compensation in respect of the same injury. Now that social security benefits have become earnings-related in totality it would appear that the state system of compensation has adopted the criteria of private insurance, that the more one contributes the more in the way of benefits one will be in receipt of. This departure in favour of earnings-related state compensation reflects a departure from the old welfare state.

Certain sections of society are less well covered by the social security system than others. The employed earner is covered by the state system as regards loss of earnings, and by disablement pensions and other benefits such as unemployability supplements, invalidity benefits, constant attendance allowances and the more recent mobility allowances for the permanently disabled, whereas the business man and the self-employed earner is rather less well covered, especially as regards loss of earnings through sickness, disablement, unemployment and loss of business or profits. For these reasons it is advisable for such persons to take out personal accident insurance covering them for loss of earnings and disablement arising from their employment and for loss of profits from reduction in turnover. In addition to these forms of first-party insurance certain kinds of liability insurance are also voluntary such as householders' insurance in respect of a householder's liability as an occupier to lawful visitors. Employers' liability and third-party and passenger motor-vehicle insurance are compulsory.

61. Liability and consequential loss insurance

It is important to understand that the cover offered by private insurance is in respect of both liability to third parties and loss consequent upon material damage suffered by the insured person, i.e. first-party insurance. Liability insurance and consequential loss insurance both offer indemnity, i.e. the exact replacement of loss.

Liability insurance

A proposer who buys liability insurance seeks to be indemnified against claims that may be made against him as a result of his legal liability to a third party. This liability is tortious. It is not possible for an insured to be indemnified against intentional conduct causing harm, as it is the policy of the law not to allow persons to insure against the consequences of anti-social conduct. This generally refers to criminal conduct but can include certain tortious conduct, such as assault.

(*a*) *Scope of cover.* Where an employee is injured by an accident or suffers a disease arising out of and in the course of his employment, tort compensation payable under a liability policy provides him with full compensation. The compensation takes the form of payment of a capital sum which he is advised to invest as prudently as possible so that the income produced by the fund can be used to the best advantage. In principle there is nothing to prevent the person squandering the money, but generally this would not happen if the injured person had a family to support. In the case of an employee killed at work a lump sum will be paid to his widow and remarriage prospects are not taken into account. In principle there is again nothing to prevent the widow squandering the money instead of applying it for the benefit of the children, though it would be hoped that the widow would create a trust. With social security the payment is not in a lump sum but rather periodical payments; moreover social security offers benefits in kind such as group care in a welfare centre, rehabilitation, meals on wheels, home help, rates rebate, invalid carriages for the disabled and discounts on the cost price of certain makes of motor vehicles. Also grants are available for the adaptation of houses for disabled persons.

(*b*) *If fault can be proved.* If fault can be proved against someone or if there is in existence a loss distributor such as an employer or nationalised industry, tort compensation or liability insurance offers a good deal more in pecuniary terms than social security. Cover is given in respect of loss of earnings, actual and calculable, medical expenses, pain and suffering, disablement, temporary or permanent, post-injury nursing expenses, loss of pleasure, change to a different, less rewarding kind of employment. Liability insurance does not compensate for injury, deformity or disablement brought about by natural causes, e.g. poliomyelitis, unless arising out of employment, as possibly in the case of a nurse. Tort compensation does offer compensation for loss of faculty, i.e. disablement irrespective of whether the victim's ability to earn his living is impaired or not. The same is true of social security. Lump sum indemnity has the built-in disadvantage that once the damages have been assessed, either in court or in a settlement out of court, they cannot normally be altered. Thus if the victim made a miraculous recovery he would be enriched to a considerable extent; whereas if the judge miscalculated and the victim's condition deteriorated more than had been expected, he will have been under-compensated.

A consequence of tort compensation is that, since it is based on liability to a third party and cannot adjust its premium rating to its pay-out, in the sense tat it cannot know in advance what the income of the injured or killed third party may be, the tort system offers earnings-related compensation based on premiums that are not earnings-related. At the time the liability insurance is bought the only earnings known are those of the insured. Yet since this insurance seeks to compensate an as yet unknown third party, the person whose income is ascertainable is the person whose income is irrelevant, because he is the person who becomes liable, not the person who may be injured and lose his income. Because the criterion of pay-out is tort liability this means that two people suffering identical injuries or disabilities will be compensated in different proportions, according to their earning-capacity prior to the injury or disablement. Liability insurance says that claimants can have earnings-related compensation in a lump sum, the criterion of premium being risk, which is not an earnings-related criterion. Conversely social security urges that claimants can only have earnings-related benefits if they make earnings-related contributions.

Consequential loss insurance

One of the chief reasons for consequential loss insurance is that, although negligence has become barely distinguishable from strict liability and the cover afforded by liability policies is wide, there are certain types of loss which are excluded. This refers to loss of an economic or pecuniary nature resulting from or consequential upon material damage, hence it is called consequential loss insurance. A typical example of this type of cover would be where a factory or a supplier's factory was destroyed by fire, as a consequence of which the factory was unable to go back into production, or the factory dependent on parts supplied by the destroyed factory was unable to assemble components with resultant loss of production and profits. In cases like this it is clear from Lord Denning's judgment in the case of *Spartan Steel and Alloys Ltd.* v. *Martin* (*Contractors*) *Ltd.* and from *Weller & Co.* v. *Foot and Mouth Disease Research Institute* that there is liability for any physical damage caused by the negligent act, but any purely pecuniary or economic loss, such as the loss of opportunity to make a profit on re-sale, would not give rise to liability. It is this latter kind of loss that will purport to be covered in a consequential loss policy, e.g. loss of profits, subject to the proviso that material damage has occurred in consequence of which the loss has eventuated. This sort of insurance is associated with property, businesses and plant, and damage thereto rather than injury to persons.

62. Profits insurance

Ever since the 1806 decision in *Lucena* v. *Craufurd* it had been established that profits, or rather loss of profits, could be insured but they had to be specifically insured. Thus in *Re Wright & Pole* an innkeeper who had insured his inn was unable to recover loss of profits consequent upon the damage caused to the inn.

Unless the contrary is stated, all policies are unvalued, i.e. profits are not recoverable. Thus fire insurance offers pure indemnity in respect of material damage caused by the fire; any loss of profits would not normally be recoverable under such a policy, unless, as indicated in *Lucena* v. *Craufurd*, their loss is specifically included in the policy and a premium paid in respect of such a contingency. Profits insurance, is concerned with the provision of indemnity for *trading losses* and before claims can be entertained in respect of such matters it is necessary that the property of the insured at the premises be insured against the same perils as the profits policy *and* that the material loss shall be paid or liability be admitted as a condition precedent to the right of recovery under the profits policy.

63. Doctrine of proximate cause

The insurer will only be liable to pay if the damage or loss is *proximately* caused by one of the insured perils. This means that the insured peril must be the proximate cause of the loss or damage to the insured's property, and this loss must be the proximate cause of the interruption of the business.

64. Rules applicable to all insurance contracts

Firstly, the premium must be paid; a mere handing over of the policy to the insured is not a waiver of this rule. Nor is issue of a renewal receipt a waiver of payment, nor the fact that the insurers had previously waived payment. Second, if there is a basis clause, answers given in the proposal form must be correct. Third, where forms are filled in by an agent, *Biggar* v. *Rock Life Assurance Co.* establishes that the agent is the agent of the insured not the insurer, with the unfortunate result that any incorrect statements or descriptions made by the agent (possibly for commission purposes) will be construed against the insured. Fourth, non-disclosure of any material facts allows the insurer to invalidate the policy. Fifth, the policy is to be regarded as void and of no effect if there is any alteration in the nature of the business or premises or property whereby the risk of damage is increased. Sixth, the claimant must take all reasonable steps to mitigate the loss otherwise the policy can be made null and void, this point having been established in the leading case of *British Westinghouse Ltd.* v. *Underground Electric Railways*.

EXAMPLES OF INSURANCE POLICIES

65. Public liability policy

This covers certain types of liability, leaving others specified in the exceptions to be borne by the insured. The main exceptions are (*a*) risks better covered under other policies, i.e. employers' liability; (*b*) motor-vehicle policies; (*c*) loss or damage necessarily and/or intentionally caused, e.g. a Gas Board investigating a gas leak excavates a footpath. In so doing it may well have damaged the footpath. An insurer would not here be expected to pay the cost of making good the damage to the footpath.

In order to qualify for indemnity under this sort of policy:

(*a*) the insured must be legally liable to pay damages; and

(*b*) the injury or damage must be accidental.

Legal liability can arise at common law or under statute. Most common law claims arise out of negligence, but some arise out of nuisance, e.g. damage to adjoining properties, as evidenced by *Gold* v. *Patman & Fotheringham Ltd.* in 1958.

Further, to qualify for indemnity the injury or damage must be accidental. The decision of the House of Lords in *Fenton* v. *Thorley & Co.* in 1903 provides the classic definition of "accident". The decision establishes that an "accident" is an "unlooked for mishap or an untoward event which is not expected or designed. Any unintended and unexpected occurrence which produces hurt or loss or such hurt or loss apart from its cause, especially if the cause is not broken."

Public liability policies also cover employers of outside labour, such as local authorities employing main contractors and sub-contractors, who can be liable at common law or under statute for injury or damage arising out of the work done for them by main and/or sub-contractors. Since damage to property gives rise to liability in nuisance it is important to see that the insurance policy seeking to indemnify the employer or local authority is jointly indorsed in the names of the employer/local authority and the contractors, both main and sub-contractors. This is particularly true of damage to adjoining property where liability is based on nuisance and the employer might not be able to claim indemnity in respect of the contractor's negligence. This sort of policy covers the liability of both employers/building owners and contractors in respect of their liabilities as occupiers under the Occupiers' Liability Act 1957.

66. Employers' liability policy

This policy provides a complete indemnity to an employer against legal liability for death, illness or disease of his employees arising out of and in the course of their employment. There is an indemnity limit of £2m. arising out of any single occurrence in respect of claims made. In the case of contract works, the contractor's employer's liability policy should be indorsed to indemnify the building owner or local authority against claims for damage made against it by employees of the contractor or sub-contractor.

67. Products' liability insurance policy

Claims arising from goods sold or supplied have increased in recent times, hence product liability insurances are not uncommon. This insurance provides indemnity against legal liability for accidental bodily injury or accidental damage to property caused by goods sold or supplied. Normally a general public liability policy excludes claims arising from goods sold or supplied. There is therefore a need for a products' liability policy. These legal liabilities in respect of products arise under the Sale of Goods Act 1893 and the Supply of Goods (Implied Terms) Act 1973 and the supplier or seller or retailer may well have to bear not only the responsibility of his own negligence (if any) but also that of the manufacturer of the goods sold or supplied, owing to the fact that makers avoid any liability by agreeing to replace parts but excluding, as the law until recently allowed them to do, liability for any consequential loss or damage. This position will change, one assumes, when a system of product liability is initiated in the U.K.

68. Making claims on insurers where there is liability

It was stated on p. 264 that approximately 2 per cent of personal injury cases go to court. The great majority are settled by negotiation out of court. Such negotiation is carried on normally between the injured party's solicitor or his trade union and the defendant's insurer. Because the great majority of personal injury cases do not go to court, it seems reasonable to assume that tort law serves as a threat to bring about a settlement.

Until fairly recently it seemed true to say that the victims of industrial accidents, as far as negotiating was concerned, were in just as unenviable a position as the victims of road accidents. Particularly because there was always a danger that physical evidence might get lost; employees often chose to remain conveniently silent for fear of recrimination. The Administration of Justice Act 1969, s. 21, however, allows the injured employee and his advisers access to premises to inspect machinery and take photographs. The difficulty of isolating the causes of accidents at work through lack of physical evidence or silence on the part of the workforce was accommodated to some extent by the adoption of the device, trotted out by courts where accidents were inexplicable, of *res ipsa loquitur.* Here proof of negligence is unnecessary. The use of this device for the injured workman's benefit may not prove so necessary in the future especially now the Safety Representatives Land Safety Committees Regulations have become operational (since October 1978). This means that the scene of an accident can be inspected by other members of the workforce representing the injured employee, these representatives being given specific powers to make periodical inspections. In this way negotiating with insurers by trade unions and the plaintiffs' solicitors will become easier.

Criteria underlying a settlement

The settlement figure is greatly influenced by what would happen if the matter went to court. (The cost of litigating the action in the High Court could be about £600 for each party and in the County Court about £250 for each party.) The unsuccessful plaintiff does not have to pay the defendant's costs, but the unsuccessful defendant does have to pay the plaintiff's costs. The basis of the out-of-court settlement is what would happen if the matter were litigated, and insurers pay 80 per cent of the probable damages that would be awarded in court if the plaintiff were to win his case. This contrasts sharply with the social security system and the system of criminal injuries compensation. In the latter two cases the adjudicators on the tribunals are concerned to see whether compensation should be paid, not how much. The average time involved in negotiating a claim is 15 months and a litigated case up to two years. This fact of delay, explicable enough in the light of the necessity of seeing whether injured workman will make a recovery, complete or partial, or is likely to deteriorate, suggests that, were it not for the system of social security, the tort system would have disappeared long ago. The fact is that an injured workman can draw industrial injury benefit from the D.H.S.S. within three days of his injury for up to about six months and claim disablement benefit afterwards, with the result that he suffers no loss of earnings, or only a marginal loss, before a settlement is made.

The consequences of going to court when a settlement has been offered but is not accepted by the injured workman

If an injured employee is not satisfied with the settlement which is made and he elects to go to court, the insurer may make a payment into court which the injured workman may accept even after the case has started, though before judgment is given. If, however, he declines the offer and judgment is finally awarded for an amount less than the insurer offered or paid into court, the injured party will have to pay both his own costs and the insurer's costs. This seems logical in view of the fact that the insurer would not have incurred the costs had the plaintiff elected to accept their paid-into-court sum. Moreover if in the course of judgment the injured workman is found contributorily negligent, the total amount finally recovered may be considerably less than the amount paid into court by the insurer. In addition, legal aid ranks as a prior charge on the damages awarded. An injured workman should, therefore, think carefully before litigating and should not do so unless his trade union advises him that he has a cast-iron case. If, however, because of his own contributory negligence and any other factors, the amount awarded in judgment is less than the amount paid into court by the insurer, and if the injured workman is a member of a trade union, normally the costs on both sides will be paid by the trade union. Where the amount claimed by the injured workman is £100 or less the insurer will normally settle, since to litigate or to allow the matter to be investigated means that the cost of legal services and the litigation costs are usually greater than the amount of the claim.

REPORT OF THE COMMITTEE ON PERSONAL INJURIES LITIGATION

This Committee was set up in 1966 under the chairmanship of Lord Justice Winn to examine and report on personal injuries litigation and to recommend certain improvements in the system, particularly regarding delay involved in obtaining compensation on the part of injured persons. Some of the observations of this Committee seem relevant here and may account for the important role which safety representatives are likely to play with respect to evidentiary matters in claims for compensation. The Report appeared in 1969.

Speaking of the delay factor in the determination of personal injury claims, which was the main object of criticism within the Committee's terms of reference, the Report stated "Nonetheless civil cases tend for many reasons not to reach conclusion so soon after the cause of action arose as the interests of justice and of the parties require" (para. 48). The Report goes on to examine a variety of reasons responsible for the delay, not least of which is the one mentioned in para. 60: "Whilst there is the general desire to avoid the risks of litigation and the hope that damages will be more speedily obtained by settlement, there is also very often the feeling that the issue of proceedings is regarded as a hostile step in what should be an amicable relationship of give and take. Thus an atmosphere is created which of itself causes delay principally in relation to the commencement of proceedings, but also in their active pursuit."

In s. 7 the Committee considers Factory Inspectors' Reports. Paragraph 160 seems to isolate the nub of a problem and bespeak the need for a vehicle for the provision of first-hand evidence relating to industrial accidents: "Accepting, as we do, that the Inspectorate is primarily and properly concerned with preventing the recurrence of accidents and therefore with seeking to control future events, and with prosecuting where they think fit for any breaches of the Act, we cannot think that it should feel no responsibility for assisting persons injured by accidents and their employers, so far as practicable, in the process of ascertaining responsibility for the injury, whether due to accident or unhealthy working conditions. In the present practice none of the information derived by Inspectors from visits paid by them for the purpose of ensuring safety or healthy conditions can be obtained by litigants, however direct their interest in it may be."

The reply of the Ministry of Labour, contained in a Further Memorandum to the Committee, is interesting and instructive. Part (B) of the Further Memorandum "concerned with the suggestion that in the interest of litigants in civil cases arising out of accidents at work H.M. Factory Inspectors should make more information available to the parties about the accidents that are investigated" states "At the present time Factory Inspectors are accepted by both sides in industry as independent and impartial persons whose work includes not only direct accident prevention but also seeking improvement in relations and better co-operation between employers and workers in safety, health and welfare matters. Their advice and help is largely accepted by both sides. If it were to be thought that their presence in a factory was in order to provide advice and information which would assist one party or another to further his case in litigation against the other it is felt that the Factory Inspector's present reputation and usefulness in this way would be greatly damaged" (para. 12).

With this statement from the Ministry, clearly correct, the Committee then went on to consider the importance of the Mines and Quarries Act 1954 and to cite evidence given to them by the National Coal Board to the effect that "in practice the site is inspected by representatives of the Board and the workmen at the earliest possible moment, since any delay could seriously interfere with production". Further on in para. 186 the Committee pointed out that the T.U.C. in their memorandum advocated that "there should be a [general] statutory requirement similar to the Mines and Quarries Act 1954 prohibiting interference with the site, plant, etc." and that the "T.U.C. is already pressing the Government for legislation which would give joint factory safety committees powers of inspection".

The appearance of the Safety Representatives' Regulations, it is thought, owes much, at least in spirit, to the above remarks of the Committee on Personal Injuries Litigation, whilst their functions related to accident prevention may well come to have considerable evidentiary value for the purposes of civil claims.

MAKING CLAIMS UNDER THE SOCIAL SECURITY SYSTEM

An injured workman who is "gainfully employed", i.e. an employed earner (the terminology of the Social Security Act 1975), equipped with a medical certificate, may submit a claim for industrial injury benefit or industrial disease benefit to the local insurance officer at the D.H.S.S. The medical certificate will say that the claimant is unfit for work. When making or submitting a claim of this nature, two separate questions are involved. Firstly, a medical question, was a particular injury caused by the accident and how should disability for the purposes of disablement pension be assessed? Second, an entitlement to benefit question, depending upon such factors as whether the injured person was an employed earner, had he got a satisfactory contribution record, etc. The medical question comes before a board of two doctors, but an appeal can be lodged to a medical appeal tribunal, consisting of two medical consultants and a lawyer. The medical appeal tribunal may confirm or reverse the decision of the medical board. The entitlement to benefit question is initially dealt with by an insurance officer at the Department of Employment, from whose decision an appeal lies to a local appeal tribunal. These local appeal tribunals consist of three members, one representing employers and employers' organisations, the others representing, respectively, the self-employed and the employed and their organisations. The tribunal has an independent chairman appointed by the Secretary of State. In both cases an appeal can be lodged beyond the medical appeal tribunal and the local entitlement appeal tribunal to the Chief National Insurance Commissioner and a number of Deputy Commissioners who must be barristers of at least 10 years' standing. The appeal may be heard by the Chief Commissioner on his own or by one of his deputies, or, where there is a point of considerable legal difficulty involved, by the Chief Commissioner and two of his deputies. Decisions of the Chief Commissioner are published by H.M.S.O. and so there is a corpus of national insurance law. Appeals of this kind are on a point of law only.

Conflict can arise between these two statutory bodies. It may well be that an insurance officer might say that there had been an industrial accident and that a medical board might feel that the injury or disability arose from a natural cause. Under the provisions of the

National Insurance Act 1972 this sort of decision on the part of the insurance officer does not bind the medical board, though prior to that Act the position had been otherwise. Appeals under the National Insurance (Industrial Injuries) Act 1965 and now the Social Security Act 1975 take less time than in the common law courts, the average time involved in an appeal reaching the Commissioner being about 12 months. No formal judgment is pronounced and the appellant is subsequently informed by letter. Legal aid is not available for representation at these tribunals. Moreover social security payments, being periodical, can be made the subject of a provisional assessment, with the result that, unlike tort, the amounts of periodical payments can subsequently be appraised in the light of changing circumstances, e.g. if an injured employee starts to make a recovery or if his condition grows worse. Lump sum compensation payable through the tort system is generally irreversible.

MAKING CLAIMS ON THE CRIMINAL INJURIES COMPENSATION BOARD

The criminal injuries compensation scheme is administered by the Criminal Injuries Compensation Board. This Board has the assistance of a permanent legal staff. Applications for compensation in respect of criminal injuries are handled by one member of the Board, but if this application is unsuccessful, a further application on appeal can be made to three members of the Board. Compensation is often in the form of a lump sum, though legal aid for representation before the Board is not available. If, however, a solicitor feels that there is a good case, he will probably represent the claimant and agree to be paid out of the award of compensation. If the claimant is a member of a trade union, it is likely that a union official will represent him.

OVERLAPPING SYSTEMS OF COMPENSATION

In the law of insurance there is a doctrine called *subrogation*. The effect of subrogation is to ensure that an injured party is not overcompensated or doubly indemnified at the expense of the tortfeasor and the insurer. This desirable result is achieved by the insurer taking over the claims, if any, of their insured third-party tortfeasors. As far as industrial injuries are concerned and actions against the employer in tort, insurers do not now normally exercise rights of subrogation against negligent employees (*see Lister* v. *Romford Ice & Cold Storage Co. Ltd*, § **21,** p. 69). A typical instance of the subrogation principle occurs with motor-vehicle accidents. In the event of a car proving to be a write-off, the insurer will indemnify the policyholder for the total loss of the car but will exercise any claim which the insured might have had against the third party. If, therefore, the facts show that the fault was substantially that of the third party, there is nothing to prevent the insurer from proceeding against the third party and claiming back in damages the part of the loss representing the third party's fault. Suppose that A and B are under separate obligations to indemnify C, C being the innocent victim of a road accident, A being the negligent motorist and B the insurer. A's liability is tortious whilst B's is contractual. Subrogation says that C cannot be doubly indemnified at the expense of A and B, the reason being that it is the purpose of indemnity to replace only the loss suffered and no more. Thus within the tort system an injured employee or road user can either sue in tort in the courts bringing his action against the actual tortfeasor or he can be compensated by the insurer. But he cannot do both.

Similarly within the social security system, since the criteria of sickness benefit and industrial injury benefit are identical, *viz.* incapacity for work, an employed earner cannot claim sickness benefit and industrial injury benefit, though he could claim sickness benefit and industrial disablement benefit since the criteria are not the same. Also when a claim is lodged for compensation from the C.I.C.B., any benefits received from social security or from tort must be brought into account in assessing the lump sum which the Board may award. This raises the question of how far subrogation operates within the parameters of the various compensation systems so as to avoid double indemnity. Or, put another way, if an employee is injured at work and after the first three days of incapacity is in receipt of industrial injury benefit, can he also at a later date be entitled to damages in tort? Indeed could he receive compensation as well from the C.I.C.B., assuming that his injury is referable to a criminal attack? Suppose further that he had bought a personal accident insurance policy (which is not likely in the case of an employee) or was entitled to sickness pay for time off work as a term of his contract of employment, or, if permanently off work, to an occupational pension, would he be entitled to receive this as well?

69. Extent of the overlap

Damages in tort and personal accident policies

Suppose that a man who had previously bought personal accident insurance to cover him for hazards arising out of and in the course of his employment was injured during his employment as the result of the negligence of a third party, e.g. an occupier. Would the injured person be able to claim compensation under his personal accident policy as well as suing the occupier under the Occupiers' Liability Act 1957? It seems that he would be able to do so. Double indemnity in respect of personal injuries goes back to the 1874 decision of *Bradburn* v. *G.W.R.* where a business man had bought personal accident insurance in case of being injured whilst travelling on one of the G.W.R. trains, the G.W.R. having encouraged passengers to buy such insurance. The plaintiff was injured whilst travelling on a train as the result of the negligence of the G.W.R. It was held that he was entitled to damages in tort *and* to insurance moneys payable under the personal accident policy. Giving judgment, Baron Pigott said that the insurer could not subrogate and recover the sum in tort damages, the reason being that the plaintiff had entered into a contract to buy insurance and that to allow the insurer the right of subrogation was tantamount to a serious breach of contract. The fact that the plaintiff had bought the insurance meant that he was entitled to receive it and liability in tort had nothing to do with the contract.

Damages in tort, sick pay and retirement pensions

Sick pay should not be confused with accident insurance. Sick pay represents a measurable loss and is an example of indemnity, *viz.* replacement of loss of earnings through incapacity to work as a result of sickness. Accident insurance, it was suggested in *Bradburn's* case, was not indemnity and was payable in respect of an immeasurable loss. This being so, Bradburn was able to recover both accident insurance moneys and tort damages. This will not be the position where sick pay has to be paid by the employer in accordance with a term of the contract of employment, in that special damages are designed to compensate the injured man for loss of earnings arising through incapacity for work, itself a measurable loss. Thus to allow an employee to claim both sick pay and special damages in tort in respect of the same incapacitating injury would be tantamount to allowing him to be in receipt of sickness benefit and industrial injury benefit, a situation which is not permitted. If, however, the injury turns out to be permanent and the sick pay period expires, since permanent injury is an immeasurable loss, the employee would be able to claim damages in tort against his employer for his injury and disablement coupled with any moneys accruing to him under his occupational pension. This was the conclusion which the House of Lords reached in 1970 in *Parry* v. *Cleaver*, on the assumption that an occupational pension as distinct from sick pay represented an immeasurable loss like industrial disablement pension.

Damages in tort and social security benefits

By virtue of s. 2(1), Law Reform (Personal Injuries) Act 1948, *half* the value of social security benefits is deductible from a subsequent award of damages in tort, unless the accident at work is fatal, in which case no benefits are deductible. Nor is a constant attendance allowance deductible. This has the consequence that widows of employees killed at work are well compensated by industrial death benefit, damages under the Fatal Accidents Act 1976 and the Law Reform (Miscellaneous Provisions) Act 1934 and by personal insurance in the form of life assurance moneys.

70. The effect of liability insurance on tort

Although the insurer keeps well out of the picture and although books on the law of tort would have us believe that the law of tort is self-supporting and does justice between plaintiff and defendant in civil actions for injury, this is a distorted picture. There are several reasons for this. Firstly, the universality of liability insurance has meant awards of greater damages and more out-of-court settlements in favour of injured persons than otherwise would have been the case. Second, certain statutory developments are inexplicable save in the light of liability insurance. This fact also accounts for a succession of case law extending the boundaries of reasonable foreseeability to such lengths as to be indistinguishable from strict liability.

The effect of insurance on statute

A series of important statutes, beginning with the Law Reform (Miscellaneous Provisions) Act 1934 and culminating in the Congenital Disabilities (Civil Liability) Act 1976, owe their emergence to the availability of third-party liability insurance. The Law Reform (Miscellaneous Provisions) Act 1934 came about because the law provided no remedy where a negligent motorist had died in a road accident in circumstances where he was insured under the provisions of the Road Traffic Act 1930, which made third-party liability insurance compulsory for motorists. The 1934 Act allowed such action to survive against the estate of the deceased motorist who had died through his own negligence, injuring the plaintiff at the same time. Similarly the Law Reform (Contributory Negligence) Act 1945 set up a *partial* defence to an action for negligence brought by an employee against an employer, the law taking the view that since employers were insuring in respect of injuries to their employees as a result of common law and statutory liability (although at that time such insurance was not compulsory), it was obtuse to deprive the employee found guilty of contributory negligence of all his compensation. Hence what had been a complete defence for the employer was turned into a partial defence, with the result that damages awarded to the injured employee would be apportioned to the extent that he was found to be contributorily negligent. The same is true of the Law Reform (Personal

Injuries) Act 1948. Consistently with the decisions of the House of Lords in *Smith* v. *Baker Ltd.* and *Wilsons & Clyde Coal Co.* v. *English* (imposing a personal liability upon the employer to counteract the doctrine of common employment not formally abolished until 1948), employers had been insuring in respect of personal liability to their employees for injuries suffered at work. To accommodate the reality that most employees are injured at work through the negligence of other employees or outside contractors and their workforce, the abolition of common employment by the Law Reform (Personal Injuries) Act 1948 and a formal recognition of vicarious liability was a forgone conclusion.

Another example of the effect of liability insurance upon industrial injury legislation occurred in the case of the various Fatal Accidents Acts. Officially liability for injury at work, whether fatal or non-fatal, depended on proof of negligence, which depended on whether the injury was reasonably foreseeable. Can it seriously be doubted that in a good many cases involving fatal injuries the injury itself might well have been foreseeable but the death following on, sometimes long after the injury, was often not foreseeable? A string of cases states that no action lies in negligence in respect of purely economic or pecuniary loss. Lord Denning was adamant about this in *Spartan Steel & Alloys Ltd.* v. *Martin (Contractors) Ltd.* Yet since the first Fatal Accidents Act in 1846 the courts have been awarding damages for purely economic or pecuniary loss. The same applies to more recent legislation. It is clear that the dominant force behind the Congenital Disabilities (Civil Liability) Act 1976 was insurance. Here an employer can be made liable for injuries which, by the most liberal stretch of the imagination, could not be regarded as foreseeable. Moreover the E.E.C. Draft Directive on Product Liability making producers strictly liable for defective products could hardly have been conceived in the absence of insurance.

The effect of insurance on tort

What better example of the effect of liability insurance on tort than the observations of Lord Denning in the 1971 case of *Nettleship* v. *Weston* when he intimated that what was important in tort was not whose fault it was but who should bear the risk. On this basis the learner driver in that case was held liable even though she could not have been expected to take the same care as a qualified driver. The existence of an employer's liability policy or a public liability policy explains decisions like *Robinson* v. *Post Office*, *Westwood* v. *Post Office*, *Chadwick* v. *British Railways Board* and *Videan* v. *British Transport Commission* (*see* § **19**, p. 44).

WHERE FAULT DETERMINES LIABILITY

Statutes relating to industrial injuries come in two categories, *viz.* fault and loss distribution. Some statutes adopt fault as the basis of liability whilst others base liability on loss distribution. An example of a statute adopting fault as the basis of an award of compensation is the Occupiers' Liability Act 1957. This statute, as we have seen, applies extensively to claims for injuries in the construction industry where fault as the basis of liability leaves much to be desired from the point of view of the accident victim. Under the Occupiers' Liability Act 1957 the contractor, building owner and to a lesser extent sub-contractor are regarded as occupiers. But what if the contractor is a fly-by-night doing sub-contract work for a large employer? It would have made sense to fix liability upon the large employer or building owner but instead, as far as lawful visitors, e.g. factory inspectors, policemen, firemen, etc., are concerned, the law has unwisely grafted liability on the contractor as occupier. Here statute has followed common law rather than reflecting social and commercial expediency.

Most statutes dealing with injuries and compensation have nailed liability to the mast of loss distribution, not fault. Obvious examples are the Employers' Liability (Compulsory Insurance) Act 1969, the Employers' Liability (Defective Equipment) Act 1969 and the Congenital Disabilities (Civil Liability) Act 1976.

71. Fault to loss distribution

Fault is alleged to be a moral concept implying that the person who is responsible for the injury will pay for it if he has been negligent. This may have the consequence that the injured person may not be compensated and raises another important moral question. In a legal system, which is higher in the order of priorities—the behaviour of the defendant (or fault) or the suffering of the victim? For many years, with one or two notable exceptions like *Rylands* v. *Fletcher*, English law took the view that the payment of compensation in the absence of fault could not be seriously entertained. Hence the unstinted growth of negligence. Now, however, the law seems to be coming round to the view that perhaps the justification of a legal system seeking to provide compensation for injury is the suffering of the victim. Unlike fault, loss distribution is a social mechanism. It aspires to place liability for injury and incapacity for work upon the shoulders of the person best able to bear it. Nevertheless it is a form of liability insurance. The employer or the local authority or the nationalised industry with its own self-insuring machinery will not pay up in the absence of liability, even though that liability is strict. We have not, therefore, reached the giddy heights of a legal system which purports to compensate for accident suffered rather than for liability. In that respect loss distribution is a halfway house between liability in tort and first-party compulsory insurance (for further development of this argument, *see* § **74**, p. 287).

72. An evaluation of fault

Some say that it is a question of time before fault is superseded by a system of accident compensation where the law recognises that the person in need of protection is not the wrongdoer but the injured party. The first steps in this direction have been taken by the establishment of the Royal Commission on Civil Liability and Compensation for Personal Injury. Hence it may not be mere speculation to suppose that England may have an accident compensation system for industrial injury and road traffic victims. Other developments might follow later—for instance, the extension of accident compensation legislation to serve the needs of those who are incapacitated and disabled through natural causes. This would cover situations comparable with the thalidomide children and young persons who have not started work who have become stricken with crippling diseases, e.g. poliomyelitis.

Eight main arguments are levelled against fault.

To concentrate liability on fault results in a multiplicity of claims

This is especially true of claims in respect of injuries arising on building sites where several parties are in occupation. Successive transfer of indemnities in this way is expensive, involving a plurality of lawsuits. Loss distribution with an affluent loss distributor somewhere in the chain would avoid all this circuity of litigation.

Fault as a deterrent

The idea that fault is a deterrent upon the wrongdoer hardly holds water, in that if a man did not foresee possible injury resulting from his conduct, for which he will be liable, it is hardly likely that the existence of the law of tort is going to make him realise. The fact is that if he did not foresee that his conduct might cause harm, he would not have any contemplation that he was in breach of the law.

Damages varying with tort

If tort is a deterrent why should not the amount of damages vary with the tort as do penalties for criminal offences? A split second's carelessness on the part of a motorist or employee may involve the motorist or the employee's employer in the payment of the same amount of damages as if the motorist had an appalling record of accidents or the employer a bad accident record. It follows that criminal law is a better form of moral censure where fines can be adjusted to meet the gravity of the offence.

Fault ignores the universality of insurance
Although actions are brought against employers personally and not the insurance company, this is no more than a façade. We all know that it is the insurer who will pay.

Fault does not take cognisance of the financial position of the defendant
One second's carelessness at the wheel of a car or lorry or a split second's inattentiveness on the part of an employee racked with the monotony of a job and both could be saddled with a debt for the rest of their lives and their children's lives as well. Is it right that such thoughtlessness, which might not even amount to guilty intent in criminal law, should deserve such a penalty?

Cost of proving fault
To prove fault can be a costly affair if the facts do not suggest the application of *res ipsa loquitur*. In *Wright* v. *Dunlop Ltd.*, for example, the court spent the best part of six weeks studying evidence in an effort to decide whether Nonox S was likely to be carcinogenic in 1945.

Contracting out of liability
If fault is consistent with tort being deterrent, why was it until the Unfair Contract Terms Act 1977 was passed always possible to contract out of the liability in negligence? This was particularly true in relation to product defects.

Fault underestimates the conduct of the plaintiff
The fact is that if no-one can be found who is at fault or if a loss distributor cannot be located, irrespective of the conduct of the plaintiff, he will not get compensated. Thus it matters little whether the plaintiff was acting chivalrously going to the rescue of another or whether he was crossing the road in a drunken stupor when he was hit by a car.

73. The future of insurance
As long as losses are distributed within a legal system on the basis of fault, insurance will protect the person at fault. This means that it is the negligent motorist or the negligent employer whom the legal system at present protects. Is this right? Should not the system protect the injured person, whether employee, road-traffic victim or the naturally disabled? Addressing himself trenchantly to this problem the American critic of fault Fleming James says "Except for Workmen's

Compensation and an occasional other forward looking statute, the main assault on the citadel of fault has not been a frontal attack but rather a boring from within. The main trends have been concerned with the enlargement of duties, the liberalisation of standards of proof; the increasing concern of society to provide a financially responsible defendant and the erosion of contributory negligence, imputation of negligence to the plaintiff, and other doctrines which impede recovery." This means that fault is a barrier denying access to the pool of insurance funds for those whose injuries are not traceable back to someone's fault. It is to that extent that liability insurance and accident compensation are mutually exclusive.

74. Some American opinions

It is instructive to examine one or two American opinions in favour of no fault liability if only because the American Insurance Association proposed the scrapping of tort liability regarding auto insurance with cover available to members on a first-party basis. In his book *Settled out of Court* Laurence Ross, himself an advocate of accident compensation, cites Keeton and O'Connel in their treatise *Basic Protection for the Traffic Victim: A Blueprint for Reforming Automobile Insurance.* In particular Keeton and O'Connell make the following criticism of the American fault system: "... the operation of the present system is excessively expensive. ... Because of the role of fault in the present system, contests over the intricate details of accidents are routine. Often these contests are also exercises in futility, since all drivers must continually make split-second judgments and many accidents are caused by slight but understandable lapses occurring at unfortunate moments...."

Later in his book Ross observes "In sum, insurance has in my opinion pretty much drawn the sting from tort law sanctions on the insured defendant, and the law in action has enormously moderated the effect of these sanctions on the claimant. The negligence system without insurance might conceivably be effective as a deterrent, but it would have to be judged grossly unfair because of a lack of proportion between the degree of fault and the degree of punishment. With insurance, I believe it is unreasonable to expect significant deterrent effects from the negligence system" (by the "law in action" Ross means the settlement out of court). In reply, the legal and insurance communities commented (cited by Ross as examples of utter nonsense): "No doubt the fundamental argument for the tort system is that the abolition of court-jury litigation in automobile accident cases would put an end to the moral and legal responsibility of individuals who inflict bodily injuries on their fellows. The abolition of responsibility is contrary to our way of life, to morality, and especially does it come with poor grace when in our time of widespread disorder we preach more individual responsibility. To embrace liability without fault would destroy the dignity of man."

Ross points out that in 1967 the A.I.A. set up a special committee to study the growing discontent involved in automobile insurance. They found that nearly all public complaints about auto insurance were centred on liability coverage, which is the heart of the expense and the delay. The outcome of this investigation was a new plan, called Complete Personal Protection. The A.I.A. plan advocates complete abolition of the tort system in automobile injury cases. Instead it proposed that compulsory insurance be bought instead of liability insurance on a first-party basis or indirectly as additional insureds or through an assigned claims plan. The insurance would provide periodic payments of medical and rehabilitation expenses, and of economic loss including wage loss plus other reasonable out-of-pocket expenses. No payments would be made for pain and suffering, but permanent disability or disfigurement would be compensated for amounts ranging up to 50 per cent of the medical expenses. In the event of death the funeral and burial expenses and survivors' benefits would be paid. There would be no limits to the degree of coverage. Disputed claims would go to a jury. Additional costs in compensating all innocent victims would be offset by two types of savings. Firstly, payments for pain and suffering would be controlled or eliminated; second, liability matters would not require investigation.

The viability of first-party insurance for accident victims

The plan proposed by A.I.A. was put into operation in New York on 1st February 1974. Could a similar experiment be attempted in the U.K.? Could the law opt for the elimination of tort liability in favour of a system of accident compensation which was first party? Could not English law allow employers and motor-vehicle users, not to mention occupiers of premises, to buy insurance directly for the benefit of other persons who might be injured in consequence of their activities? In this way insurance would directly benefit the accident victim rather than indemnify the tortfeasor.

Privity of contract

Contract law says that a contractual benefit can only extend to the parties who made the contract. This is known as privity of contract and it prevents an employer from buying insurance for the benefit of his employees. The decisions in *Re Schebsman* and *Green* v. *Russell* show that it is impossible to do this because the contract was made by an employer and an insurer for the benefit of a third party who was a stranger to the consideration of a contract. One might have thought, however, that in our car-orientated, machine-orientated society insurance might have devised machinery for eliminating and obviating the tiresome technicality of privity. How such insurance would be funded, whether from a central government source resembling our present social security system or by private insurance, is a political rather than legal matter.

75. The New Zealand approach: an Accident Compensation Act

New Zealand passed its Accident Compensation Act in 1972 and it was amended shortly afterwards in 1973. This Act ushered in a comprehensive system of social insurance. By adopting the view that accidents and disability, whether caused at work or arising from natural causes, are the responsibility of society, it has gone further than North America in affording financial cover for road and work accident victims. This Act has set up a system of compensation which is payable in case of accidents arising at work, on the road or through criminal attack. Sickness and congenital disability are still covered by the less remunerative social security system. New Zealand therefore has an integrated accident compensation system which compares favourably with our own fragmented accident compensation system. Self-inflicted injuries and suicides are not covered by the scheme set up in 1972. In this way practically all personal injuries caused by accidents are covered by national insurance.

AN ACCIDENT COMPENSATION ACT FOR BRITAIN?

The question which follows is—is Britain likely to pass an Accident Compensation Act, the probable effect of which would be to integrate the present media of accident compensation, so that compensation for industrial injuries, road accident injuries, sickness and congenital disability, would be dealt with by one centralised system? Here we must look to the recommendations of the Pearson Commission.

76. The purpose of the Royal Commission

The Royal Commission on Civil Liability and Compensation for Personal Injury came about as a result of certain statements made in the House of Commons by the Prime Minister on 19th December 1972 when he said "The Government have been considering proposals made from time to time in the past, which are now particularly relevant in the light of the Report of the Robens Committee on Safety and Health at Work and in connection with the recent concern over the thalidomide cases, that there should be an inquiry into the basis of civil liability in the U.K. for causing death or personal injury. It is the Government's view that a wide-ranging inquiry is required into the basis on which compensation should be recoverable."

Following on from this the brief of the Royal Commission was expressed to be: "To consider to what extent, in what circumstances and by what means compensation should be payable in respect of death or personal injury (including ante-natal injury) suffered by any person—

(*a*) in the course of employment;

(*b*) through the use of a motor vehicle or other means of transport (but only road-transport injuries);

(*c*) through the manufacture, supply or use of goods or services (product liability);

(*d*) on premises belonging to or occupied by another (occupiers' liability); or

(*e*) otherwise through the act or omission of another where compensation under the present law is recoverable only on proof of fault or under the rules of strict liability;

having regard to the cost and other implications of the arrangements for the recovery of compensation, whether by way of compulsory insurance or otherwise." In para. 7 the Commission's First Circular made it clear that one of the most important matters was whether there should be a system of compensation not based on liability.

77. Factors leading to the reform of compensation law

Later the First Circular posed the question whether, if a system of liability without fault was ushered or partly ushered in, it should be in substitution for or in addition to the existing system of tort liability. Or should tort liability be scrapped or retained so as to operate in conjunction with the new system? Assuming that it was retained, should benefits received or receivable under the new system be taken into account either totally or partially when damages in tort were being assessed? Further, if a system of no fault liability were rejected in favour of the retention of the present fault system, perhaps it might be advisable to alter the burden of proof in civil cases so that the employer would have to show that the injury was not caused by him or his employees. This would be tantamount to the introduction of the *res ipsa loquitur* doctrine.

78. Possible changes in the law affecting compensation

In the light of a system of compensation for injuries at work which is already partially automatic under the Social Security Act 1975, the system could become totally automatic with benefits being paid as a right on a periodic basis. Owing to the long association of the state with industrial injuries this would not amount to a complete break with the past. In certain other areas a change would be more revolutionary. Thus the introduction of a system of product liability analogous with American product law, which seems likely, would mean that producers and suppliers would become strictly liable for defective products, whether causing injury or not. This would mean more insurance for manufacturers, the cost of which would be absorbed in the product. The conclusions of the Pearson Commission may well have been accelerated to this extent by the E.E.C. Draft Directive on Product Liability coupled with the European Convention on Product Liability in regard to Personal Injury and Death. Moreover the law relating to occupiers might have changed. The Commission considered the introduction of a system of strict liability whereby the occupier would have to prove that the injury to an entrant on premises was not caused by the occupier's fault or that of others for whom he might be responsible, but rejected the idea in favour of a retention of the present law, with an extension of liability in favour of the "uninvited entrant".

Overlapping systems

Persons injured at work may be doubly compensated by state benefits and tort damages. The Pearson Commission has recommended that the no fault social security system for injuries at work be retained with higher benefits for the first six months of incapacity followed by a long-term invalidity pension calculated on the same basis as the new state pensions, something approximating to earnings-related benefit for the duration of incapacity for work. Moreover the tort system of liability is to be retained with no recommendation for a reversal in the burden of proof in cases of negligence. The Royal Commission also recommends that the present industrial disablement benefit be retained.

Options open to the Commission

The following were the options open to the Royal Commission:

(*a*) retention of the present system of tort liability without any major modification;

(*b*) introduction of a system of strict liability, the consequence of which would be that the burden of proof in civil actions would always be on the defendant not the plaintiff;

(*c*) introduction of a system of no fault insurance, i.e. no fault liability underwritten by compulsory insurance. This would be operated by insurance companies and underwriters and supplemented by tort liability with modifications;

(*d*) introduction of a system of no fault insurance operated by insurance companies, with tort liability for the categories of injuries so underwritten being abolished;

(*e*) introduction of a system of no fault insurance operated by a governmental agency and supplemented by tort liability with modifications;

(*f*) introduction of a system of no fault insurance operated by a governmental agency, with liability in tort in respect of injuries so underwritten being abolished.

79. Main recommendations of the Royal Commission

The main recommendations of the Royal Commission were as follows:

Firstly, that tort liability should be retained, the liability continuing to be based on negligence except where there are special reasons for introducing strict liability. There are certain areas where the introduction of strict liability is justified, as in the case of defective products and vaccine damage. The notion that compensation is designed to make full reparation of loss should be retained, but tort damages should be reduced by taking into account the full amount of social security benefits, thus avoiding double compensation. A system of periodic payments should be introduced for pecuniary loss caused by death or by serious and lasting injury, although the parties should still be at liberty to settle for an agreed lump sum.

Second, that as regards work injuries, the industrial injuries scheme administered by the D.H.S.S. should provide the basis for improved provision for those injured at work. Higher benefits should be paid for the first six months of incapacity or widowhood, followed by pensions calculated in the same way as in the new state pensions scheme, but in all cases at the improved levels that would accrue if contributions to the scheme had been paid for 20 years. The scheme should be extended to cover the self-employed and commuting accidents. Moreover the conditions for compensation for occupational diseases should be less restrictive, and the increased compensation costs should be met by the employer. With respect to industrial and occupational diseases, it might be enough, as the trade unions suggested, to compensate a claimant for an occupational disease even though it was not prescribed but merely occupational or, alternatively, on the basis of individual proof that a particular occupation caused his disease.

Third, as regards product liability, there is no justification for introducing a no fault scheme for injuries caused by defective products. Strict liability would meet most of the objections to the present law. And a scheme based broadly on the Strasbourg Convention and the E.E.C. Draft Directive should be introduced.

Fourth, as regards the liability of occupiers, no change in the present law is recommended except that the Law Commission's proposals relating to uninvited entrants should be implemented to supplement the Occupiers' Liability Act. This is to the effect that an occupier of premises owes a duty to an uninvited entrant upon the premises in respect of a danger if, but only if, the danger is one against which, in all the circumstances of the case, the occupier can reasonably be expected to offer him some protection. The duty owed by an occupier in accordance with the above sub-section is to take such care as is reasonable to see that the entrant does not suffer personal injury or death by reason of the danger. On this basis the occupier would not necessarily owe a particular trespasser a duty at all.

Fifth, a state scheme for criminal injuries is desirable in principle.

Sixth, as regards ante-natal injury, the tort action under the Congenital Disabilities (Civil Liability) Act 1976 should be retained, but its operation should be retricted because there are grave objections to the tort action within the family in the sphere of ante-natal injury.

Appendix I

New Cases and Legislation

NEW CASES

The sequel to the case involving the Procurator Fiscal and Strathclyde Regional Council's Director of Roads (*see* §**30**, p. 198) appeared in 1977 as *Armour* v. *Skeen* (Procurator Fiscal, Glasgow) before the High Court of Justiciary. This case concerned s. 37, H.A.S.A.W.A., and a conviction under that section by the lower court. In the lower court the success of the prosecution by the Procurator Fiscal hinged on the expression "neglect" in subs. 37(1), the court finding that the neglect on the appellant's part related to his failure to fulfil the responsibilities for health and safety placed on those in his position by the Regional Council. This included having a sound safety policy for his department, informing his employees of the implications and requirements of H.A.S.A.W.A., and training and instructing employees in safe working practice. The appellant appealed against the lower court's decision on the ground that he personally was not under a statutory duty to provide a safe system of work and therefore the accident could not be attributable to any neglect of such duty. Moreover he argued that he could not be prosecuted personally as he did not fall within the ambit of subs. 37(1). It was held, dismissing the appeal, that the Director of Roads was personally liable for neglect within the meaning of s. 37(1), H.A.S.A.W.A. The court felt that s. 2(1) imposed on employers the duty to ensure the safety at work of employees and that this did not mean that members of the Council as policymakers were alone responsible for the safety policy, and that there was no duty on the appellant to carry out the policy. The neglect on the part of a person covered by s. 37(1) need not be a neglect only in relation to a duty which the legislation has placed on that person. Section 37(1) refers to *any* neglect in duty which means any neglect in duty, however constituted, to which the contravention of the safety provision was attributable.

NEW REGULATIONS

Offshore oil and gas industries

As from 1st September 1977, as a result of the Health and Safety at Work Act (Application outside Great Britain) Order 1977, the provisions of the Health and Safety at Work Act now cover the workforce in the offshore oil and gas industries. The instrument is issued under subs. 84(3), H.A.S.A.W.A. Hitherto health and safety aspects of the offshore industries had been covered by the Mineral Workings (Offshore Installations) Act 1971 and the regulations made under it. These regulations make no provision for safety representatives and safety committees. The position is, therefore, that although existing legislation remains in force there is a uniform set of health and safety standards offshore (within territorial waters) as well as onshore. H.A.S.A.W.A. therefore applies, as from 1st September, to offshore installations, pipeline work, offshore construction work, loading and unloading ships, shipbuilding and repairing and diving operations.

Increase of penalties for breaches of H.A.S.A.W.A.: the Criminal Law Act 1977

On 29th July the Criminal Law Act 1977 went on to the statute book, bringing with it an updating of certain parts of the criminal law machine. In particular, by virtue of the Criminal Law Act 1977 Commencement No. 1 Order, which came into effect on 8th September, fines for breach of certain sections of H.A.S.A.W.A. have been increased. The offences in respect of which the penalties have been increased are enumerated in Sched. 6, H.A.S.A.W.A., and are as follows:

(*a*) contravention of a requirement imposed under s. 14 or obstructing any person in the exercise of his powers under s. 14 (power of the Commission to direct inquires and investigations):—£400 to £1,000;

(*b*) contravening a requirement imposed by a factory inspector:—£400 to £1,000;

(*c*) preventing anyone from appearing before an inspector:—£400 to £1,000;

(*d*) intentionally obstructing an inspector:—£400 to £1,000;

(*e*) falsely pretending to be an inspector:—£400 to £1,000.

Increase of penalties for breaches of the Employers' Liability (Compulsory Insurance) Act 1969: the Criminal Law Act 1977

The reform of the law is found in Sched. 6, Criminal Law Act 1977. The Employers' Liability (Compulsory Insurance) Act 1969 created two offences, the first of failing to buy compulsory insurance on the part of an employer (with exceptions, *see* p. 112) in respect of his direct and vicarious liability towards his employees injured at work. The penalty for failure to do this was £200. This has now been increased to £500. In addition it was an offence not to display a certificate of insurance in respect of these matters. This was punishable by a maximum fine of £50. This has been increased to £200.

Appendix II

List of Prescribed Industrial Diseases and Injuries and Occupations for which they are prescribed

THE SOCIAL SECURITY (INDUSTRIAL INJURIES) (PRESCRIBED DISEASES) REGULATIONS 1975 NO. 1537

SCHEDULE 1

Regulations 2 and 4

PART I

Description of disease or injury	Nature of occupation
Poisoning by:	Any occupation involving:
1. Lead or a compound of lead.	The use or handling of, or exposure to the fumes, dust or vapour of, lead or a compound of lead, or a substance containing lead.
2. Manganese or a compound of manganese.	The use or handling of, or exposure to the fumes, dust or vapour of, manganese or a compound of manganese, or a substance containing manganese.
3. Phosphorous or phosphine or poisoning due to the anti-cholinesterase action of organic phosphorus compounds.	The use or handling of, or exposure to the fumes, dust or vapour of, phosphorus or a compound of phosphorus, or a substance containing phosphorus.
4. Arsenic or a compound of arsenic.	The use or handling of, or exposure to the fumes, dust or vapour of, arsenic or a compound of arsenic, or a substance containing arsenic.
5. Mercury or a compound of mercury.	The use or handling of, or exposure to the fumes, dust or vapour of, mercury or a compound of mercury, or a substance containing mercury.
6. Carbon bisulphide.	The use or handling of, or exposure to the fumes or vapour of, carbon bisulphide or a compound of carbon bisulphide, or a substance containing carbon bisulphide.
7. Benzene or a homologue.	The use or handling of, or exposure to the fumes of, or vapour containing, benzene or any of its homologues.

Description of disease or injury	Nature of occupation
8 A nitro- or amino- or chloro-derivative of benzene or of a homologue of benzene, or poisoning by nitrochlorbenzene.	The use or handling of, or exposure to the fumes of, or vapour containing, a nitro- or amino- or chloro-derivative of benzene or of a homologue of benzene or nitrochlorbenzene.
9. Dinitrophenol or a homologue or by substituted dinitrophenols or by the salts of such substances.	The use or handling of, or exposure to the fumes of, or vapour containing, dinitrophenol or a homologue or substituted dinitrophenols or the salts of such substances.
10. Tetrachloroethane.	The use or handling of, or exposure to the fumes of, or vapour containing, tetrachloroethane.
11. Tri-cresyl phosphate.	The use or handling of, or exposure to the fumes of, or vapour containing, tricresyl phosphate.
12. Tri-phenyl phosphate.	The use or handling of, or exposure to the fumes of, or vapour containing, triphenyl phosphate.
13. Diethylene dioxide (dioxan).	The use or handling of, or exposure to the fumes of, or vapour containing, diethylene dioxide (dioxan).
14. Methyl bromide.	The use or handling of, or exposure to the fumes of, or vapour containing, methyl bromide.
15. Chlorinated naphthalene.	The use or handling of, or exposure to the fumes of, or dust or vapour containing, chlorinated naphthalene.
16. Nickel carbonyl.	Exposure to nickel carbonyl gas.
17. Nitrous fumes.	The use or handling of nitric acid or exposure to nitrous fumes.
18. Gonioma kamassi (African boxwood).	The manipulation of gonioma kamassi or any process in or incidental to the manufacture of articles therefrom.
19. Anthrax.	The handling of wool, hair, bristles, hides or skins or other animal products or residues, or contact with animals infected with anthrax.
20. Glanders.	Contact with equine animals or their carcases.
21. (*a*) Infection by leptospira icterohaemorrhagiae.	Work in places which are, or are liable to be, infested by rats.
(*b*) Infection by leptospira canicola.	Work at dog kennels or the care or handling of dogs.

Description of disease or injury	Nature of occupation
22. Ankylostomiasis.	Work in or about a mine.
23. (*a*) Dystrophy of the cornea (including ulceration of the corneal surface) of the eye, (*b*) Localised new growth of the skin, papillomatous or keratotic, (*c*) Squamous-celled carcinoma of the skin, due in any case to arsenic, tar, pitch, bitumen, mineral oil (including paraffin), soot or any compound, product (including quinone or hydroquinone), or residue of any of these substances.	The use or handling of, or exposure to, arsenic, tar, pitch, bitumen, mineral oil (including paraffin), soot or any compound, product (including quinone or hydroquinone), or residue of any of these substances.
24.	
25. Inflammation, ulceration or malignant disease of the skin or subcutaneous tissues or of the bones, or blood dyscrasia, or cataract, due to electro-magnetic radiations (other than radiant heat), or to ionising particles.	Exposure to electro-magnetic radiations other than radiant heat, or to ionising particles.
26. Heat cataract.	Frequent or prolonged exposure to rays from molten or red-hot material.
27. Decompression sickness.	Subjection to compressed or rarefied air.
28. Cramp of the hand or forearm due to repetitive movements.	Prolonged periods of handwriting, typing or other repetitive movements of the fingers, hand or arm.
29.	
30.	
31. Subcutaneous cellulitis of the hand (Beat hand).	Manual labour causing severe or prolonged friction or pressure on the hand.
32. Bursitis or subcutaneous cellulitis arising at or about the knee due to severe or prolonged external friction or pressure at or about the knee (Beat knee).	Manual labour causing severe or prolonged external friction or pressure at or about the knee.
33. Bursitis or subcutaneous cellulitis arising at or about the elbow due to severe or prolonged external friction or pressure at or about the elbow (Beat elbow).	Manual labour causing severe or prolonged external friction or pressure at or about the elbow.
34. Traumatic inflammation of the tendons of the hand or forearm, or of the associated tendon sheaths.	Manual labour, or frequent or repeated movements of the hand or wrist.

Description of disease or injury	Nature of occupation
35. Miner's nystagmus.	Work in or about a mine.
36. Poisoning by beryllium or a compound of beryllium.	The use or handling of, or exposure to the fumes, dust or vapour of, beryllium or a compound of beryllium, or a substance containing beryllium.
37. (*a*) Carcinoma of the mucous membrane of the nose or associated air sinuses. (*b*) Primary carcinoma of a bronchus or of a lung.	Work in a factory where nickel is produced by decomposition of a gaseous nickel compound which necessitates working in or about a building or buildings where that process or any other industrial process ancillary or incidental thereto is carried on.
38. Tuberculosis.	Close and frequent contact with a source or sources of tuberculous infection by reason of employment— (*a*) in the medical treatment or nursing of a person or persons suffering from tuberculosis, or in a service ancillary to such treatment or nursing; (*b*) in attendance upon a person or persons suffering from tuberculosis, where the need for such attendance arises by reason of physical or mental infirmity; (*c*) as a research worker engaged in research in connection with tuberculosis; (*d*) as a laboratory worker, pathologist or person taking part in or assisting at post-mortem examinations of human remains where the occupation involves working with material which is a source of tuberculous infection.
39. Primary neoplasm of the epithelial lining of the urinary bladder (Papilloma of the bladder) or of the renal pelvis or of the ureter or of the urethra.	(*a*) Work in a building in which any of the following substances is produced for commercial purposes:— (i) alpha-naphthylamine or beta-naphthylamine; (ii) diphenyl substituted by at least one nitro or primary amino group or by at least one nitro and primary amino group; (iii) any of the substances mentioned in sub-paragraph (ii) above if further ring substituted by halogeno, methyl or methoxy groups, but not by other groups; (iv) the salts of any of the substances mentioned in sub-paragraphs (i) to (iii) above; (v) auramine or magenta;

Description of disease or injury	Nature of occupation
	(*b*) the use or handling of any of the substances mentioned in sub-paragraphs (i) to (iv) of paragraph (*a*), or work in a process in which any such substance is used or handled or is liberated; (*c*) the maintenance or cleaning of any plant or machinery used in any such process as is mentioned in paragraph (*b*), or the cleaning of clothing used in any such building as is mentioned in paragraph (*a*) if such clothing is cleaned within the works of which the building forms a part or in a laundry maintained and used solely in connection with such works.
40. Poisoning by cadmium.	Exposure to cadmium fumes.
41. Inflammation or ulceration of the mucous membrane of the upper respiratory passages or mouth produced by dust, liquid or vapour.	Exposure to dust, liquid or vapour.
42. Non-infective dermatitis of external origin (including chrome ulceration of the skin but excluding dermatitis due to ionising particles or electro-magnetic radiations other than radiant heat).	Exposure to dust, liquid or vapour or any other external agent capable of irritating the skin (including friction or heat but excluding ionising particles or electro-magnetic radiations other than radiant heat).
43. Pulmonary disease due to the inhalation of the dust of mouldy hay or of other mouldy vegetable produce, and characterised by symptoms and signs attributable to a reaction in the peripheral part of the broncho-pulmonary system, and giving rise to a defect in gas exchange (Farmer's lung).	Exposure to the dust of mouldy hay or other mouldy vegetable produce by reason of employment:— (*a*) in agriculture, horticulture or forestry; or (*b*) loading or unloading or handling in storage such hay or other vegetable produce; or (*c*) handling bagasse.
44. Primary malignant neoplasm of the mesothelium (diffuse mesothelioma) of the pleura or of the peritoneum.	(*a*) The working or handling of asbestos or any admixture of asbestos; (*b*) the manufacture or repair of asbestos textiles or other articles containing or composed of asbestos; (*c*) the cleaning of any machinery or plant used in any of the foregoing operations and of any chambers, fixtures and appliances for the collection of asbestos dust; (*d*) substantial exposure to the dust arising from any of the foregoing operations.
45. Adeno-carcinoma of the nasal cavity or associated air sinuses.	Attendance for work in or about a building where wooden furniture is manufactured.

Description of disease or injury	Nature of occupation
46. Infection by brucella abortus.	Contact with bovine animals infected by brucella abortus, their carcases or parts thereof or their untreated products, or with laboratory specimens or vaccines of or containing brucella abortus, by reason of employment— (*a*) as a farm worker; (*b*) as a veterinary worker; (*c*) as a slaughterhouse worker; (*d*) as a laboratory worker; or (*e*) in any other work relating to the care, treatment, examination or handling of such animals, carcases or parts thereof, or products.
47. Poisoning by acrylamide monomer.	The use or handling of, or exposure to, acrylamide monomer.
48. Substantial permanent sensori-neural hearing loss due to occupational noise amounting to at least 50dB in the better ear, being the average, after exclusion of hearing losses not due to occupational noise, of pure tone losses measured by audiometry over the 1, 2 and 3 kHz frequencies (occupational deafness).	(*a*) The use of pneumatic percussive tools or high-speed grinding tools in the cleaning, dressing or finishing of cast metal or of ingots, billets or blooms, or (*b*) the use of pneumatic percussive tools on metal in the shipbuilding or ship repairing industries, or (*c*) work wholly or mainly in the immediate vicinity of drop-forging plant or forging press plant engaged in the shaping of hot metal.

THE SOCIAL SECURITY (INDUSTRIAL INJURIES) (PRESCRIBED DISEASES) AMENDMENT REGULATIONS 1975 NO. 2241

49. Viral hepatitis	Any occupation involving: (*a*) Close and frequent contact with human blood or human blood products; or (*b*) close and frequent contact with a source of viral hepatitis infection by reason of employment in the medical treatment or nursing of a person or persons suffering from viral hepatitis, or in a service ancillary to such treatment or nursing.

THE SOCIAL SECURITY (INDUSTRIAL INJURIES) (PRESCRIBED DISEASES) AMENDMENT REGULATIONS 1977 NO. 250

50. (*a*) Angiosarcoma of the liver; and (*b*) Osteolysis of the terminal phalanges of the fingers	Any occupation involving—work in or about machinery or apparatus used for the polymerization of vinyl chloride monomer, a process which, for the purposes of this provision, comprises all operations up to and including the drying of the slurry produced by the polymerization and the packaging of the dried product; or work in a building or structure in which any part of the aforementioned process takes place.

Appendix III

The Structure of the Health and Safety Commission

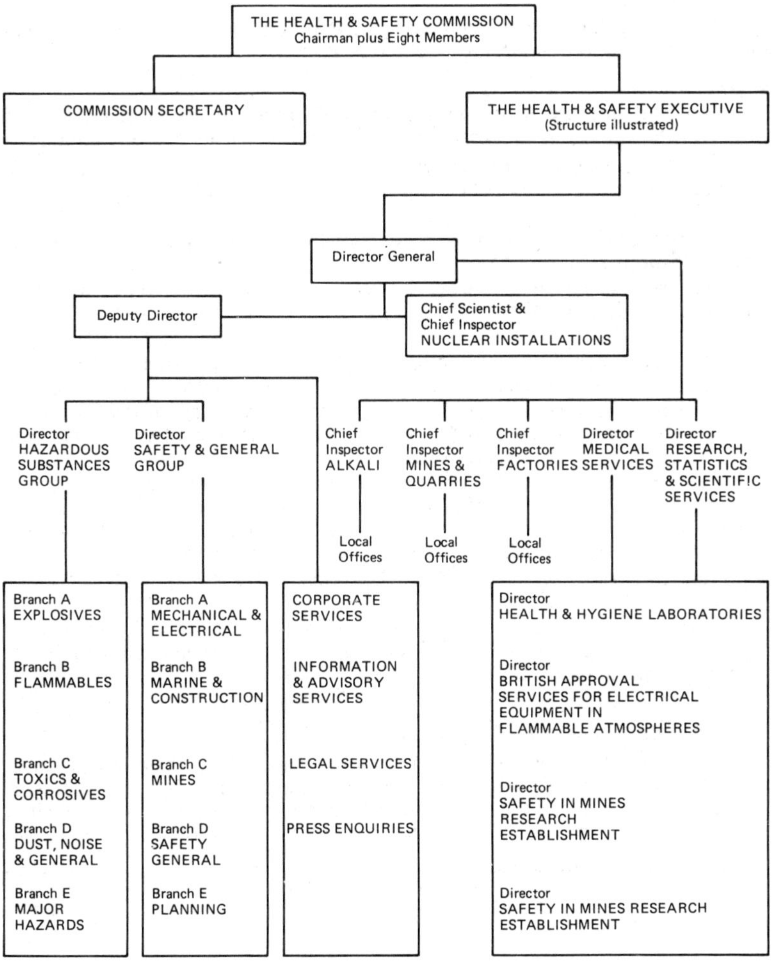

Appendix IV

The Power Presses Regulations 1965 No. 1441

Citation and commencement

1. These Regulations may be cited as the Power Presses Regulations 1965 and shall come into operation on 20th July 1966.

Interpretation

2.—(1) The Interpretation Act 1889**(a)** shall apply to the interpretation of these Regulations as it applies to the interpretation of an Act of Parliament.

(2) In these Regulations, unless the context otherwise requires, the following expressions have the meanings hereby assigned to them respectively, that is to say—

" approved " means approved for the time being by certificate of the Chief Inspector;

" clutch mechanism " means in relation to a power press a device designed to impart when required the movement of the flywheel to any tool;

" factory " includes any premises to which these Regulations apply;

" fixed fencing " means fencing provided for the tools of a power press being fencing which has no moving parts associated with or dependent upon the mechanism of a power press and includes that part of a closed tool which acts as a guard;

" power press " means a press or a press brake which in either case is used wholly or partly for the working of metal by means of tools or for the purpose of die proving, being a press or a press brake which is power driven and which embodies a flywheel and a clutch mechanism;

" the principal Act " means the Factories Act 1961 as amended by or under any other Act; and

" safety device " means the fencing and any other safeguard provided for the tools of a power press in pursuance of requirements imposed by or under the principal Act.

Application and operation of Regulations

3.—(1) Subject to paragraph (2) of this Regulation, these Regulations apply to the following premises in which any power press is used, that is to say, all factories and all premises to which the provisions of the principal Act are applied by section 123(1) of that Act (which relates to electrical stations).

(a) 52 & 53 Vict. c. 63.

(2) Nothing in these Regulations shall apply as respects any power press when used for working hot metal or which is not being used for working cold metal and is intended next to be used for working hot metal.

(3) The provisions of these Regulations are in addition to and not in substitution for or in diminution of other requirements imposed by or under the principal Act.

Appointment of persons to prepare power presses for use

4.—(1) Except as provided in paragraph (4) of this Regulation, no persons shall set, re-set, adjust or try out the tools on a power press or install or adjust any safety device thereon, being installation or adjustment preparatory to production or die proving, or carry out an inspection and test of any safety device thereon required by Regulation 7, unless he—

(*a*) has attained the age of eighteen;

(*b*) has been trained in accordance with the Schedule to these Regulations;

(*c*) is competent to carry out those duties; and

(*d*) has been appointed by the occupier of the factory to carry out those duties in respect of the class or description of power press or the class or description of safety device to which the power press or the safety device (as the case may be) belongs; and every such appointment shall be made by signed and dated entry in, or by signed and dated certificate attached to, a register kept for the purposes of this Regulation.

(2) Any appointment made under paragraph (1)(*d*) of this Regulation may be made in respect of all the duties specified in paragraph (1) of this Regulation or may be limited to certain of those duties, and where the said appointment is so limited, particulars of the limitation shall be set out in the entry or certificate (as the case may be) of the appointment. Every person appointed under the said paragraph (1)(*d*) shall be furnished by the occupier with a copy of the entry or certificate of his appointment and of any entry revoking the same.

(3) Any appointment made under paragraph (1)(*d*) of this Regulation may be revoked by the occupier at any time by signed and dated entry in the said register.

(4) Paragraph (1) of this Regulation shall not apply to and shall not prevent the employment on setting, re-setting, adjusting or trying out of the tools on a power press or on installing or adjusting as aforesaid a safety device thereon, or on the carrying out of an inspection and test of any safety device thereon required by Regulation 7, of a person who (whether or not he has attained the age of eighteen) is undergoing training (whether or not in accordance with the Schedule to these Regulations) for those duties or any of them and who is working under the immediate supervision of a competent person appointed under the said paragraph (1) in respect of the duties on which the first-mentioned person is employed, provided that an entry has been made in, or a certificate has been attached to, the register kept for the purposes of this Regulation, specifying that the said first-mentioned person is undergoing such training. Every such entry or certificate shall be signed by the said competent person and shall specify the date on which the entry was made or the certificate was given (as the case may be).

Examination and testing of power presses and safety devices

5.—(1) No power press or safety device shall be taken into use in any factory for the first time in that factory, or in the case of a safety device for the first time on any power press, unless it has been thoroughly examined and tested, in the case of a power press, after installation in the factory, or in the case of a safety device, when in position on the power press in connection with which it is to be used, by a competent person.

(2) No power press shall be used unless it has been thoroughly examined and tested by a competent person—

(*a*) in the case of a power press on which the tools are fenced exclusively by means of fixed fencing within the immediately preceding period of twelve months; or

(*b*) in any other case, within the immediately preceding period of six months.

(3) No power press shall be used unless every safety device (other than fixed fencing) thereon has within the immediately preceding period of six months when in position on that power press, been thoroughly examined and tested by a competent person.

(4) The competent person carrying out an examination and test under the foregoing provisions of this Regulation shall make a report of the examination and test in the approved form and containing the approved particulars and every such report shall within fourteen days of the completion of the examination and test be entered in or attached to a register kept for the purposes of this Regulation.

Defects disclosed during a thorough examination and test

6.—(1) Where any defect is disclosed in any power press or in any safety device by any examination and test under Regulation 5 and, in the opinion of the competent person carrying out the examination and test, either—

(*a*) the said defect is a cause of danger to employed persons and in consequence the power press or safety device (as the case may be) ought not to be used until the said defect has been remedied; or

(*b*) the said defect may become a cause of danger to employed persons and in consequence the power press or safety device (as the case may be) ought not to be used after the expiration of a specified period unless the said defect has been remedied,

such defect shall, as soon as possible after the completion of the examination and test, be notified in writing by the competent person to the occupier of the factory and, in the case of a defect falling within sub-paragraph (*b*) of this paragraph such notification shall include the period within which, in the opinion of the competent person, the defect ought to be remedied.

(2) Where any such defect is notified to the occupier in accordance with the foregoing provisions of this Regulation the power press or safety device (as the case may be) having the said defect shall not be used—

(*a*) in the case of a defect falling within sub-paragraph (*a*) of paragraph (1) of this Regulation until the said defect has been remedied; and

(*b*) in the case of a defect falling within sub-paragraph (*b*) of paragraph (1) of this Regulation, after the expiration of the said specified period unless the said defect has been remedied.

(3) As soon as is practicable after any defect of which notification has been given under paragraph (1) of this Regulation has been remedied, a record shall be made by or on behalf of the occupier stating the measures by which and the date on which the defect was remedied. The said record, where practicable, shall be made in an approved form on the document containing the report of the examination and test under Regulation 5

during which the defect was disclosed, or, in any other case, shall be made in writing and attached as soon as is practicable to the said report.

Inspection and test of safety devices

7.—(1) No power press shall be used after the setting, re-setting or adjustment of the tools thereon unless a person appointed or authorised for the purpose under Regulation 4 has inspected and tested every safety device thereon while it is in position on the said power press and has given a certificate in accordance with paragraph (3) of this Regulation:

Provided that an inspection, test and certificate as aforesaid shall not be required where any adjustment of the tools has not caused or resulted in any alteration to or disturbance of any safety device on the power press and if, after the adjustment of the tools, the safety devices remain, in the opinion of such a person as aforesaid, in efficient working order.

(2) No power press shall be used after the expiration of the fourth hour of any working period unless between the time when the said power press last ceased to be used before the commencement of that period and the expiration of the said fourth hour a person appointed or authorised for the purpose under Regulation 4 has inspected and tested (whether or not in pursuance of paragraph (1) of this Regulation) every safety device thereon while it is in position on the said power press and has given a certificate in accordance with paragraph (3) of this Regulation.

In this paragraph the expression "working period" means the daily hours worked in a factory or, in the case of a factory where a shift system is in operation, a shift.

(3) Every certificate of inspection and test made under this Regulation—

(*a*) shall contain sufficient particulars to identify every safety device inspected and tested and the power press on which that safety device was fitted at the time of the inspection and test;

(*b*) shall state the date and time of the inspection and test;

(*c*) shall be signed by the person carrying out the inspection and test; and

(*d*) shall state that every safety device on the said power press is in efficient working order.

(4) In determining, for the purpose of giving a certificate of an inspection and test made under this Regulation or of forming an opinion for the purposes of the proviso to paragraph (1) of this Regulation, whether or not any safety device is in efficient working order, the person carrying out the inspection and test or forming the opinion shall not treat as a defect any defect in a safety device which at the time of the inspection and test is the subject of a notification in writing under Regulation 6(1) by virtue of which the use of the safety device may be continued during a specified period which has not then expired unless he is of the opinion that the said defect has then become a cause of danger to employed persons.

(5) Every certificate of an inspection and test under this Regulation shall be kept available at or near the power press to which it relates for inspection by the persons employed until the expiration of the period during which the said power press may be used in reliance upon that certificate and shall be preserved in the factory and kept available for a period of six months thereafter for inspection by any inspector.

Defects disclosed during an inspection and test

8.—(1) Where it appears to any person as a result of any inspection and test carried out by him under Regulation 7 that any necessary safety device is not in position or is not properly in position on a power press or that any safety device which is in position on a power press is not in his opinion suitable, he shall notify the occupier or his agent forthwith.

(2) Except as provided in paragraph (3) of this Regulation, where any defect is disclosed in a safety device by any inspection and test under Regulation 7, the person carrying out the inspection and test shall notify the occupier or his agent forthwith.

(3) Where any defect in a safety device is the subject of a notification in writing under Regulation 6 by virtue of which the use of the safety device may be continued during a specified period without the said defect having been remedied, the requirement in paragraph (2) of this Regulation shall not apply to the said defect until the said period has expired.

Identification of power presses and safety devices

9. For the purpose of identification in the registers required to be kept by these Regulations, every power press and every safety device provided for the same shall be distinctively and plainly marked.

Indication of speed and direction of flywheel

10. The maximum permissible flywheel speed as certified by the makers and the direction of rotation of the flywheel shall be conspicuously marked on every power press constructed after the coming into operation of these Regulations and where such a power press is in use the flywheel shall not be driven at a greater speed than that so marked, or in the direction opposite to that so marked; but nothing in this paragraph shall prevent in any emergency the reversal of the direction of rotation of the flywheel.

Preservation of registers

11. Every register kept for the purposes of Regulation 4 or 5 shall be preserved in the factory and kept available for inspection by any inspector for a period of two years after the date of the last entry in the register.

Exemption certificates

12. The Chief Inspector may (subject to such conditions as may be specified therein) by certificate in writing (which he may in his discretion revoke at any time) exempt from all or any of the requirements of these Regulations—

(*a*) any power press or safety device or any class or description of power presses or safety devices; or

(*b*) any operation or process or any class or description of operations or processes,

if he is satisfied that the requirements in respect of which the exemption is granted are not necessary for the protection of persons employed or are impracticable or inappropriate. Where such exemption is granted a legible copy of the certificate, showing the conditions (if any) subject to which it has been granted, shall be kept posted in any factory in which the exemption applies in a position where it may be conveniently read by the persons employed.

SCHEDULE

Regulation 4

Particulars of training required by paragraph (1) *of Regulation* 4

The training shall include suitable and sufficient practical instruction in the following matters in relation to each type of power press and safety device in respect of which it is proposed to appoint the person being trained, that is to say—

1. *Power press mechanisms* with particular reference to their bearing on safety. The nature and function of clutch mechanisms, flywheels, brakes and ancillary equipment.

2. *Safety devices.* Types of devices. Functions of each type of device. Methods of installing, inspecting and testing including detection of defects.

3. *Accident causation and prevention.* The causes and prevention of accidents with special reference to power presses.

4. *The work of the tool setter.* Safe methods of working. Lubrication. Co-operation with the press operator.

5. *Tool design*—in relation to safe methods of working.

EXPLANATORY NOTE

(*This Note is not part of the Regulations.*)

These Regulations contain requirements in relation to the use of power presses (other than when used or to be used for working hot metal) in factories and electrical stations to which the Factories Act 1961 applies. Such requirements are additional to those imposed by the said Act of 1961. The Regulations provide that only competent and properly trained persons shall prepare power presses for use; for the periodical examination of power presses and their safety devices; for the inspection and test of safety devices after the tools have been set or re-set and before or during each working shift; for the marking of power presses and safety devices for purposes of identification; and for the marking on every power press constructed after the coming into operation of these Regulations of the maximum permissible speed and the direction of rotation of the flywheel.

Appendix V

The Asbestos Regulations 1969
No. 690

PART I

APPLICATION, INTERPRETATION AND GENERAL

Citation, commencement and revocation

1.—(1) These Regulations may be cited as the Asbestos Regulations 1969 and shall come into operation on 14th May 1970.

(2) The Asbestos Industry Regulations 1931**(a)** are hereby revoked.

(3) The Shipbuilding and Ship-repairing Regulations 1960**(b)** shall have effect as if the following provisions were omitted—

(*a*) in Regulation 3(2), the definition of the word "asbestos";
(*b*) in Regulation 76(1), sub-paragraphs (*a*) to (*d*); and
(*c*) in Regulation 77, paragraphs (*a*) to (*d*).

Interpretation

2.—(1) The Interpretation Act 1889**(c)** shall apply to the interpretation of these Regulations as it applies to the interpretation of an Act of Parliament, and as if these Regulations and the Regulations hereby revoked were Acts of Parliament.

(2) In these Regulations, unless the context otherwise requires, the following expressions have the meanings hereby assigned to them respectively, that is to say—

"approved" means approved for the time being by certificate of the Chief Inspector;

"asbestos" means any of the following minerals, that is to say, crocidolite, amosite, chrysotile, fibrous anthophyllite and any mixture containing any of the said minerals;

"factory" includes any premises and place to which these Regulations apply;

"protective clothing" means overalls and headgear, which (in either case) will when worn exclude asbestos dust;

"the principal Act" means the Factories Act 1961 as amended by or under any other Act.

(3) References in these Regulations to asbestos dust shall be taken to be references to dust consisting of or containing asbestos to such an extent as is liable to cause danger to the health of employed persons.

(a) S.R. & O. 1931/1140 (Rev. VII, p. 181: 1931 p. 410).
(b) S.I. 1960/1932 (1960 II, p. 1427). **(c)** 1889 c. 63.

Application of Regulations

3.—(1) These Regulations shall apply to the following premises and places in which any process to which these Regulations apply is carried on, that is to say, factories and all premises and places to which the provisions of Part IV of the principal Act with respect to special regulations for safety and health are applied by the following provisions of that Act, namely, section 123 (which relates to electrical stations), section 124 (which relates to institutions), so much of section 125 as relates to warehouses other than warehouses within any dock or forming part of any wharf or quay, section 126 (which relates to ships) and section 127 (which relates to building operations and works of engineering construction).

(2) These Regulations apply to every process involving asbestos or any article composed wholly or partly of asbestos, except a process in connection with which asbestos dust cannot be given off.

(3) The provisions of these Regulations are in substitution for the provisions of so much of section 1(2) of the principal Act as relates to the cleaning of parts of factories, being parts in which any process to which these Regulations apply is carried on or into which any asbestos dust from any such process carried on in the factory is liable to escape, and of so much of section 63(1) of that Act as relates to the measures to be taken against asbestos dust, but save as aforesaid are in addition to and not in substitution for or in diminution of other requirements imposed by or under the principal Act.

Certificates of exemption

4. The Chief Inspector may (subject to such conditions, if any, as may be specified therein) by certificate in writing (which he may in his discretion revoke at any time) exempt from all or any of the requirements of these Regulations—

(*a*) any factory or part of any factory; or

(*b*) any class or description of factories or of parts thereof; or

(*c*) any machine, plant, apparatus or process or any class or description of machines, plant, apparatus or processes; or

(*d*) the employment of any person or class or description of persons,

if he is satisfied that the requirements in respect of which the exemption is granted are not necessary for the protection of persons employed or are impracticable.

Obligations under Regulations

5.—(1) Except in the case of any place where building operations or works of engineering construction are carried on it shall be the duty—

(*a*) of every employer (whether or not an occupier) who is undertaking in a factory any process to which these Regulations apply to comply with the requirements of these Regulations in relation to any person employed by him, any plant or material under his control, and any part of the factory in which he is carrying on the process or into which asbestos dust in connection with that process is liable to escape;

(*b*) of every occupier of any factory (whether or not such an employer as aforesaid) where any such process is being carried on to comply with the requirements of Regulations 8, 18 and 19 in relation to any person employed in the factory (other than any person employed by an employer who is undertaking in the factory any process to which these Regulations apply) and with the requirements of Part III of these Regulations; and

(*c*) of every person who is undertaking in a factory any process to which these Regulations apply and who employs no other person in the carrying on of that process to comply, for the protection of employed persons working there, with the requirements of these Regulations (other than Regulations 8, 14 and 18 to 20) with which he would have to comply if he were the employer of a person doing there the work which he does himself.

(2) In the case of places where building operations or works of engineering construction are carried on it shall be the duty of every contractor and of every employer who is undertaking any process to which these Regulations apply to comply with the requirements of these Regulations in relation to any person employed by him and any person employed there who is liable to be exposed to asbestos dust in connection with any process carried on by the contractor or employer (as the case may be), and in relation to any plant or material under his control and any part of the site where he is carrying on the process or where asbestos dust in connection with that process is liable to escape.

(3) It shall be the duty of every person employed to comply with the requirements of such of these Regulations as relate to the performance of or refraining from an act by him.

Notifications

6.—(1) Where at the date of the coming into operation of these Regulations a person is undertaking in a factory any process involving crocidolite (including the removal from any surface of lagging consisting of or containing crocidolite), he shall within 28 days of that date give written notice to the inspector for the district that he is undertaking such a process and of the address of the factory at which he is doing so.

(2) Before a person first begins after the coming into operation of these Regulations to undertake in a factory any such process as aforesaid, he shall give not less than 28 days written notice to the inspector for the district, or such shorter notice as the inspector may agree to accept, of the date on which and of the address of the factory at which he will begin to do so.

PART II

EXHAUST VENTILATION AND PROTECTIVE EQUIPMENT

Exhaust ventilation

7.—(1) Subject to the provisions of the next following Regulation, no process to which these Regulations apply shall be carried on in any factory unless equipment is provided, maintained and used which produces an exhaust draught which prevents the entry into the air of any workplace of asbestos dust:

Provided that the foregoing requirements of this Regulation shall not apply where any such process is carried on in such a manner as to be as safe as it would be if the said requirements were complied with.

(2) Exhaust ventilation equipment provided in accordance with this Regulation shall while any work of maintenance or repair to the machinery, apparatus or other plant or equipment in connection with which it is provided is being carried on, be kept in use so as to produce an exhaust draught which prevents the entry into the air of any workplace of asbestos dust.

(3) Exhaust ventilation equipment provided in accordance with this Regulation shall be inspected at least once in every seven days and shall be thoroughly examined and tested by a competent person at least once in every period of fourteen months, and a report of the results of every such examination and test containing approved particulars and signed by the person making or responsible for the carrying out of the examination and test shall be made within fourteen days after the examination and test.

(4) Every such report as aforesaid shall be attached to the general register and be preserved and kept available for inspection by any inspector for a period of two years after it is made.

Protective equipment

8.—(1) Where in any factory the requirements of paragraph (1) or (2) of the last foregoing Regulation apply, but it is impracticable to comply with those requirements, there shall be provided for the use of each person employed in any part of the factory, being a part into which asbestos dust from a process to which these Regulations apply is liable to escape—

(*a*) approved respiratory protective equipment; and

(*b*) protective clothing.

(2) All respiratory protective equipment and protective clothing provided in pursuance of the foregoing paragraph of this Regulation shall be maintained.

(3) All respiratory protective equipment and protective clothing so provided shall be used by the persons for whom they are provided while employed in any such part of the factory as aforesaid.

(4) No respiratory protective equipment so provided which has been worn by a person shall be provided for the use of another person unless it has been thoroughly cleaned and disinfected since last being worn.

(5) No person shall be employed to perform any work for which respiratory protective equipment is provided in pursuance of this Regulation unless he has been fully instructed in the proper use of that equipment.

Part III

Cleanliness of Premises and Plant

Duty to keep clean

9. All machinery, apparatus, work benches and other plant and equipment used in any factory for the purposes of any process to which these Regulations apply, all external surfaces of exhaust ventilation equipment provided in pursuance of these Regulations and all floors, inside walls, ceilings, ledges and other internal surfaces of any part of a building or of any ship, being a part in which any process to which these Regulations apply is carried on or into which asbestos dust from any such process carried on in the factory is liable to escape, shall, so far as is practicable, be kept in a clean state and free from asbestos waste and dust in accordance with the provisions of this Part of these Regulations.

Cleaning by dustless method

10. The cleaning required by the last foregoing Regulation shall, so far as is practicable, be carried out—

(*a*) by means of vacuum cleaning equipment so designed and constructed and so used; or

(*b*) by some other suitable method such,

that asbestos dust neither escapes nor is discharged into the air of any workplace.

Cleaning where cleaning by dustless method is impracticable

11.—(1) In so far as it is impracticable to carry out the cleaning required by Regulation 9 in accordance with the provisions of the last foregoing Regulation, such cleaning shall be carried out in accordance with the provisions of this Regulation.

(2) Protective clothing and approved respiratory protective equipment shall be provided for and used by every person present in any part of a factory while such cleaning is being carried out in that part otherwise than in accordance with Regulation 10.

(3) No respiratory protective equipment so provided which has been worn by a person shall be provided for the use of another person unless it has been thoroughly cleaned and disinfected since last being worn.

(4) No person shall be employed to perform any work for which respiratory protective equipment is provided in pursuance of this Regulation unless he has been fully instructed in the proper use of that equipment.

(5) Any surface on which asbestos waste or dust is deposited as a result of the carrying out of cleaning in pursuance of this Regulation, not being a surface of the cleaning equipment used, shall be cleaned forthwith.

Vacuum cleaning equipment

12. Vacuum cleaning equipment used for cleaning in accordance with the provisions of this Part of these Regulations shall be maintained and have all external surfaces kept in a clean state and free from asbestos waste and dust.

Requirements as to construction of buildings before first use for certain processes

13. After the date of the coming into operation of these Regulations no process specified in the Schedule to these Regulations shall be carried on in a building by the occupier of the building unless any part of such building where any such process is carried on and has not previously been carried on—

(*a*) is so constructed that all its interior surfaces are smooth and impervious and is so designed and constructed as to have in it as few surfaces as is practicable on which asbestos dust can settle; and

(*b*) has a vacuum cleaning system which has filters and a source of suction located at a fixed central point with pipes throughout the building fitted with suitably placed inlets to which portable cleaning implements can be attached and is a system which is so designed and constructed that asbestos dust cannot escape or be discharged from it into the air of any workplace:

Provided that the requirements of this Regulation shall not apply where a process or processes so specified are carried on in a factory for a total number of hours not exceeding eight in any week.

Cleanliness of accommodation for clothing

14. Where there is provided for the use of employed persons under section 59 (accommodation for clothing) of the principal Act or under any regulations made under that Act accommodation for clothing not worn during working hours, the accommodation so provided shall be such or shall be so situated that asbestos dust is not deposited on any such clothing in the said accommodation.

PART IV

STORAGE AND DISTRIBUTION

Storage

15. All loose asbestos in a factory shall when not in use be kept in suitable closed receptacles which prevent the escape of asbestos dust therefrom, and all asbestos waste in a factory shall, when stored, be kept in such receptacles.

Distribution, etc.

16. No loose asbestos or asbestos waste shall be despatched from or as far as is practicable received into a factory except in suitable closed receptacles which prevent the escape of asbestos dust therefrom or shall be distributed within a factory except in such receptacles or in a totally enclosed system of conveyance.

Marking of receptacles

17. All such receptacles as aforesaid shall, where they contain crocidolite, be clearly and boldly marked with the words "Blue asbestos—do not inhale dust".

PART V

ACCOMMODATION FOR AND CLEANING OF PROTECTIVE EQUIPMENT

Accomodation, etc. for protective clothing and respiratory protective equipment

18.—(1) Suitable accomodation in a conveniently accessible position shall be provided for the use of persons employed when putting on or taking off respiratory protective equipment and protective clothing provided in pursuance of these Regulations and for the storage of such clothing and equipment when it is not in use.

(2) All protective clothing and respiratory protective equipment which has been provided in pursuance of these Regulations for the use of persons employed shall when not required for use be kept in the accommodation for storage provided in pursuance of paragraph (1) of this Regulation, except when it is necessary for it to be removed for cleaning or replacement.

(3) So far as practicable, the same protective clothing shall (until it becomes necessary to replace it) be available for use by a person employed on each occasion on which he requires to wear it in pursuance of these Regulations.

Cleaning of protective clothing

19.—(1) All protective clothing provided in pursuance of these Regulations shall at suitable intervals be cleaned.

(2) The said cleaning shall be carried out within the factory where the clothing is provided:

Provided that this requirement shall not apply in the case of protective clothing which before despatch from the factory for cleaning is properly packed in suitable containers which prevent the escape of asbestos dust therefrom and the containers in which it is packed are clearly and boldly marked with the words "Asbestos contaminated clothing".

Part VI

Young Persons

Employment of young persons

20. No young person shall be employed—

(*a*) in any process to which these Regulations apply or in any part of a factory into which asbestos dust from any such process carried on in the factory would otherwise be liable to escape unless the requirements of Regulation 7 are complied with in respect of that process; or

(*b*) in carrying out any cleaning in accordance with Part III of these Regulations otherwise than by means of vacuum cleaning equipment or other suitable method in accordance with the provisions of Regulation 10.

SCHEDULE

Regulation 13

Processes specified for the purposes of Regulation 13

1. Breaking, crushing, disintegrating, opening and grinding of asbestos, and the mixing of asbestos (including the mixing of asbestos with any other material) or the sieving of asbestos, and all processes involving manipulation of asbestos incidental thereto.

2. All processes in the manufacture of yarn or cloth composed of asbestos or asbestos mixed with any other material, including preparatory and finishing processes.

3. The making of insulation slabs or sections, composed wholly or partly of asbestos, and processes incidental thereto.

4. The making and repairing of insulating mattresses, composed wholly or partly of asbestos, and processes incidental thereto.

5. Any other process in the manufacture of articles composed wholly or partly of asbestos in connection with which asbestos dust is given off.

EXPLANATORY NOTE

(*This Note is not part of the Regulations.*)

These Regulations impose requirements for the safety and welfare of persons employed in factories and in other premises and places to which the Factories Act 1961 applies, where processes are carried on which involve asbestos or any article composed wholly or partly of asbestos and are capable of giving rise to asbestos dust. The Regulations require that the processes be carried on under an exhaust draught or in some other equally safe way, or that protective clothing and respiratory protective equipment be provided for the use of persons employed. The Regulations also impose requirements as to the cleaning of premises and plant, the construction of buildings newly used for the processes, the storage and distribution of asbestos, the accommodation to be provided for protective equipment and the cleaning of protective clothing. The employment of young persons in connection with processes to which the Regulations apply and in cleaning work is prohibited except in specified circumstances. The Regulations also require that notification of processes involving crocidolite be sent to the inspector for the district.

The Asbestos Industry Regulations 1931 and certain provisions concerning asbestos of the Shipbuilding and Ship-repairing Regulations 1960 are revoked.

Appendix VI

The Abrasive Wheels Regulations 1970 No. 535

Citation, commencement and revocation

1.—(1) These Regulations may be cited as the Abrasive Wheels Regulations 1970 and shall come into operation on 2nd April 1971.

(2) Regulation 9 of the Grinding of Metals (Miscellaneous Industries) Regulations 1925(**a**) and Regulation 13 of the Grinding of Cutlery and Edge Tools Regulations 1925(**b**) are hereby revoked.

Interpretation

2.—(1) The Interpretation Act 1889(**c**) shall apply to the interpretation of these Regulations as it applies to the interpretation of an Act of Parliament and as if these Regulations and the Regulations hereby revoked were Acts of Parliament.

(2) For the purposes of these Regulations, unless the context otherwise requires, the following expressions have the meanings hereby assigned to them respectively, that is to say :—

"abrasive wheel" means—

(*a*) a wheel, cylinder, disc or cone which, whether or not any other material is comprised therein, consists of abrasive particles held together by mineral, metallic or organic bonds whether natural or artificial ;

(*b*) a mounted wheel or point and a wheel or disc having in either case separate segments of abrasive material ;

(*c*) a wheel or disc made in either case of metal, wood, cloth, felt, rubber or paper and having any surface consisting wholly or partly of abrasive material ; and

(*d*) a wheel, disc or saw to any surface of any of which is attached a rim or segments consisting in either case of diamond abrasive particles,

and which is, or is intended to be, power driven and which is for use in any grinding or cutting operation ;

"approved" means approved for the time being for the purposes of these Regulations by certificate of the Chief Inspector ;

"factory" includes any place to which these Regulations apply ;

(**a**) S.R. & O. 1925/904 (Rev. VII, p. 328: 1925, p. 356).
(**b**) S.R. & O. 1925/1089 (Rev. VII, p. 323: 1925, p. 362).
(**c**) 1889 c. 63.

"mounted wheel or point" means a wheel or point consisting in either case of abrasive particles held together by mineral, metallic or organic bonds whether natural or artificial and securely and permanently mounted on the end of a mandrel or quill ;

"overhang" means in relation to a mounted wheel or point that part of the mandrel or quill which is exposed between the collet in which the mandrel or quill is held and the part of the abrasive material nearest to the said collet ;

"principal Act" means the Factories Act 1961 as amended by or under any other Act.

Application and operation of Regulations

3.—(1) These Regulations, other than Regulation 19, shall apply to any abrasive wheel used for any grinding or cutting operation in any of the following, that is to say, factories and any premises, places, processes, operations and works to which the provisions of Part IV of the principal Act with respect to special Regulations for safety and health are applied by any of the following provisions of that Act, namely, section 123 (which relates to electrical stations), section 124 (which relates to institutions), section 125 (which relates to certain dock premises and certain warehouses), section 126 (which relates to ships) and section 127 (which relates to building operations and works of engineering construction).

(2) In relation to abrasive wheels, the provisions of these Regulations are in substitution for the provisions of section 14(1) of the principal Act and accordingly the provisions of that subsection shall not apply in relation to any abrasive wheel.

(3) Regulation 19 shall apply to any abrasive wheel (other than an abrasive wheel of any of the kinds referred to in Regulation 4(1), (4) and (6) and to any machine of the kind referred to in Regulation 7(1).

(4) Regulation 67(2) of the Shipbuilding and Ship-repairing Regulations 1960(**a**) and Regulation 42 of the Construction (General Provisions) Regulations 1961(**b**) shall not apply in relation to any abrasive wheel.

Exceptions

4.—(1) Regulations 6, 7, 11(*a*) and (*b*), 12 and 19 shall not apply to any abrasive wheel manufactured of metal, wood, cloth, felt, rubber or paper and having any surface consisting wholly or partly of abrasive material.

(2) Regulations 11(*a*) and (*b*) and 12 shall not apply to any abrasive wheel which consists wholly of abrasive particles held together by natural bonds.

(3) Regulation 6 shall not apply to any abrasive wheel having separate segments of abrasive material.

(4) Regulations 6, 7, 10(1), 11, 12, 18 and 19 shall not apply to any abrasive wheel which does not exceed 235 millimetres in diameter, is manufactured of cloth, felt, rubber or paper and has any surface consisting wholly or partly of abrasive material, when that abrasive wheel is used, or in the case of Regulation 19, is for use, in a portable machine.

(5) Regulations 7, 10(1), 11 and 12 shall not apply to any abrasive wheel when it is used for the grinding of glass.

(a) S.I. 1960/1932 (1960 II, p. 1427).
(b) S.I. 1961/1580 (1961 II, p. 3207).

(6) Regulations 6, 7, 12 and 19 shall not apply to any wheel, disc or saw being a wheel, disc or saw of a kind specified in Regulation 2(2)(*d*).

Exemptions

5. The Chief Inspector may (subject to such conditions, if any, as may be specified therein) by certificate in writing (which he may in his discretion revoke at any time) exempt from all or any of the requirements of these Regulations—

(*a*) any particular abrasive wheel or any type of abrasive wheel ; or

(*b*) any machine or part of a machine or any class or description of machines or parts of machines ; or

(*c*) any operation or process or any class or description of operations or processes,

if he is satisfied that the said requirements in respect of which the exemption is granted are impracticable or inappropriate or are not necessary for the protection of persons employed.

Speeds of abrasive wheels

6.—(1) No abrasive wheel having a diameter of more than 55 millimetres shall be taken into use in any factory for the first time in that factory after the coming into operation of these Regulations unless the abrasive wheel or its washer is clearly marked with the maximum permissible speed in revolutions per minute specified by the manufacturer for that abrasive wheel.

(2) No abrasive wheel having a diameter of 55 millimetres or less shall be taken into use in any factory for the first time in that factory after the coming into operation of these Regulations unless there is kept permanently fixed in the room in which grinding is ordinarily carried out with that abrasive wheel a notice clearly stating the maximum permissible speed in revolutions per minute specified by the manufacturer for that abrasive wheel or for abrasive wheels of the class to which that abrasive wheel belongs and, in the case of mounted wheels and points, the overhang permissible at that speed :

Provided that when grinding with such an abrasive wheel is not ordinarily carried out in any one room the said notice shall be kept posted at a place and in a position where it may easily be read by persons employed in grinding with that abrasive wheel.

(3) No abrasive wheel shall be operated at a speed in excess of the maximum permissible speed in revolutions per minute specified in accordance with the foregoing provisions of this Regulation for that abrasive wheel :

Provided that where the diameter of an abrasive wheel has been reduced its said maximum permissible speed may be increased to that speed which bears the same proportion to the said maximum permissible speed as the original diameter of the abrasive wheel bears to its reduced diameter.

Speeds of spindles

7.—(1) There shall be securely affixed to every power driven machine having any spindle on which an abrasive wheel is, or is intended to be, mounted a notice specifying—

(*a*) in the case of each such spindle (other than a spindle to which sub-paragraph (*b*) or (*c*) of this paragraph applies) its maximum working speed ;

(*b*) in the case of any such spindle for which there are provided arrangements for operating the spindle at more than one specific working speed, each such speed ; and

(*c*) in the case of any such spindle for which there are provided arrangements for operating the spindle at an infinite number of working speeds within a specified range, the maximum and minimum working speeds of the spindle.

(2) No such spindle shall while an abrasive wheel is mounted on it be operated at a speed in excess of the maximum working speed specified in accordance with the foregoing provisions of this Regulation for that spindle.

(3) The occupier shall, when so required by an inspector, provide him with all such facilities and information as are necessary to enable him to determine the working speed of any spindle, shaft, pulley, or other appliance, which is used to operate an abrasive wheel.

(4) The speed of every air driven spindle on which an abrasive wheel is mounted shall be controlled by a governor or other device so that the speed of the spindle does not at any time exceed the maximum working speed specified for that spindle in accordance with paragraph (1) of this Regulation.

(5) Every governor and other device used for controlling the speed of an air driven spindle on which an abrasive wheel is mounted shall be properly maintained.

Mounting

8. Every abrasive wheel shall be properly mounted.

Training and appointment of persons to mount abrasive wheels

9.—(1) Except as provided in paragraphs (4) and (5) of this Regulation, no person shall mount any abrasive wheel unless he—

(*a*) has been trained in accordance with the Schedule to these Regulations;

(*b*) is competent to carry out that duty ; and

(*c*) has been appointed by the occupier of the factory to carry out that duty in respect of the class or description of abrasive wheel to which the abrasive wheel belongs ; and every such appointment shall be made by a signed and dated entry in, or signed and dated certificate attached to, a register kept for the purposes of this Regulation.

(2) Particulars of the class or description of abrasive wheel to which any appointment made in accordance with paragraph (1)(*c*) of this Regulation relates shall be set out in the entry or certificate (as the case may be) of the appointment. Every person appointed under the said paragraph (1)(*c*) shall be furnished by the occupier with a copy of the entry or certificate of his appointment and of any entry revoking the same.

(3) Any appointment made under paragraph (1)(*c*) of this Regulation may be revoked by the occupier at any time by signed and dated entry in the said register.

(4) Paragraph (1) of this Regulation shall not apply to a person who is undergoing training in the work of mounting abrasive wheels and working under the immediate supervision of a competent person appointed under the said paragraph (1).

(5) Paragraph (1) of this Regulation shall not apply to any person who mounts any mounted wheel or point.

Provision of guards

10.—(1) Except as provided in paragraph (2) of this Regulation, a guard shall be provided and kept in position at every abrasive wheel in motion.

(2) Paragraph (1) of this Regulation shall not apply to an abrasive wheel where, by reason of the work being done thereat, or of the work which is ordinarily done thereat (or, in the case of a new wheel, is intended ordinarily to be done thereat), or the nature of the wheel, the use of a guard is impracticable.

Construction, maintenance, etc., of guards

11. Every guard provided in pursuance of these Regulations shall—

(*a*) so far as is reasonably practicable, be of such a design and so constructed as to contain every part of the abrasive wheel in the event of any fracture of the abrasive wheel or of any part thereof occurring while the abrasive wheel is in motion ;

(*b*) be properly maintained and so secured as to prevent its displacement in the event of any such fracture as aforesaid ; and

(*c*) enclose the whole of the abrasive wheel except such part thereof as is necessarily exposed for the purpose of any work being done at that abrasive wheel or, where a guard which is not adjustable is used, for the purpose of the work which is ordinarily done thereat (or, in the case of a new abrasive wheel, is intended ordinarily to be done thereat).

Tapered wheels and protection flanges

12.—(1) Where the work which is ordinarily done (or, in the case of a new abrasive wheel, is intended ordinarily to be done) at any abrasive wheel requires that the exposed arc of the wheel shall exceed 180 degrees measured at the centre of the wheel then, where practicable, the wheel used shall be tapered from its centre towards its periphery by at least 6 per cent on each side and shall be mounted between suitable protection flanges.

(2) Protection flanges between which any abrasive wheel is mounted in accordance with this Regulation shall be of substantial construction and properly maintained and shall have the same degree of taper as the wheel and shall—

(*a*) in the case of a wheel of 300 millimetres or less in diameter, be of a diameter equal to at least half the diameter of the wheel ;

(*b*) in the case of a wheel of more than 300 millimetres but not exceeding 750 millimetres in diameter, be of a diameter equal to at least the diameter of the wheel less 150 millimetres ; and

(*c*) in the case of a wheel of more than 750 millimetres in diameter, be of a diameter equal to at least the diameter of the wheel less 200 millimetres.

Selection of abrasive wheels

13. All practicable steps shall be taken to ensure that any abrasive wheel used is suitable for the work for which it is used to the extent necessary to reduce the risk of injury to persons employed.

Machine controls

14. No abrasive wheel shall be used in a machine unless the said machine is provided with an efficient device or efficient devices for starting, and cutting off, the power to the machine and the control or controls of the device or devices shall be in such a position and of such design and construction as to be readily and conveniently operated by the person operating the machine.

Rests

15.—(1) Where at any abrasive wheel there is a rest for supporting a workpiece that rest shall at all times while the wheel is in motion be—

(*a*) properly secured ; and

(*b*) adjusted so as to be as close as practicable to the exposed part of the abrasive wheel.

(2) Every such rest as aforesaid shall be of substantial construction and properly maintained.

Cautionary notice

16. The approved cautionary notice as to the dangers arising from the use of abrasive wheels and the precautions to be observed shall be affixed in every room in which grinding or cutting by means of abrasive wheels is ordinarily carried out, or, where such grinding or cutting is not ordinarily carried out in a room, at a place and in such a position where it may be easily read by persons employed in grinding or cutting.

Condition of floors

17.—(1) The floor immediately surrounding every fixed machine on which an abrasive wheel is, or is intended to be, mounted shall be maintained in good and even condition and shall, so far as practicable, be kept clear of loose material and prevented from becoming slippery.

(2) All reasonably practicable steps shall be taken to ensure that the floor of any room or place in which any portable machine on which an abrasive wheel is mounted is used is in good and even condition, is kept clear of loose material and is prevented from becoming slippery.

Duties of employed persons

18.—(1) No employed person using an abrasive wheel shall wilfully misuse or remove any guard, or wilfully misuse any protection flanges or other appliance provided in pursuance of these Regulations or any rest for a workpiece.

(2) Every employed person shall make full and proper use of guards, protection flanges and other appliances provided in pursuance of these Regulations and of rests for workpieces and if he discovers any defect in the same shall forthwith report such defect to the occupier, manager or other appropriate person.

Sale or hire of machinery

19. The provisions of section 17(2) of the principal Act (which prohibits the sale or letting on hire of certain machines which do not comply with the requirements of that section) shall extend to—

(*a*) any abrasive wheel to which this Regulation applies having a diameter of more than 55 millimetres which is for use in a factory and which is not marked in the manner required by Regulation 6(1) ; and

(*b*) any machine supplied with its prime mover as a unit which is for use in a factory and is of a kind specified in Regulation 7(1) which does not comply with the requirements of that paragraph.

Regulation 9

SCHEDULE

Particulars of Training required by paragraph (1) *of Regulation* 9

The training shall include suitable and sufficient instruction in the following matters in relation to each class or description of abrasive wheel in respect of which it is proposed to appoint the person being trained, that is to say—

(1) Approved advisory literature relating to the mounting of abrasive wheels;

(2) Hazards arising from the use of abrasive wheels and precautions which should be observed;

(3) Methods of marking abrasive wheels as to type and speed;

(4) Methods of storing, handling and transporting abrasive wheels;

(5) Methods of inspecting and testing abrasive wheels to check for damage;

(6) The functions of all components used with abrasive wheels, including flanges, washers, bushes and nuts used in mounting and including knowledge of the correct and incorrect methods of assembling all components and correct balancing of abrasive wheels;

(7) The proper method of dressing an abrasive wheel;

(8) The adjustment of the rest of an abrasive wheel;

(9) The requirements of these Regulations.

EXPLANATORY NOTE

(*This Note is not part of the Regulations.*)

These Regulations impose requirements in relation to abrasive wheels used in factories and in other premises and places to which the Factories Act 1961 applies. As respects abrasive wheels, the Regulations are in substitution for section 14(1) of the Factories Act 1961, for Regulation 67(2) of the Shipbuilding and Ship-repairing Regulations 1960 and for Regulation 42 of the Construction (General Provisions) Regulations 1961 (all of which require that dangerous parts of machinery shall be securely fenced).

The Regulations provide for the maximum permissible speeds of abrasive wheels to be specified and for the control of such speeds; for the proper mounting of abrasive wheels; for the training and appointment of persons permitted to mount abrasive wheels; for the provision and maintenance of guards and protection flanges for abrasive wheels; for the controls to be provided for machines in which abrasive wheels are used; for the securing and adjustment of rests for workpieces at abrasive wheels; for the affixing of cautionary notices as to the dangers from the use of abrasive wheels and the precautions to be observed; and for the condition of floors where there are machines on which abrasive wheels are mounted.

The Regulations also prohibit the sale or letting on hire for use in a factory of certain abrasive wheels which are not marked with their maximum permissible speeds and of certain machines on which abrasive wheels are or are intended to be mounted to which the notice as to working speeds is not affixed as required by the Regulations.

Appendix VII

The Protection of Eyes Regulations 1974 No. 1681

Citation, commencement and revocation

1.—(1) These Regulations may be cited as the Protection of Eyes Regulations 1974 and shall come into operation on 10th April 1975.

(2) The Regulations specified in columns 1 and 2 of Schedule 3 to these Regulations are hereby revoked to the extent respectively specified in relation thereto in column 3 of that Schedule.

Interpretation

2.—(1) The Interpretation Act 1889**(a)** shall apply to the interpretation of these Regulations as it applies to the interpretation of an Act of Parliament and as if these Regulations and the Regulations hereby revoked were Acts of Parliament.

(2) In these Regulations, unless the context otherwise requires, the following expressions have the meanings hereby assigned to them respectively, that is to say—

"approved" means approved for the time being for the purposes of these Regulations by certificate of the Chief Inspector;

"eye protectors" means any of the following (being equipment made to be worn by a person), that is to say, goggles, visors, spectacles and face screens;

"factory" includes any premises and place to which these Regulations apply;

"fixed shield" means a screen which is free-standing or which is, or is made to be, attached to machinery, plant or other equipment, or to a building or structure;

"foundry" means an iron foundry or a steel foundry as defined in the Iron and Steel Foundries Regulations 1953**(b)** and those parts of factories in which is carried on any operation or process specified in Regulation 3(1) or (2) of the Non-ferrous Metals (Melting and Founding) Regulations 1962**(c)**;

"person occasionally employed", in relation to any week, means a person who in that week is to be employed in either or both of the following, that is to say—

(*a*) in any one or more of the specified processes;

(a) 1889 c. 63.

(b) S.I. 1953/1464 (1953 I, p. 735).

(c) S.I. 1962/1667 (1962 II, p. 2021).

(*b*) in a place where there is a reasonably foreseeable risk of injury to his eyes from the carrying on of the process specified in paragraph 3 of Schedule 2 to these Regulations,

on not more than two days and for a total of not more than 15 minutes on any one day;

"possession" does not include ownership;

"shield" means a helmet or hand shield, being equipment made to be worn or held by a person;

"specified process" means any process specified in Schedule 1 to these Regulations.

Application of Regulations

3.—(1) These Regulations shall apply to—

(*a*) all factories; and

(*b*) all premises, places, processes, operations and works to which the provisions of Part IV of the Factories Act 1961 with respect to special regulations for safety and health are applied by any of the following provisions of that Act, namely, section 123 (which relates to electrical stations), section 124 (which relates to institutions), section 125 (which relates to certain dock premises and certain warehouses), section 126 (which relates to ships) and section 127 (which relates to building operations and works of engineering construction),

where any specified process is carried on.

(2) The provisions of these Regulations shall be in addition to and not in substitution for or diminution of other requirements imposed by or under the Factories Act 1961.

Exemption certificates

4. The Chief Inspector may (subject to such conditions as may be specified therein) by certificate in writing (which he may in his discretion revoke at any time) exempt from all or any of the requirements of these Regulations—

(*a*) any factory or part of any factory; or

(*b*) any class or description of factories or parts thereof; or

(*c*) any machine, plant, apparatus, process, operation or work, or any class or description of machines, plant, apparatus, processes, operations or works,

if he is satisfied that the requirements in respect of which the exemption is granted are not necessary for the protection of persons employed or are not reasonably practicable. Where such exemption is granted a legible copy of the certificate, showing the conditions (if any) subject to which it has been granted, shall be kept posted in any factory to which the exemption applies in a position where it may be conveniently read by the persons employed.

Protection of persons employed in the specified processes

5.—(1) The employer of every person who is employed in a factory for any of his time in any of the specified processes shall provide for his use—

(*a*) in the case of a person employed in a process specified in Part I of Schedule 1 to these Regulations, eye protectors;

(*b*) in the case of a person employed in a process specified in Part II of the said Schedule, a shield or a sufficient number of fixed shields; and

(*c*) in the case of a person employed in a process specified in Part III or Part IV of the said Schedule, eye protectors or a shield or a sufficient number of fixed shields.

(2) In addition, the employer shall, where necessary and so far as reasonably practicable, provide fixed shields between adjacent positions in each of which persons employed in a process specified in paragraph 6, 7 or 32 of Schedule 1 to these Regulations are working independently of the persons so employed in an adjacent position.

Protection in certain cases of persons at risk from, but not employed in, a specified process

6. Where any process specified in Schedule 2 to these Regulations is being carried on in a factory and persons other than those employed in that process are employed in a place where there is a reasonably foreseeable risk of injury to their eyes from the carrying on of the process—

(*a*) in the case of the process specified in paragraph 3 of the said Schedule, eye protectors or a shield or a sufficient number of fixed shields shall be provided by their employer for their use; and

(*b*) in the case of any other process, a sufficient number of fixed shields shall be provided for their protection by the employer of the persons employed in that process.

Eye protectors—issue and availability

7. Eye protectors provided in pursuance of Regulation 5(1)(*a*) or (*c*) or 6(*a*) shall be given by the employer into the possession of each person for whom they are provided except in the case of persons occasionally employed for whose use a sufficient number of eye protectors shall be provided, maintained and kept readily available by their employer.

Replacement of eye protectors

8.—(1) Where consequent upon a report under Regulation 11 it is confirmed that any eye protectors given into the possession of a person in accordance with the last foregoing Regulation have been lost or destroyed or are so defective as to be unsuitable for the purpose for which they were provided,the employer shall forthwith give into the possession of the person for whom they were provided (unless he has ceased to be a person within the protection provided for by Regulation 5(1)(*a*) or (*c*) or 6(*a*) or has become a person occasionally employed or unless these Regulations are being complied with in his case, in accordance with these Regulations, by the provision of a shield or fixed shields) eye protectors to replace those lost, destroyed or defective.

(2) The employer shall, in addition to any eye protectors provided in accordance with the last foregoing Regulation, keep available as many eye protectors as will be sufficient, so far as can reasonably be foreseen, to enable him to comply with the requirements of the foregoing paragraph.

Eye protectors and shields—construction and marking

9. Eye protectors and shields provided in pursuance of these Regulations shall—

(*a*) be suitable for the person for whose use they are provided;

(*b*) be made in conformity with an approved specification for eye protectors or shields, as the case may be, being an approved specification for eye protectors or shields which are appropriate to the specified process in which the said person is employed or from the carrying on of which there is a reasonably foreseeable risk of injury to the eyes of the said person, as the case may be; and

(*c*) be marked in such a manner and accompanied by such particulars as may be approved in order to indicate the purpose or purposes for which the eye protectors or shields were designed.

Fixed shields—construction, maintenance and positioning

10. Fixed shields provided in pursuance of these Regulations shall—

(*a*) when provided in pursuance of Regulation 5(1) for the use of a person employed in a process specified in Part II or Part III of Schedule 1 to these Regulations, be made in conformity with an approved specification for fixed shields;

(*b*) be properly maintained and, if it is necessary that they should be transparent, be kept clean; and

(*c*) whether or not required by these Regulations to be made in conformity with an approved specification for fixed shields, be so constructed and kept in position as to protect, so far as practicable, the eyes of the persons for whose protection they are provided while those persons are carrying on a specified process, or while those persons are employed in a place where there is a reasonably foreseeable risk of injury to their eyes from the carrying on of such a process, as the case may be.

Duties of employed persons

11.—(1) Every person who is provided with eye protectors or a shield in pursuance of these Regulations shall use the said eye protectors or shield at all times while he is employed in a specified process or while he is employed in a place where there is a reasonably foreseeable risk of injury to his eyes from the carrying on of the process specified in paragraph 3 of Schedule 2 to these Regulations, as the case may be, except when Regulations 5(1), 6(*a*) and 10 are being complied with in his case, in accordance with these Regulations, by the provision of fixed shields.

(2) Every person who is provided with eye protectors or a shield in pursuance of these Regulations shall—

(*a*) take reasonable care of the said eye protectors or shield and not wilfully misuse them or it; and

(*b*) report forthwith to the employer or his agent the loss or destruction of, or any defect in, the said eye protectors or shield.

(3) Every person for whose use a fixed shield is provided in pursuance of these Regulations shall make full and proper use of it.

SCHEDULE 1

Regulations 2(2), 5 and 10

Specified Processes

Part I

Processes in which approved eye protectors are required

1. The blasting or erosion of concrete by means of shot or other abrasive materials propelled by compressed air.

2. The cleaning of buildings or structures by means of shot or other abrasive materials propelled by compressed air.

3. Cleaning by means of high-pressure water jets.

4. The striking of masonry nails by means of a hammer or other hand tool or by means of a power driven portable tool.

5. Any work carried out with a hand-held cartridge operated tool, including the operation of loading and unloading live cartridges into such a tool, and the handling of such a tool for the purpose of maintenance .repair or examination when the tool is loaded with a live cartridge.

6. The chipping of metal, and the chipping, knocking out, cutting out or cutting off of cold rivets, bolts, nuts, lugs, pins, collars, or similar articles from any structure or plant, or from part of any structure or plant, by means of a hammer, chisel, punch, or similar hand tool, or by means of a power driven portable tool.

7. The chipping or scurfing of paint, scale, slag, rust or other corrosion from the surface of metal and other hard materials by means of a hand tool or by means of a power driven portable tool or by applying articles of metal or such materials to a power driven tool.

8. The use of a high-speed metal cutting saw or an abrasive cutting-off wheel or disc, which in either case is power driven.

9. The pouring or skimming of molten metal in foundries.

10. Work at a molten salt bath when the molten salt surface is exposed.

11. The operation, maintenance, dismantling or demolition of plant or any part of plant, being plant or part of plant which contains or has contained acids, alkalis, dangerous corrosive substances, whether liquid or solid, or other substances which are similarly injurious to the eyes, and which has not been so prepared (by isolation, reduction of pressure, emptying or otherwise), treated or designed and constructed as to prevent any reasonably foreseeable risk of injury to the eyes of any person engaged in any such work from any of the said contents.

12. The handling in open vessels or manipulation of acids, alkalis, dangerous corrosive materials, whether liquid or solid, and other substances which are similarly injurious to the eyes, where in any of the foregoing cases there is a reasonably foreseeable risk of injury to the eyes of any person engaged in any such work from drops splashed or particles thrown off.

13. The driving in or on of bolts, pins, collars or similar articles to any structure or plant or to part of any structure or plant by means of a hammer, chisel, punch or similar hand tool or by means of a power driven portable tool, where in any of the foregoing cases there is a reasonably foreseeable risk of injury to the eyes of any person engaged in the work from particles or fragments thrown off.

14. Injection by pressure of liquids or solutions into buildings or structures or parts thereof where in the course of any such work there is a reasonably foreseeable risk of injury to the eyes of any person engaged in the work from any such liquids or solutions.

15. The breaking up of metal by means of a hammer, whether power driven or not, or by means of a tup, where in either of the foregoing cases there is a reasonably foreseeable risk of injury to the eyes of any person engaged in the work from particles or fragments thrown off.

16. The breaking, cutting, cutting into, dressing, carving or drilling by means of a power driven portable tool or by means of a hammer, chisel, pick or similar hand tool other than a trowel, of any of the following, that is to say—

(*a*) glass, hard plastics, concrete, fired clay, plaster, slag or stone (whether natural or artificial);

(*b*) materials similar to any of the foregoing;

(*c*) articles consisting wholly or partly of any of the foregoing;

(*d*) stonework, brickwork or blockwork;

(*e*) bricks, tiles or blocks (except blocks made of wood),

where in any of the foregoing cases there is a reasonably foreseeable risk of injury to the eyes of any person engaged in the work from particles or fragments thrown off.

17. The use of compressed air for removing swarf, dust, dirt or other particles, where in the course of any such work there is a reasonably foreseeable risk of injury to the eyes of any person engaged in the work from particles or fragments thrown off.

18. Work at a furnace containing molten metal, and the pouring or skimming of molten metal in places other than foundries, where there is a reasonably foreseeable risk of injury to the eyes of any person engaged in any such work from molten metal.

19. Processes in foundries where there is a reasonably foreseeable risk of injury to the eyes of any person engaged in any such work from hot sand thrown off.

20. Work in the manufacture of wire rope where there is a reasonably foreseeable risk of injury to the eyes of any person engaged in the work from particles or fragments thrown off or from flying ends of wire.

21. The operation of coiling wire, and operations connected therewith, where there is a reasonably foreseeable risk of injury to the eyes of any person engaged in any such work from particles or fragments thrown off or from flying ends of wire.

22. The cutting of wire or metal strapping under tension, where there is a reasonably foreseeable risk of injury to the eyes of any person engaged in any such work from flying ends of wire or flying ends of metal strapping.

23. Work in the manufacture of glass and in the processing of glass and the handling of cullet, where in any of the foregoing cases there is a reasonably foreseeable risk of

injury to the eyes of any person engaged in the work from particles or fragments thrown off.

PART II

Processes in which approved shields or approved fixed shields are required

24. Any process involving the use of an exposed electric arc or an exposed stream of arc plasma.

PART III

Processes in which approved eye protectors or approved shields or approved fixed shields are required

25. The welding of metals by means of apparatus to which oxygen or any flammable gas or vapour is supplied under pressure.

26. The hot fettling of steel castings by means of a flux-injected burner or air carbon torch, and the de-seaming of metal.

27. The cutting, boring, cleaning, surface conditioning or spraying of material by means of apparatus (not being apparatus mechanically driven by compressed air) to which air, oxygen or any flammable gas or vapour is supplied under pressure excluding any such process elsewhere specified, where in any of the foregoing cases there is a reasonably foreseeable risk of injury to the eyes of any person engaged in the work from particles or fragments thrown off or from intense light or other radiation.

28. Any process involving the use of an instrument which produces light amplification by the stimulated emission of radiation, being a process in which there is a reasonably foreseeable risk of injury to the eyes of any person engaged in the process from radiation.

PART IV

Processes in which approved eye protectors or approved shields or fixed shields are required

29. Truing or dressing of an abrasive wheel where in either of the foregoing cases there is a reasonably foreseeable risk of injury to the eyes of any person engaged in the work from particles or fragments thrown off.

30. Work with drop hammers, power hammers, horizontal forging machines, and forging presses, other than hydraulic presses, used in any case for the manufacture of forgings.

31. The dry grinding of materials or articles by applying them by hand to a wheel, disc or band which in any such case is power driven or by means of a power driven portable tool, where in any of the foregoing cases there is a reasonably foreseeable risk of injury to the eyes of any person engaged in the work from particles or fragments thrown off.

32. The fettling of metal castings, involving the removal of metal, including runners, gates and risers, and the removal of any other material during the course of such fettling, where in any of the foregoing cases there is a reasonably foreseeable risk of injury to the eyes of any person engaged in the work from particles or fragments thrown off.

33. The production of metal castings at pressure die casting machines where there is a reasonably foreseeable risk of injury to the eyes of any person engaged in any such work from molten metal thrown off.

34. The machining of metals, including any dry grinding process not elsewhere specified, where there is a reasonably foreseeable risk of injury to the eyes of any person engaged in any such work from particles or fragments thrown off.

35. The welding of metals by an electric resistance process or a submerged electric arc, where there is a reasonably foreseeable risk of injury to the eyes of any person engaged in any such work from particles or fragments thrown off.

SCHEDULE 2

Regulations 2(2) and 6

CASES IN WHICH PROTECTION IS REQUIRED FOR PERSONS AT RISK FROM, BUT NOT EMPLOYED IN THE SPECIFIED PROCESS

1. The chipping of metal, and the chipping, knocking out, cutting out or cutting off of cold rivets, bolts, nuts, lugs, pins, collars, or similar articles from any structure or plant, or from any part of any structure or plant, by means of a hammer, chisel, punch or similar hand tool, or by means of a power driven portable tool where in any of the foregoing cases there is a reasonably foreseeable risk of injury to the eyes of any person not engaged in any such work from particles or fragments thrown off.

2. Any process involving the use of an exposed electric arc or an exposed stream of arc plasma.

3. Work with drop hammers, power hammers, horizontal forging machines and forging presses other than hydraulic presses used in any case for the manufacture of forgings.

4. The fettling of metal castings, involving the removal of metal, including runners, gates and risers, and the removal of any other material during the course of such fettling, where in any of the foregoing cases there is a reasonably foreseeable risk of injury to the eyes of any person not engaged in any such work from particles or fragments thrown off.

5. Any process involving the use of an instrument which produces light amplification by the stimulated emission of radiation, where in any such process there is a reasonably foreseeable risk of injury to the eyes of any person not engaged in the process from radiation.

SCHEDULE 3

Regulation 1(2)

Column 1 Regulations revoked	Column 2 References	Column 3 Extent of revocation
The Chemical Works Regulations 1922.	S.R. & O. 1922/731 (Rev. VII, p.246: 1922, p. 255).	Regulations 10(*b*) and 25(*a*)(vi) in so far as they require the provision of goggles.
The Protection of Eyes Regulations 1938.	S.R. & O. 1938/654 (Rev. VII, p. 124: 1938 I, p. 1222).	The whole regulations.
The Iron and Steel Foundries Regulations 1953.	S.I. 1953/1464 (1953 I, p. 735).	Regulation 8(1)(*c*).

Column 1 Regulations revoked	Column 2 References	Column 3 Extent of revocation
The Shipbuilding and Ship-repairing Regulations 1960.	S.I. 1960/1932 (1960 II, p. 1427).	Regulations 74(1), (2)(*a*) and (3) and 75.
The Construction (General Provisions) Regulations 1961.	S.I. 1961/1580 (1961 II, p. 3207).	Regulation 52 and the Schedule.
The Non-ferrous Metals (Melting and Founding) Regulations 1962.	S.I. 1962/1667 (1962 II, p. 2021).	Regulation 13(1)(*c*).

EXPLANATORY NOTE

(*This Note is not part of the Regulations.*)

These Regulations make provision for the protection of the eyes of persons employed in factories and certain other premises and places to which the Factories Act 1961 applies, while any of the processes specified in Schedule 1 to the Regulations is being carried on. They require the provision of eye protectors or of shields (either worn by a person, held in the hand or independently mounted) for the protection of persons employed in (and in one case, at risk from) such processes. Eye protectors and, with certain exceptions, shields are required to conform with specifications approved by the Chief Inspector of Factories. Independently mounted shields are required to be provided for other persons employed in positions where their eyes are at risk of injury from particles or fragments thrown off or radiation produced in the carrying on of certain specified processes.

The Regulations supersede the Protection of Eyes Regulations 1938, which are revoked, and certain provisions requiring the provision of goggles or other means of protection contained in the Chemical Works Regulations 1922, the Iron and Steel Foundries Regulations 1953, the Shipbuilding and Ship-repairing Regulations 1960, the Construction (General Provisions) Regulations 1961 and the Non-ferrous Metals (Melting and Founding) Regulations 1962, which provisions are also revoked.

Appendix VIII

The Construction (Working Places) Regulations 1966 No. 94

Part I

Application and Interpretation

Citation, commencement and revocation

1.—(1) These Regulations may be cited as the Construction (Working Places) Regulations 1966 and shall come into operation on 1st August 1966.

(2) The following provisions are hereby revoked, that is to say,

(*a*) Regulations 6 to 33 of the Building (Safety, Health and Welfare) Regulations 1948(**a**);

(*b*) in Regulation 3(1) of the Construction (General Provisions) Regulations 1961(**b**) the words from "every contractor" (in the first place where those words occur after sub-paragraph (*b*)) to "and of"; and

(*c*) Regulations 7 and 54 of the said Regulations of 1961.

(3) In Regulation 3(1)(*a*) of the said Regulations of 1961 "8" shall be substituted for "7", and in Regulation 3(1)(*b*) thereof the reference to Regulation 54 shall be omitted.

Application of Regulations

2.—(1) These Regulations apply—

(*a*) to building operations; and

(*b*) to works of engineering construction;

undertaken by way of trade or business, or for the purpose of any industrial or commercial undertaking, or by or on behalf of the Crown or any municipal or other public authority, and to any line or siding which is used in connection therewith and for the purposes thereof and is not part of a railway or tramway.

(2) The provisions of these Regulations shall be in addition to and not in substitution for or in diminution of other requirements imposed by or under the principal Act.

Obligations under Regulations

3.—(1) It shall be the duty of every contractor, and every employer of workmen, who is undertaking any of the operations or works to which these Regulations apply—

(*a*) to comply with such of the requirements of the following Regulations as affect him or any workman employed by him, that is to say, Regula-

(**a**) S.I. 1948/1145 (Rev. VII, p. 191: 1948 I, p. 953).
(**b**) S.I. 1961/1580 (1961 II, p. 3207).

tions 6 to 23, 25, 30, 31, 32, 36 and 38 and, in so far as they relate to the falling or slipping of persons, Regulations 24, 26, 27, 28, 29, 33, 34 and 35: Provided that the requirements of the said Regulations shall be deemed not to affect any workman if and so long as his presence in any place is not in the course of performing any work on behalf of his employer and is not expressly or impliedly authorised or permitted by his employer; and

(*b*) to comply with such of the requirements of the following Regulations as relate to any work, act or operation performed or about to be performed by any such contractor or employer of workmen, that is to say, Regulations 37 and 39 and, in so far as they relate to the falling of materials and articles, Regulations 24, 26, 27, 28, 29, 33, 34 and 35,

and it shall be the duty of every contractor and every employer of workmen who erects or alters any scaffold to comply with such of the requirements of these Regulations as relate to the erection or alteration of scaffolds having regard to the purpose or purposes for which the scaffold is designed at the time of erection or alteration; and of every contractor and every employer of workmen who erects, installs, works or uses any other plant or equipment to which any of the provisions of these Regulations applies, to erect, install, work or use any such plant or equipment in a manner which complies with those provisions.

(2) It shall be the duty of every person employed to comply with the requirements of such of these Regulations as relate to the performance of or the refraining from an act by him and to co-operate in carrying out these Regulations and if he discovers any defect in the plant or equipment to report such defect without unreasonable delay to his employer or foreman, or to a person appointed by the employer under Part II of the Construction (General Provisions) Regulations 1961.

Interpretation

4.—(1) The Interpretation Act 1889**(a)** shall apply to the interpretation of these Regulations as it applies to the interpretation of an Act of Parliament, and as if these Regulations and the Regulations hereby revoked were Acts of Parliament.

(2) In these Regulations, unless the context otherwise requires, the following expressions have the meanings hereby assigned to them respectively, that is to say—

"ladder" does not include a folding step-ladder;

"ladder scaffold" means a scaffold with a working platform which is supported directly on a ladder or by means of a crutch or bracket on a rung or rungs of a ladder;

"lifting appliance" means a crab, winch, pulley block or gin wheel used for raising or lowering, and a hoist, crane, sheer legs, excavator, dragline, piling frame, aerial cableway, aerial ropeway or overhead runway;

"lifting gear" means a chain sling, rope sling, or similar gear, and a ring, link, hook, plate clamp, shackle, swivel or eye-bolt;

"plant or equipment" includes any plant, equipment, gear, machinery, apparatus or appliance, or any part thereof;

"the principal Act" means the Factories Act 1961 as amended by or under any other Act;

"scaffold" means any temporarily provided structure on or from which persons perform work in connection with operations or works to which

(a) 52 & 53 Vict. c. 63.

these Regulations apply, and any temporarily provided structure which enables persons to obtain access to or which enables materials to be taken to any place at which such work is performed, and includes any working platform, gangway, run, ladder or step-ladder (other than an independent ladder or step-ladder which does not form part of such a structure) together with any guard-rail, toe-board or other safeguards and all fixings, but does not include a lifting appliance or a structure used merely to support such an appliance or to support other plant or equipment;

"sloping roof" has the meaning assigned to it by Regulation 35;

"slung scaffold" means a scaffold suspended by means of lifting gear, ropes or chains or rigid members and not provided with means of raising or lowering by a lifting appliance or similar device;

"suspended scaffold" means a scaffold (not being a slung scaffold) suspended by means of ropes or chains and capable of being raised or lowered but does not include a boatswain's chair or similar appliance;

"trestle scaffold" includes a scaffold in which the supports for the platform are any of the following which are self-supporting that is to say, split heads, folding step-ladders, tripods or movable contrivances similar to any of the foregoing; and

"working platform" includes a working stage.

Part II

Exemptions

Certificates of exemption

5. The Chief Inspector may (subject to such conditions, if any, as may be specified therein) by certificate in writing (which he may in his discretion revoke at any time) exempt from all or any of the requirements of these Regulations—

(*a*) any particular plant or equipment or any class or description of plant or equipment; or

(*b*) any particular work or any class or description of work;

if he is satisfied that the requirements in respect of which the exemption is granted are not necessary for the protection of persons employed or are not reasonably practicable.

Part III

Safety of Working Places and Access and Egress

General

6.—(1) Without prejudice to the other provisions of these Regulations, there shall, so far as is reasonably practicable, be suitable and sufficient safe access to and egress from every place at which any person at any time works, which access and egress shall be properly maintained.

(2) Without prejudice to the other provisions of these Regulations, every place at which any person at any time works shall, so far as is reasonably practicable, be made and kept safe for any person working there.

Provision of scaffolds, etc.

7. Without prejudice to the other provisions of these Regulations, where work cannot safely be done on or from the ground or from part of a building or other permanent structure, there shall be provided, placed and kept in position for use and properly maintained either scaffolds or where appropriate ladders or other means of support, all of which shall be sufficient and suitable for the purpose.

Supervision of work and inspection of material

8. No scaffold shall be erected or be substantially added to or altered or be dismantled except under the immediate supervision of a competent person and so far as possible by competent workmen possessing adequate experience of such work. All material for any scaffold shall be inspected by a competent person on each occasion before being taken into use.

Construction and material

9.—(1) Every scaffold and every part thereof shall be of good construction, of suitable and sound material and of adequate strength for the purpose for which it is used.

(2) Sufficient material shall be provided for and shall be used in the construction of scaffolds.

(3) Timber used for scaffolds shall be of suitable quality, be in good condition, and have the bark completely stripped off.

(4) Timber used for scaffolds, trestles, ladders, and folding step-ladders shall not be so painted or treated that defects cannot easily be seen.

(5) Metal parts used for scaffolds shall be of suitable quality and be in good condition and free from corrosion or other patent defect likely to affect their strength materially.

Defective material

10.—(1) No defective material or defective part shall be used for a scaffold.

(2) No rope or bond which is defective whether through contact with an acid or other corrosive substance or otherwise shall be used.

(3) All material and parts for scaffolds shall when not in use be kept under good conditions and apart from any materials or parts unsuitable for scaffolds.

Maintenance of scaffolds

11. Every scaffold shall be properly maintained, and every part shall be kept so fixed, secured or placed in position as to prevent so far as is practicable accidental displacement.

Partly erected or dismantled scaffolds

12. No scaffold or part of a scaffold shall be partly erected or dismantled and remain in such a condition that it is capable of being used unless—

- (*a*) the scaffold as so erected or dismantled complies with these Regulations; or
- (*b*) a prominent warning notice indicating that the scaffold or part as the case may be, is not to be used is affixed near any point at which the scaffold or part, as the case may be, is liable to be approached for the purpose of use; or
- (*c*) access to the scaffold or part, as the case may be, is as far as reasonably practicable effectively blocked.

Standards or uprights, ledgers and putlogs

13.—(1) Standards or uprights of scaffolds shall—

(*a*) where practicable, be either vertical or slightly inclined towards the building or other structure; and

(*b*) be fixed sufficiently close together to secure the stability of the scaffold having regard to all the circumstances.

(2) The foot or base of any standard or upright shall be placed on an adequate base plate in a manner to prevent slipping or sinking, or its displacement shall be prevented in some other sufficient way.

(3) Ledgers shall be as nearly as possible horizontal and shall be securely fastened to the standards or uprights by efficient means.

(4) Putlogs or other supports on which a platform rests shall be securely fastened to the ledgers or to the standards or uprights, or their movement shall be prevented by other efficient means. Where one end of a putlog is supported by a wall that end shall extend into or on to the wall sufficiently to provide a supporting surface of sufficient area.

(5) The distance between two consecutive putlogs and other supports on which a platform rests shall be fixed with due regard to the anticipated load and the nature of the platform flooring. The distance with single planking shall not as a general rule exceed 3 feet 3 inches with planks of 1¼ inches in thickness, 5 feet with planks 1½ inches in thickness, or 8 feet 6 inches with planks 2 inches in thickness.

Ladders used in scaffolds

14.—(1) Ladders serving as uprights or scaffolds shall—

(*a*) be of adequate strength;

(*b*) be placed so that the two stiles or sides of each ladder are evenly supported or suspended; and

(*c*) be secured to prevent slipping.

(2) Ladder scaffolds shall be used only if the work is of such a light nature and the material required for the work is such that this type of scaffold can be used safely.

Stability of scaffolds

15.—(1) Every scaffold shall be securely supported or suspended and shall where necessary be sufficiently and properly strutted or braced to prevent collapse, and shall be rigidly connected with the building or other structure unless the scaffold is so designed and constructed as to ensure stability without such connection.

(2) Every structure and appliance used as a support for a scaffold shall be of sound construction, have a firm footing or be firmly supported, and shall where necessary be sufficiently and properly strutted or braced to prevent collapse and to ensure stability.

(3) Every scaffold which can be moved on wheels or skids (not being a suspended scaffold or slung scaffold) shall—

(*a*) be constructed with due regard to stability and, if necessary for stability, be adequately weighted at the base;

(*b*) be used only on a firm and even surface not so sloping as to involve risk of instability of the scaffold or any load thereon;

(*c*) be adequately secured to prevent movement when any person is working upon it or upon any ladder or other plant or equipment, being a ladder, plant or equipment which is supported by the scaffold; and

(*d*) be moved only by the application of force at or near the base.

(4) Loose bricks, drain pipes, chimneys pots or other unsuitable material shall not be used for the construction or support of scaffolds save that bricks or small blocks may, if they provide a firm support, be used to support a platform not more than 2 feet above the ground or floor.

Slung scaffolds

16.—(1) No chain, wire rope, lifting gear, metal tube or other means of suspension for slung scaffolds shall be used unless the following requirements (in so far as they are applicable) are observed, (in addition as respects chains, ropes and lifting gear to satisfying the requirements of the Construction (Lifting Operations) Regulations 1961**(a)**), that is to say—

(*a*) it is suitable and of adequate strength for the purpose for which it is used;

(*b*) it is properly and securely fastened to safe anchorage points and to the scaffold ledgers or other main supporting members;

(*c*) it is so placed as to ensure stability of the scaffold;

(*d*) it is as nearly vertical as is reasonably practicable; and

(*e*) it is kept taut.

(2) No rope other than a wire rope shall be used for the suspension of a slung scaffold.

(3) Where chains or wire ropes are used for the suspension of a slung scaffold, steps shall be taken to prevent such chains or wire ropes coming into contact at points of suspension with edges where this would cause danger.

(4) Every slung scaffold shall be secured to prevent undue horizontal movement while it is used as a working platform.

Cantilever, jib, figure and bracket scaffolds

17.—(1) No cantilever scaffold or jib scaffold shall be used unless it is adequately supported, fixed and anchored, has outriggers of adequate length and strength and is where necessary sufficiently and properly strutted or braced to ensure rigidity and stability.

(2) No figure scaffold or bracket scaffold supported or held by dogs, spikes, or similar fixings liable to pull out of the stone-work, brickwork or other surface in which they are gripped or fixed shall be used.

Support for scaffolds, etc.

18. No part of a building or other structure shall be used as support for a scaffold, ladder, folding step-ladder or crawling ladder or for part of a scaffold, ladder, folding step-ladder or crawling ladder, unless the part of the building or other structure is of sound material and sufficiently stable and of sufficient strength to afford safe support. Gutters shall not be used as such supports

(a) S.I. 1961/1581 (1961 II, p. 3223).

unless they and their fixings are suitable and are of adequate strength and in the case of overhanging eaves gutters shall not be so used unless in addition they have been specially designed as walkways.

Suspended scaffolds (Not power operated)

19.—(1) The requirements of this Regulation (in addition, as respects lifting appliances, chains, ropes and lifting gear used in connection therewith, to the requirements of the Construction (Lifting Operations) Regulations 1961 shall be observed as respects—

(*a*) every suspended scaffold; and

(*b*) plant or equipment which is permanent plant or equipment of a building and which, but for the fact that it is permanently provided, would be a suspended scaffold;

being in any case a suspended scaffold, plant or equipment which is not raised or lowered by a power-driven lifting appliance or power-driven lifting appliances and no such suspended scaffold, plant or equipment shall be used unless it complies with the requirements of this Regulation.

(2) In the application of the succeeding paragraphs of this Regulation, references therein to suspended scaffolds shall be construed as references to suspended scaffolds to which this Regulation applies and as including references to plant or equipment of the kind referred to in sub-paragraph (*b*) of the foregoing paragraph of this Regulation.

(3) Every suspended scaffold shall be provided with adequate and suitable chains or ropes and winches or other lifting appliances or similar devices and shall be suspended from suitable outriggers, joists, runways, rail tracks or other equally safe anchorage.

(4) The winches or other lifting appliances or similar devices of a suspended scaffold shall be—

(*a*) provided with a brake or similar device which comes into operation when the operating handle or lever is released; and

(*b*) adequately protected against the effects of weather, dust or material likely to cause damage.

(5) The outriggers for a suspended scaffold shall be of adequate length and strength and properly installed and supported and, subject to paragraph (15) of this Regulation, shall be installed horizontally and provided with adequate stops at their outer ends. The outriggers shall be properly spaced having regard to the construction of the scaffold and of the runway, joist or rail track on which the scaffold is carried.

(6) Where counterweights are used with outriggers the counterweights shall be securely attached to the outriggers and shall be not less in weight than three times the weight which would counterbalance the weight suspended from the outrigger including the weight of the runway, joist or rail track, the suspended scaffold and persons and other load thereon.

(7) The points of suspension of every suspended scaffold shall be an adequate horizontal distance from the face of the building or other structure.

(8) Every runway, joist and rail track supporting a suspended scaffold shall be of suitable and sound material, adequate strength for the purpose for which it is used and free from patent defect, shall be provided with adequate stops at each end and shall be properly secured to the building or other structure or,

where outriggers are used, to the outriggers.

(9) The suspension ropes or chains of a suspended scaffold—

(*a*) shall be securely attached to the outriggers or other supports and to the platform framework or to any lifting appliance or other device attached thereto, as the case may be; and

(*b*) shall be kept in tension.

(10) Where winches are used with suspended scaffolds the suspension ropes shall be of such a length that at the lowest position at which the scaffold is intended to be used there are not less than two turns of rope remaining on each winch drum and the length of each rope shall be clearly marked on its winch.

(11) Every part of a suspended scaffold and all plant and equipment used for the purposes thereof shall be of good construction, suitable and sound material, of adequate strength for the purpose for which it is used and shall be properly maintained, and, where constructed of metal, shall be free from corrosion and other patent defects, being corrosion and defects likely materially to affect its strength. Adequate arrangements shall be made to prevent undue tipping, tilting or swinging of a suspended scaffold and to secure it to prevent undue horizontal movement while it is being used as a working platform.

(12) No rope other than a wire rope shall be used for the raising, lowering and suspension of a suspended scaffold, except that the raising lowering and suspension may be carried out by means of fibre ropes and pulley blocks in the case of work to which paragraph (15) of this Regulation applies.

(13) The platform of every suspended scaffold shall—

(*a*) except to the extent necessary for drainage, be closely boarded, planked or plated; and

(*b*) subject to paragraph (15) of this Regulation, be of adequate width to afford adequate working space at every working point and shall, in any event,

(i) be at least 25 inches wide if used as a footing only and not for the deposit of any material;

(ii) be at least 34 inches wide if used for the deposit of material; and

(iii) not be used for the support of any higher scaffold;

and shall be so arranged or secured that at each working position the space between the face of the building or other structure and the platform is as small as reasonably practicable; so however, that where workmen sit at the edge of the platform to work there may be a space not exceeding 12 inches. Where necessary, devices shall be provided and used to keep the platform a sufficient distance from the wall when persons have to work in a sitting position.

(14) If a suspended scaffold is carried on fibre ropes and pulley blocks the ropes shall be spaced not more than 10 feet 6 inches apart.

(15) Where the work to be carried out from a suspended scaffold is of such a light nature and the material required for the work is such that a cradle or similar light-weight suspended scaffold can be used with safety and where such suspended scaffold is used, the following requirements of this Regulation shall not apply, that is to say—

(*a*) the requirement of paragraph (5) that the outriggers shall be installed horizontally and that stops shall be provided; and

(*b*) the requirements of paragraph (13)(*b*) as to the width of the platform.

The platform of a suspended scaffold to which this paragraph applies shall be not less than 17 inches wide.

Boatswain's chairs, cages, skips, etc. (Not power operated)

20.—(1) No boatswain's chair, cage, skip or similar plant or equipment (not being a boatswain's chair, cage, skip or similar plant or equipment which is raised or lowered by a power-driven lifting appliance) shall be used unless—

(*a*) it is of good construction, suitable and sound material, adequate strength, free from patent defect and properly maintained;

(*b*) the outriggers or other supports are of adequate strength and properly installed and supported;

(*c*) the chains, ropes, lifting gear or other means of suspension used therewith (in addition to satisfying the requirements of the Construction (Lifting Operations) Regulations 1961) are securely attached to the outriggers or other supports and to the chair, cage, skip or similar plant or equipment or to any lifting appliance or other device attached thereto, as the case may be;

(*d*) suitable means are provided to prevent any occupant falling out;

(*e*) it is free of materials or articles liable to interfere with the occupant's handhold or foothold or otherwise endanger him;

(*f*) suitable measures are taken to prevent spinning or tipping in a manner dangerous to any occupant;

(*g*) in the case of any skip or other receptacle it is at least 3 feet deep; and

(*h*) its installation has been, and its use is, supervised by a competent person.

(2) No boatswain's chair, cage, skip or similar plant or equipment (not being a boatswain's chair, cage, skip or similar plant or equipment which is raised or lowered by a power-driven lifting appliance) shall be used as a working place in circumstances in which a suspended scaffold could be used unless the work is of such short duration as to make the use of a suspended scaffold unreasonable or the use of a suspended scaffold is not reasonably practicable.

Trestle scaffolds

21.—(1) All trestles and supports used for the construction of any trestle scaffold shall be of good construction, suitable and sound material, adequate strength for the purposes for which they are used and free from patent defect and shall be properly maintained.

(2) A trestle scaffold shall not be used—

(*a*) if the scaffold is so situated that a person would be liable to fall from its working platform a distance of more than 15 feet; or

(*b*) if constructed with more than one tier where folding supports are used.

(3) No trestle scaffold shall be erected on a scaffold platform unless—

(*a*) the width of the said platform is such as to leave sufficient clear space for the transport of materials along the platform; and

(*b*) the trestles or supports are firmly attached to the said platform and adequately braced to prevent displacement.

Inspection of scaffolds, boatswain's chairs, etc.

22.—(1) Subject to the provisions of this Regulation, no scaffold (including any boatswain's chair, cage, skip or similar plant or equipment) and no plant or equipment used for the purposes of any of the foregoing shall be used unless (in addition to satisfying the requirements of the Construction (Lifting Operations) Regulations 1961)—

(*a*) it has been inspected by a competent person within the immediately preceding 7 days;

(*b*) it has been inspected by a competent person since exposure to weather conditions likely to have affected its strength or stability or to have displaced any part; and

(*c*) a report has been made of the results of every such inspection in the form set out in the Schedule to these Regulations and containing the particulars therein specified and signed by the person making the inspection:

Provided that sub-paragraph (*a*) shall not apply in the case of a scaffold no part of which has been erected for more than 7 days, and sub-paragraph (*c*) shall not apply to a ladder scaffold, a trestle scaffold or a scaffold from no part of which a person is liable to fall a distance of more than 6 feet 6 inches.

(2) Paragraph (1) of this Regulation shall not require a scaffold to be inspected by reason only that it has been added to, altered or partly dismantled.

(3) In the case of a site where the employer for whom the inspection was carried out has reasonable grounds for believing that the operations or works will be completed in a period of less than 6 weeks, the provision in this Regulation requiring that a report shall have been made and signed shall be deemed to have been satisfied if the person in charge of the operations or works carried on by that employer at that site has himself carried out the inspection and is a competent person and if, within 1 week of the date of the inspection, he reports to his employer in writing that the scaffold, boatswain's chair, cage, skip or similar plant or equipment (as the case may be) and any plant and equipment used for the purposes of any of the foregoing was inspected by him and that he found it in good order or observed certain defects, as the case may be, and the date of such inspection and the results thereof together with the name of the person making the inspection are entered by the employer in the said form together with the said particulars.

Scaffolds used by workmen of different employers

23. Where a scaffold or part of a scaffold is to be used by or on behalf of an employer other than the employer for whose workmen it was first erected, the first-mentioned employer shall, before such use, and without prejudice to any other obligations imposed upon him by these Regulations, take express steps, either personally or by a competent agent, to satisfy himself that the scaffold or part thereof is stable, that the materials used in its construction are sound and that the safeguards required by these Regulations are in position.

Construction of working platforms, gangways and runs

24.—(1) Every working platform, gangway and run from any part of which a person is liable to fall a distance of more than 6 feet 6 inches shall be closely boarded, planked or plated: Provided that this requirement shall not apply to—

(*a*) a platform, gangway or run consisting of open metal work having interstices none of which exceeds 6 square inches in area, if there is no risk of persons below any such platform, gangway or run being struck by materials or articles falling through the platform, gangway or run; or

(*b*) a platform, gangway or run, the boards, planks or plates of which are so secured as to prevent their moving and so placed that the space between adjacent boards, planks or plates does not exceed 1 inch, if there is no risk of persons below any such platform, gangway or run being struck by materials or articles falling through the platform.

(2) No gangway or run shall be used the slope of which exceeds 1 vertical to 1½ horizontal.

(3) Where the slope of a gangway or run renders additional foothold necessary, and in every case where the slope is more than 1 vertical to 4 horizontal there shall be provided proper stepping laths which shall—

(*a*) be placed at suitable intervals; and

(*b*) be the full width of the gangway or run, except that where necessary they may be interrupted over widths of not more than 4 inches to facilitate the movement of barrows.

Boards and planks in working platforms, gangways and runs

25.—(1) Every board or plank forming part of a working platform, gangway or run shall be—

(*a*) of a thickness which is such as to afford adequate security having regard to the distance between the putlogs or other supports; and

(*b*) not less than 8 inches wide, or, in the case of boards or planks exceeding 2 inches in thickness, not less than 6 inches wide.

(2) No board or plank which forms part of a working platform, gangway or run shall project beyond its end support to a distance exceeding four times the thickness of the board or plank unless it is effectively secured to prevent tipping, or to a distance which, having regard to the thickness and strength of the plank, renders the projecting part of the plank an unsafe support for any weight liable to be upon it.

(3) Suitable measures shall be taken by the provision of adequate bevelled pieces or otherwise to reduce to a minimum the risk of tripping and to facilitate the movement of barrows where boards or planks which form part of a working platform, gangway or run overlap each other or are not of reasonably uniform thickness where they meet each other or owing to warping or for some other reason do not provide an even surface.

Provided that this paragraph shall not apply to a working platform, gangway or run one side of which is contiguous to a curved surface of any cylindrical or spherical structure forming part of a work of engineering construction.

(4) Every board or plank which forms part of a working platform, gangway or run shall—

(*a*) rest securely and evenly on its supports; and

(*b*) rest on at least 3 supports unless, taking into account the distance between the supports and the thickness of the board or plank, the conditions are such as to prevent undue or unequal sagging.

(5) Where work has to be done at the end of a wall or working face the working platform at such wall or face shall, wherever practicable, extend at least 24 inches beyond the end of the wall or face.

Widths of working platforms

26.—(1) Subject to paragraphs (2) and (4) of this Regulation, every working platform (other than working platforms of suspended scaffolds and working platforms referred to in Regulations 28(6)(*c*) and 35) from which a person is liable to fall a distance of more than 6 feet 6 inches shall—

(*a*) if used as a footing only and not for the deposit of any material, be at least 25 inches wide;

(*b*) if used for the deposit of material, be at least 34 inches wide and have a clear passage way between one side of the working platform and the deposited material adequate in width for the passage of persons being a passage way which is in any case at least 17 inches wide;

(*c*) if used for the passage of materials, afford a clear passage way which is adequate in width for the passage of the materials without removal of the guard-rails and toe-boards being a passage way which is in any case at least 25 inches wide;

(*d*) if used for the support of any higher platform, be at least 42 inches wide;

(*e*) if used to dress or roughly shape stone be at least 51 inches wide;

(*f*) if used for the support of any higher platform and is one upon which stone is dressed or roughly shaped, be at least 59 inches wide;

and in every case be of sufficient width to afford adequate working space at every part.

(2) Subject to paragraph (4) of this Regulation, the following working platforms to which this Regulation applies shall be at least 17 inches wide—

(*a*) a platform of a ladder scaffold or a platform supported directly by folding trestles or folding step-ladders or a platform under a roof used for work on or in the vicinity of the roof being a platform which is supported by or suspended from roof members or the roof, where in any such case the work thereon is of a light nature and of short duration in any one position and a platform less than 25 or 34 inches wide (as the case may be) can be used with safety; and

(*b*) a platform which is used for work in connection with cylindrical or spherical metal structures.

(3) Where work at the face of a building or other structure is done from a working platform to which this Regulation applies the space between such face and the working platform shall be as small as practicable; so, however, that where workmen sit at the edge of the platform to work, there may be a space not exceeding 12 inches.

(4) The provisions of paragraphs (1) and (2) of this Regulation shall not apply to a working platform to which this Regulation applies where it is impracticable by reason of limitations of space to provide a platform of the width required by the said paragraphs; so, however, that in any such case the platform shall be as wide as is reasonably practicable.

Widths of gangways and runs

27.—(1) Subject to paragraph (2) of this Regulation, every gangway and run from any part of which a person is liable to fall a distance of more than 6 feet 6 inches shall,

(*a*) if used for the passage of persons only, be at least 17 inches wide;

(*b*) if used for the passage of materials, be adequate in width for the passage of the materials and in any case be at least 25 inches wide.

(2) The foregoing provisions of this Regulation shall not apply to a gangway or run where it is impracticable by reason of limitations of space to provide a gangway or run of the width required by the said provisions; so, however, that in any such case the gangway or run shall be as wide as is reasonably practicable.

Guard-rails and toe-boards at working platforms and places

28.—(1) Every side of a working platform or working place, being a side thereof from which a person is liable to fall a distance of more than 6 feet 6 inches, shall, subject to paragraphs (3) to (6) and except as provided in paragraph (7) of this Regulation, be provided with a suitable guard-rail or guard-rails of adequate strength to a height of between 3 feet and 3 feet 9 inches above the platform or place and above any raised standing place on the platform, and with toe-boards or other barriers up to a sufficient height which shall in no case be less than 6 inches. Such guard-rails and toe-boards or other barriers shall be so placed as to prevent so far as possible the fall of persons, materials and articles from such platform or place.

(2) Without prejudice to the provisions of Regulation 11, the outward movement of guard-rails and toe-boards or barriers shall (unless they are so designed and used as to prevent such movement) be prevented by placing them on the inside of the uprights or by other equally effective means.

(3) Where guard-rails are required to be provided, the distance between any toe-board or other barrier and the lowest guard-rail above it shall not exceed 30 inches.

(4) Guard-rails, toe-boards and barriers required by paragraph (1) of this Regulation may be removed or remain unerected for the time and to the extent necessary for the access of persons or the movement of materials or other purposes of the work; but guard-rails, toe-boards and barriers removed or remaining unerected for any of those purposes shall be replaced or erected as soon as practicable.

(5) On the side of a suspended scaffold next to the wall or working face—

(*a*) guard-rails where required by this Regulation need not extend to a height of more than 27 inches above the platform if the work is impracticable with a guard-rail at a greater height; and

(*b*) guard-rails and toe-boards or other barriers shall not be required if the workers sit at the edge of the platform to work and ropes or chains affording all the workers a safe and secure handhold are provided.

(6) None of the requirements of paragraphs (1) and (2) of this Regulation shall apply to—

(*a*) the platform of a ladder scaffold if a secure handhold is provided along the full length of such platform;

(*b*) the platform of a trestle scaffold when the platform is supported on folding trestles, split heads or similar devices or folding step-ladders;

(*c*) a platform which is used only in the course of erecting any framework or prefabricated unit forming part of a building or other permanent structure for the purposes of jointing, bolting-up, riveting or welding work and which is used for such a short period as to make the provision of guard-rails and toe-boards or barriers unreasonable if—

(i) the platform is at least 34 inches wide;

(ii) there is adequate handhold; and

(iii) the platform is not used for the deposit of materials or articles otherwise than in boxes or receptacles suitable to prevent the fall of the materials or articles from the platform;

(*d*) a temporary platform passing between two adjacent glazing bars of a roof with a sloping surface if those bars or the roof framework afford secure handhold along the full length of the platform:

Provided that toe-boards or barriers shall be provided in accordance with paragraphs (1) and (2) of this Regulation except where the provision of toe-boards or barriers is impracticable on account of the nature or circumstances of the work;

(*e*) a platform under a roof being a platform which is supported by or suspended from roof members or the roof and which is used only for work on or in the vicinity of the roof and of a light nature and of such short duration as to make the provision of guard-rails and toe-boards or barriers unreasonable if—

(i) there is adequate handhold at every working position; and

(ii) the material required for the work is such that the platform can be used with safety;

(*f*) a working platform or working place one side of which is contiguous to the concave surface of a cylindrical or spherical structure so long as reasonably practicable steps are being taken to prevent persons working thereon from falling a distance of more than 6 feet 6 inches.

(7) Except as provided in Regulation 35, the provisions of this Regulation shall not apply to working platforms and working places being working platforms and working places to which that Regulation applies.

Guard-rails, etc. for gangways, runs and stairs

29.—(1) Except for the time and to the extent necessary for the access of persons or the movement of materials, stairs shall be provided throughout their length with hand-rails or other efficient means to prevent the fall of persons. If necessary to prevent danger to any person the hand-rails or other means shall be continued beyond the end of the stairs.

(2) Every side of any gangway, run or stairs from which a person is liable to fall a distance of more than 6 feet 6 inches shall be provided—

(*a*) with a suitable guard-rail or guard-rails of adequate strength to a height of between 3 feet and 3 feet 9 inches above the gangway, run or stairs; and

(*b*) except in the case of stairs, with toe-boards or other barriers up to a sufficient height which shall in no case be less than 6 inches and so placed as to prevent as far as possible the fall of persons, materials and articles. The space between any such toe-board or barrier and the lowest guard-rail above it shall not exceed 30 inches.

Provided that the provisions of this paragraph shall not apply to a temporary gangway which is used only in the course of erecting any framework forming part of a building or other permanent structure for work of such short duration as to make the provision of a gangway with guard-rails and toe-boards or other barriers unreasonable.

(3) Guard-rails, toe-boards and barriers required by paragraph (2) of this Regulation may be removed or remain unerected for the time and to the extent necessary for the access of persons or the movement of materials or other purposes of the work; but guard-rails, toe-boards and barriers removed or remaining unerected for any of those purposes shall be replaced or erected as soon as practicable.

Platforms, gangways, runs and stairs, etc. to afford safe foothold

30.—(1) If a platform, gangway, run or stair becomes slippery, appropriate steps shall as soon as reasonably practicable be taken by way of sanding, cleaning or otherwise to remedy the condition.

(2) Every platform, gangway, run or stair shall be kept free from any unnecessary obstruction and material and free from rubbish and any projecting nails.

Construction and maintenance of ladders and folding step-ladders

31.—(1) Every ladder and folding step-ladder shall be of good construction, of suitable and sound material and of adequate strength for the purpose for which it is used and shall be properly maintained.

(2) No ladder shall be used in which a rung is missing or is defective.

(3) Every rung of a ladder shall be properly fixed to the stiles or sides. No ladder shall be used in which any rung depends for its support solely on nails, spikes or other similar fixing. Where in the case of a wooden ladder the tenon joints are not secured by wedges, reinforcing ties shall be used. Wooden stiles or sides and wooden rungs of ladders shall have the grain running lengthwise.

Provided that the requirements of this paragraph shall not apply to ladders to which Regulation 35 applies.

Use of ladders and folding step-ladders

32.—(1) This Regulation shall—

(*a*) apply to ladders and folding step-ladders being ladders and folding step-ladders which afford a means of access, egress, communication or support to a person or persons employed; and

(*b*) not apply to any ladder lying upon a roof or to any crawling board or crawling ladder.

(2) Subject to the provisions of paragraph (4) of this Regulation, no ladder standing on a base shall be used unless—

(*a*) except as provided in the next following paragraph of this Regulation, it is securely fixed near to its upper resting place, or, in the case of a vertical ladder near to its upper end:

Provided that where such fixing is impracticable the ladder shall be securely fixed at or near to its lower end;

(*b*) it has a level and firm footing and is not standing on loose bricks or other loose packing;

(*c*) it is secured where necessary to prevent undue swaying or sagging; and (*d*) it is equally and properly supported on each stile or side.

(3) Subject to the provisions of paragraph (4) of this Regulation, where it is impracticable in the case of a ladder standing on a base to comply with either of the requirements of sub-paragraph (*a*) of the last foregoing paragraph, a person shall be stationed at the foot of the ladder when in use to prevent it slipping.

(4) Paragraphs (2) and (3) of this Regulation shall not apply to a ladder which is not more than 10 feet in length and which is not used as a means of communication, if the ladder is securely placed so as to prevent it from slipping or falling.

(5) No ladder shall be used unless—

(*a*) (i) it extends to a height of at least 3 feet 6 inches above the place of landing or the highest rung to be reached by the feet of any person using the ladder, as the case may be, or, if this is impracticable, to the greatest practicable height; or

(ii) there is other adequate handhold; and

(*b*) there is sufficient space at each rung to provide adequate foothold.

(6) Every ladder, other than ladders to which paragraphs (2) and (3) of this Regulation apply, shall before being used—

(*a*) be securely suspended;

(*b*) be secured where necessary to prevent undue swinging or swaying; and

(*c*) be equally and properly suspended by each stile or side.

(7) No folding step-ladder shall be used unless it has a level and firm footing or while it is standing on loose bricks or other loose packing.

(8) No ladder or run of ladders rising a vertical distance of over 30 feet shall be used unless it is, if practicable, provided with an intermediate landing place or intermediate landing places so that the vertical distance between any two successive landing places shall not exceed 30 feet. Every landing place shall be of adequate dimensions and, if a person is liable to fall therefrom a distance of more than 6 feet 6 inches, shall, except in so far as it is not reasonably practicable, be provided with sufficient and suitable guard-rails to a height of between 3 feet and 3 feet 9 inches above the landing place and with toe-boards or other barriers up to a sufficient height which shall in no case be less than 6 inches, so placed as to prevent as far as possible the fall of persons, materials and articles and so that the space between any toe-board or other barrier and the lowest guard-rail above it shall not exceed 30 inches. Where a ladder passes through an opening in the floor of a landing place, the opening shall be as small as is reasonably practicable.

Openings, corners, breaks, edges and open joisting

33.—(1) Paragraphs (1) to (3) of this Regulation apply to every opening, corner, break or edge being an opening, corner, break or edge which any person employed is liable to approach or near or across which any person is liable to pass—

(*a*) in or of a roof (other than a roof to which Regulation 35 applies), floor, wall or other similar part either of a building or of any other structure whether the roof, floor, wall or other similar part of the building or of the other structure is complete or only partly complete or is in course of construction, maintenance, repair or demolition; or (*b*) in or of a working platform, gangway or run.

(2) Subject to Regulation 34, in the case of any such opening, corner, break or edge through or from which any person is liable to fall a distance of more than 6 feet 6 inches or to fall into any liquid or material so as to involve risk of drowning or serious injury there shall be provided either—

(*a*) a suitable guard-rail or guard-rails of adequate strength to a height of between 3 feet and 3 feet 9 inches above the surface across which persons are liable to pass together with toe-boards or other barriers up to a sufficient height which shall in no case be less than 6 inches, so placed as to prevent as far as possible the fall of persons, materials and articles and so that the space between any toe-board or other barrier and the lowest guard-rail above it shall not exceed 30 inches; or

(*b*) a covering so constructed as to prevent the fall of persons, materials and articles; any such covering shall be clearly and boldly marked to show its purpose or be securely fixed in position.

(3) Subject to Regulation 34, in the case of any such opening, corner, break or edge (not being an opening, corner, break or edge to which the last foregoing paragraph of this Regulation applies) through or from which materials or articles are liable to fall so as to endanger persons employed, suitable precautions by way of the erection of toe-boards, secure covering or otherwise shall be taken to prevent materials and articles so falling.

(4) Subject to Regulation 34, when work is done on or immediately above any open joisting through which a person is liable to fall a distance of more than 6 feet 6 inches, such joisting shall be securely covered by boards or other temporary covering to the extent necessary to afford safe access to or foothold for the work, or other effective measures shall be taken to prevent persons from falling.

Exceptions from Regulation 33

34.—(1) Guard-rails, toe-boards, barriers and coverings required by Regulation 33 may be removed or remain unerected—

(*a*) where and when this is necessary in order to proceed with any permanent filling in, covering or enclosure; or

(*b*) for the time and to the extent necessary for the access of persons or the movement of materials or other purposes of the work;

but guard-rails, toe-boards, barriers and coverings removed or remaining unerected for any of those purposes shall be replaced or erected as soon as practicable.

(2) Without prejudice to Regulation 28, Regulation 33 shall not apply to any opening, corner, break or edge created in the course of demolition operations to which Part X of the Construction (General Provisions) Regulations 1961 applies, or to any opening, corner, break or edge created in the course of any other demolition operation, if in the course of such last mentioned demolition operation the opening, corner, break or edge is not left unattended.

Sloping roofs

35.—(1) In this regulation "sloping roof" means a roof or part of a roof being a roof or part having a pitch of more than 10 degrees which is covered either wholly or partly and—

(*a*) which is in the course of construction, maintenance, repair or demolition; or

(*b*) which is used as a means of access to or egress from operations or works on a roof or part of a roof being operations or works to which these Regulations apply.

(2) Except as provided in paragraphs (6) and (7) of this Regulation, where any sloping roof has—

(*a*) a pitch of more than 30 degrees; or

(*b*) a pitch of 30 degrees or less and a surface on or from which a person is by reason of the nature or condition of the surface or of the weather liable to slip or fall to such an extent that he is liable to fall from the edge of the roof;

work thereon or therefrom shall only be carried out by workmen who are suitable for such work and the requirements of paragraphs (3) and (4) of this Regulation shall be complied with.

(3) Where any sloping roof is used as a means of access to or egress from operations or works on a roof or a part of a roof being operations or works to which these Regulations apply, sufficient and suitable crawling ladders or crawling boards shall be provided on that sloping roof.

(4) Where any work is done on or from any sloping roof sufficient and suitable crawling ladders or crawling boards shall be provided on that sloping roof and (except where the work is not extensive) either—

(*a*) a barrier shall be provided at the lower edge of the sloping roof, other than the upper surface of a tank or similar structure of metal construction, of such a design and so constructed as to prevent any person falling from that edge; or

(*b*) the work shall be done from a securely supported working platform not less than 17 inches wide which complies with the requirements of Regulation 28(1) to (4).

(5) Crawling ladders and crawling boards provided in pursuance of paragraph (3) or (4) of this Regulation shall be—

(*a*) of good construction, suitable and sound material, adequate strength for the purposes for which they are used, free from patent defect and properly maintained;

(*b*) properly supported; and

(*c*) securely fixed or anchored to the sloping surface or over the roof ridge or securely fixed in some other effective way, so as, in every case, to prevent slipping.

(6) The provision of crawling ladders or crawling boards shall not be required in the case of any sloping roof where the handhold and foothold afforded by the battens or other similar members of the structure are such that the said sloping roof is as safe for every person thereon as it would be if the said provision had been made.

(7) The requirements of paragraph (4)(*a*) and (*b*) of this Regulation shall apply only in the case of any sloping roof from the eaves of which a person is liable to fall a distance of more than 6 feet 6 inches.

(8) Suitable and sufficient means shall be provided to prevent the fall of materials or articles from a sloping roof.

Work on or near fragile materials

36.—(1) Without prejudice to the provisions of Regulations 33 and 35, no person shall pass across, or work on or from, material which would be liable to fracture if his weight were to be applied to it and so situated that if it were to be so fractured he would be liable to fall a distance of more than 6 feet 6 inches unless such one or more of all or any of the following, that is to say, suitable and sufficient ladders, crawling ladders, crawling boards and duck-boards (which shall in any case be securely supported and, if necessary, secured so as to prevent their slipping) as are necessary are provided and so used that the weight of any person so passing or working is wholly or mainly supported by such ladders or boards unless his weight is supported by other equally safe and sufficient means.

(2) Without prejudice to the provisions of Regulations 33 and 35, no person shall pass or work near material of the kind and situated as specified in paragraph (1) of this Regulation unless provision is made by means of such one or more of all or any of the following, that is to say, suitable guard-rails, suitable coverings and other suitable means as are necessary for preventing, so far as reasonably practicable, any person so passing or working from falling through the said material.

(3) Where any person passes across or near or works on or near material of the kind and situated as specified in paragraph (1) of this Regulation prominent warning notices shall, except where the material consists wholly of glass, be affixed at the approaches to the place where the material is situated.

(4) References in this Regulation to a person's weight shall be construed as references to the aggregate of his weight and that of anything he may for the time being be supporting by his person.

(5) All plant and equipment provided in pursuance of this Regulation shall be of good construction, suitable and sound material, adequate strength for the purpose for which it is used, free from patent defect and properly maintained.

Loads on scaffolds

37.—(1) A scaffold shall not be overloaded and so far as practicable the load thereon shall be evenly distributed.

(2) When any material is transferred on or to a scaffold it shall be moved or deposited without imposing any violent shock. Materials shall not be kept upon a scaffold unless needed for work within a reasonable time.

Prevention of falls and provision of safety nets and belts

38.—(1) Where by reason of the special nature or circumstances of any part of the work or of the access thereto or the egress therefrom it is impracticable to comply with all or any of the requirements of the provisions of Regulations 6, 7, 24 to 30, 33, 35 and 36, so far as they relate to the falls of persons, the requirements of those provisions shall be complied with so far as practicable and in any such case, except as provided in paragraphs (2) and (3) of this Regulation, there shall in addition where practicable be provided and so erected and kept in such positions as to be effective to protect persons carrying on that part of the work or using the said access or egress suitable safety nets or safety sheets of such a design and so constructed and installed as to prevent so far as practicable injury to persons falling on to them:

Provided that such safety nets or safety sheets may be removed or remain unerected for the time and to the extent necessary for the access of persons or the movement of materials or other purposes of the work, but shall be replaced or erected as soon as practicable.

(2) Where by virtue of the preceding paragraph of this Regulation safety nets or safety sheets would be required to be provided for the protection of the persons carrying on any part of the work or using the access thereto or the egress therefrom but all such persons are able to carry on that work or use the said access or egress while making use of safety belts attached continuously to a suitable and securely fixed anchorage, such safety nets or safety sheets shall not be required to be provided if there are provided and so used by those persons while carrying on that work and using the said access and egress suitable and sufficient safety belts or other suitable and sufficient equipment.

(3) In any of the following cases, that is to say—

(*a*) where it is impracticable to provide all such safety nets or safety sheets as would be required to comply with the requirements of paragraph (1) of this Regulation;

(*b*) where it is not reasonably practicable so to provide all such nets or sheets by reason of the frequent movement of materials or other purposes of the work; and

(*c*) where the work is of such short duration as to make the provision of all such nets or sheets unreasonable;

safety nets or safety sheets in accordance with the provisions of that paragraph shall be provided to the extent to which it is reasonably practicable to provide them and in any such case as aforesaid there shall also be provided (together with suitable and sufficient anchorages) suitable and sufficient safety belts or other suitable and sufficient equipment having suitable fittings and being of such a design and so constructed as to prevent serious injury in the event of a fall to persons using them.

(4) All safety nets, safety sheets, safety belts and other equipment provided in pursuance of this Regulation shall be properly maintained.

Part IV

Keeping of Records

Reports, etc.

39.—(1) The reports required by Regulation 22 shall be kept on the site of the operations or works and when there are no relevant operations or works shall be kept at an office of the contractor or employer for whom the examination was carried out:

Provided that in the case of a site where the contractor or employer has reasonable grounds for believing that the operations or works will be completed in a period of less than 6 weeks, the contractor or employer may keep the said reports at his office.

(2) All reports and other documents required for the purposes of these Regulations shall at all reasonable times be open to inspection by any inspector and the person keeping any such report or other document shall send to any such inspector such extracts therefrom or copies thereof as the inspector may from time to time require for the purpose of the execution of his duties under the principal Act.

SCHEDULE

Regulation 22

FACTORIES ACT 1961

CONSTRUCTION (WORKING PLACES) REGULATIONS 1966

SCAFFOLD INSPECTIONS

FORM OF REPORTS OF RESULTS OF INSPECTIONS UNDER REGULATION 22 OF SCAFFOLDS, INCLUDING BOATSWAIN'S CHAIRS, CAGES, SKIPS AND SIMILAR PLANT OR EQUIPMENT (AND PLANT OR EQUIPMENT USED FOR THE PURPOSES THEREOF)

Name or title of Employer or Contractor..................................

Address of Site..

Work Commenced—Date..

Location and Description of Scaffold, etc. and other Plant or Equipment Inspected (1)	Date of Inspection (2)	Result of Inspection. State whether in good order (3)	Signature (or, in case where signature is not legally required, name) of person who made the inspection (4)

EXPLANATORY NOTE

(This Note is not part of the Regulations.)

For the protection of persons employed on building operations and on works of engineering construction, these Regulations impose requirements as to the safety of the working places and of the means of access and egress to and from those places. In the case of building operations these Regulations replace similar requirements in the Building (Safety, Health and Welfare) Regulations 1948.

Appendix IX

Regulations to which Regulation 3(1)(*a*), 3(1)(*b*) Construction (Working Places) Regulations 1966 applies

Regulation 3(1)(*a*) extends coverage to regulations as follows:

Regulation

6. Working places and access to work;
7. Provision and maintenance of scaffolds, ladders or other means of support;
8. Supervision of scaffolding and inspection of material;
9. Construction and material of scaffolds;
10. Defective scaffold material;
11. Maintenance of scaffolds;
12. Partly erected or dismantled scaffolds;
13. Standards or uprights, ledgers and putlogs;
14. Ladders used in scaffolds;
15. Stability of scaffolds;
16. Slung scaffolds;
17. Cantilever, jib, figure and bracket scaffolds;
18. Support for scaffolds;
19. Suspended scaffolds (not power operated);
20. Boatswain's chairs, cages, skips, etc.;
21. Trestle scaffolds;
22. inspection of scaffolds, boatswain's chairs, etc.;
23. Scaffolds used by workmen of different employers;
24. Construction of working platforms, gangways and runs;

Regulation

25. Boards and planks in working platforms, gangways and runs;
26. Width of working platforms;
27. Widths of gangways and runs;
28. Guard-rails and tow-boards at working platforms and places;
29. Guard-rails, etc., for gangways, runs and stairs;
30. Platforms, gangways, runs and stairs, etc.;
31. Construction and maintenance of ladders and folding step-ladders;
32. Use of ladders and folding step-ladders;
33. Openings, corners, breaks, edges and open joisting;
34. Exceptions from reg. 33;
35. Sloping roofs.
36. Work on or near fragile materials;
38. Prevention of falls and provision of safety nets and belts.

Regulation 3(1)(*b*) extends coverage to regulations as follows:

In addition to applying to reg. 24, 26, 27, 28, 29, 33, 34 and 35 this regulation applies to:

Regulation

37. Loads on scaffolds;
39. Reports.

Appendix X

Fire Certificates (Special Premises) Regulations 1976 No. 2003

Citation and commencement

1. These Regulations may be cited as the Fire Certificates (Special Premises) Regulations 1976 and shall come into operation on 1st January 1977.

Interpretation

2.—(1) In these Regulations unless the context otherwise requires—

(*a*) "fire certificate" means a certificate issued under these Regulations by the Health and Safety Executive;

"responsible person" in relation to any premises means the person having control of those premises, whether as occupier or otherwise, in connection with the carrying on by him of a trade, business or other undertaking (whether for profit or not) except that in the case of a mine within the meaning of the Mines and Quarries Act 1954(a) the responsible person shall be the manager of that mine;

(*b*) any reference to a numbered Regulation is a reference to the Regulation in these Regulations bearing that number.

(2) The Interpretation Act 1889(b) shall apply to the interpretation of these Regulations as it applies to the interpretation of an Act of Parliament.

Premises for which a fire certificate is required

3.—(1) A fire certificate shall be required in respect of any premises of a description specified in Part I of Schedule 1 to these Regulations unless either—

(*a*) an exemption is granted in respect of them under Regulation 15, or

(*b*) in the case of premises to which paragraph 15 of Part I of that Schedule applies, each condition in Part II of that Schedule is satisfied.

(2) Where any premises to which paragraph (1) of this Regulation applies ("specified premises") are within the close curtilage or precincts of larger premises, any part of those larger premises used for the purposes of the specified premises (including use as a means of access) shall for the purposes of these Regulations, be treated as forming part of the specified premises.

(a) 1954 c. 70.

(b) 1889 c. 63.

Application for and issue of a fire certificate

4.—(1) An application for a fire certificate shall be made by the responsible person to the Health and Safety Executive and shall contain the particulars specified in Schedule 2 to these Regulations.

(2) On receipt of an application for a fire certificate the Executive may require the applicant within such time as it may specify:—

(*a*) to furnish such plans of the premises to be covered by the certificate as it may specify; and

(*b*) to furnish, in so far as it is reasonably practicable for him to do so, such plans of adjoining premises as it may specify;

and if the applicant fails to furnish the required plans within that time or such further time as the Executive may allow, the application shall be deemed to have been withdrawn at the end of that time or further time, as the case may be.

(3) Where an application for a fire certificate has been duly made and all such plans (if any) as are required to be furnished under paragraph (2) of this Regulation have been duly furnished, it shall be the duty of the Executive to cause an inspection to be carried out; and if it is satisfied that—

(*a*) the means of escape which are provided in case of fire;

(*b*) the means (other than means for fighting fire) which are provided for securing that such means of escape can be safely and effectively used at all material times;

(*c*) the means which are provided whether on the premises or elsewhere for fighting fire; and

(*d*) the means which are provided whether on the premises or elsewhere for giving warning in case of fire,

are such as may reasonably be required in the circumstances of the case in connection with the premises, the Executive shall issue a fire certificate.

(4) Where the Executive, after causing such an inspection to be carried out is not satisfied as aforesaid, it shall by notice served on the applicant—

(*a*) inform the applicant of that fact and of the steps which would have to be taken to satisfy it; and

(*b*) notify him that it will not issue a fire certificate unless those steps are taken (whether by the applicant or otherwise) within such time as may be specified in the notice;

and if at the end of that time or such further time as may be allowed by the Executive or by any order made on, or in proceedings arising out of, an appeal under Regulation 12 against the notice those steps have not been taken, the fire certificate shall be deemed to have been refused.

Contents of a fire certificate

5.—(1) Every fire certificate shall specify—

(*a*) the address of the premises;

(*b*) the name of the responsible person;

(*c*) the description of the premises by reference to the class or classes listed in Schedule 1 to these Regulations;

(*d*) the means of escape which are provided in case of fire;

(*e*) the means (other than means for fighting fire) which are provided for securing that such means of escape can be safely and effectively used at all material times;

(*f*) the type, number and location of the means which are provided (whether in the premises or elsewhere) for fighting fire; and

(*g*) the type and location of the means which are provided (whether in the premises or elsewhere) for giving warning in case of fire,

and may do so by means of or by reference to a plan or photograph.

(2) A fire certificate may impose such conditions as the Executive consider appropriate in the circumstances—

(*a*) for securing that the means of escape in case of fire which are provided are properly maintained and kept free from obstruction;

(*b*) for securing that the means which are provided as mentioned in paragraph (1)(*e*) to (*g*) of this Regulation are properly maintained;

(*c*) for securing that the means which are provided as mentioned in paragraph (1)(*f*) and (*g*) of this Regulation are tested and examined at regular intervals and that records are kept of such tests and examinations;

(*d*) for securing that persons at work on the premises receive appropriate instruction and training in what to do in case of fire, and that records are kept of instruction and training given for that purpose;

(*e*) for limiting the number of persons who may be on the premises at any one time or at such times as the Executive may specify;

(*f*) for limiting the quantity and disposition of any substance or article which may be on the premises at any one time or at such times as the Executive may specify; and

(*g*) as to other precautions to be observed in relation to the risk to persons in case of fire.

(3) Any conditions imposed by virtue of paragraph (2) of this Regulation may be framed so as to apply to the whole of the premises or so as to apply to one or more parts of them and different conditions may be imposed in relation to different parts.

(4) A fire certificate shall be sent or delivered to the responsible person and where he is not the occupier of the premises a copy of it shall be sent to the occupier.

(5) An occupier of premises (whether or not he is the responsible person) shall keep the fire certificate, or the copy as the case may be, on the premises and shall make it available for inspection at reasonable times by any person who might be affected by any of its provisions.

(6) Where a fire certificate has been issued, the occupier of the premises (whether or not he is also the responsible person) shall keep a notice posted at a suitable place on or about the premises to which it relates in such characters and in such a position as to be easily seen and read by any person who might be affected by any of the provisions of the certificate; the notice shall state—

(*a*) that a fire certificate has been issued;

(*b*) where it, or a copy of it, may be inspected, and

(*c*) the date of posting of the notice.

(7) Where the responsible person is not the occupier he shall keep the fire certificate and shall make it available for inspection at reasonable times by any person who might be affected by any of its provisions.

Changes by the responsible person affecting adequacy of matters specified in a fire certificate

6.—(1) Where the responsible person proposes to make any material change which will affect any of the matters specified or referred to in a fire certificate which is in force, he shall, before commencing to make that change, notify the Executive in writing of the proposal.

(2) The provisions of Regulation 4(2)(*a*) and (*b*) shall apply to a notice under paragraph (1) of this Regulation as they apply to an application for a fire certificate.

(3) If the Executive is satisfied that if the proposed change were made, all of the matters mentioned in Regulation 5(1)(*d*) to (*g*) and 5(2) would still be adequate, it shall on production of the fire certificate cause a copy of the notice to be attached to it together with a statement that it is so satisfied or it may vary the certificate or cancel it and issue a new one.

(4) If the Executive is satisfied that if the change were made any of the matters mentioned in Regulation 5(1)(*d*) to (*g*) and 5(2) would be inadequate it shall—

(*a*) inform the responsible person of that fact and of the steps which would have to be taken to prevent their becoming inadequate in the event of the change being made;

(*b*) notify him that the fire certificate will be cancelled if the change is made without those steps being taken.

(5) Where a notice has been given under paragraph (4)(*b*) of this Regulation the Executive may cancel the certificate if the changes are made without the steps referred to in paragraph (4)(*a*) having been taken; and where the steps are taken it may vary the certificate or cancel it and issue a new one.

Changes by persons other than the responsible person affecting adequacy of matters specified in a fire certificate

7.—(1) Where the responsible person becomes aware that some other person proposes to make or has made any material change which affects any of the matters specified or referred to in a fire certificate which is in force, the responsible person shall notify the Executive in writing of the proposal or change as the case may be.

(2) The provisions of Regulation 4(2)(*a*) and (*b*) shall apply to a notice under paragraph (1) of this Regulation as they apply to an application for a fire certificate.

(3) If the Executive is satisfied that notwithstanding the change all of the matters mentioned in Regulation 5(1)(*d*) to (*g*) and 5(2) would still be adequate, it shall on production of the fire certificate cause a copy of the notice to be attached to it together with a statement that it is so satisfied or it may vary the certificate or cancel it and issue a new one.

(4) If the Executive is satisfied that any of the matters mentioned in Regulation 5(1)(*d*) to (*g*) and 5(2) would become inadequate if the proposed change were made, or where the change has been made it is satisfied that any of those matters have become inadequate, it shall—

(*a*) inform the responsible person of that fact and of the steps which would have to be taken to prevent their becoming or, as the case may be, remaining inadequate;

(*b*) notify him that the fire certificate will be cancelled if those steps are not taken within such time as may be specified in the notice.

(5) Where a notice has been given under paragraph (4)(*b*) of this Regulation, the Executive may cancel the certificate if the steps are not taken within the time specified in the notice, or within such further time as may be allowed by the Executive or by any order made on, or in proceedings arising out of, an appeal under Regulation 12 against the notice; and where the steps are taken it may vary the certificate or may cancel it and issue a new one.

Other cases where a fire certificate becomes inadequate

8. If the Executive considers that a fire certificate should be varied or cancelled because any of the matters mentioned in Regulation 5(1)(*d*) to (*g*) and 5(2) is or has become inadequate it shall—

(*a*) inform any person to whom the certificate, or a copy of it, was given under Regulation 5(4) of that fact and the steps, if any, which would have to be taken for those matters to become adequate;

(*b*) notify him that it will cancel the certificate unless those steps are taken within such time as may be specified in the notice; and if at the end of that time, or such further time as may be allowed by the Executive or by any order made on, or in any proceedings arising out of, an appeal under Regulation 12 against the notice, those steps have been taken, the Executive may vary the certificate accordingly and if they have not been taken may, subject to any such order, cancel it.

Applications for variation of a fire certificate

9.—(1) Where a fire certificate is in force, the responsible person may apply to the Executive for the certificate to be varied; any such application shall state the grounds on which it is made and the provisions of Regulation 4(2) shall apply.

(2) If the Executive is satisfied that the variation should be made it shall vary the certificate accordingly or may cancel it and issue a new one; and if it is not so satisfied it shall notify the responsible person that it refuses to vary the certificate.

Changes of responsible person

10. Where a fire certificate is in force in respect of any premises and a person becomes the responsible person in respect of those premises he shall notify the Executive and the Executive shall vary the certificate accordingly.

Persons not to work on specified premises unless a fire certificate is in force

11.—(1) Subject to paragraph (2) of this Regulation a responsible person shall not work or permit any other person to work on any premises for which a fire certificate is required unless such a certificate is in force and the conditions attached to it are being complied with or an application has been made for a certificate and it has not been refused or withdrawn.

(2) Paragraph (1) of this Regulation shall not apply where the work is solely for the purpose of—

(*a*) taking any steps required under Regulation 4(4), 6(4), 7(4) or 8;

(*b*) complying with a condition imposed under Regulation 5(2); or

(*c*) otherwise making the premises safe.

Right of appeal

12.—(1) A person who is aggrieved by—

(*a*) anything mentioned in a notice served under Regulation 4(4) as a step which would have to be taken as a condition of the issue of a fire certificate or by the period allowed by such a notice for the taking of any steps mentioned in it;

(*b*) a refusal to issue a fire certificate;

(*c*) the inclusion of anything in, or the omission of anything from, a fire certificate;

(*d*) a refusal to vary a fire certificate;

(*e*) anything mentioned in a notice served under Regulation 6(4), 7(4) or 8 with respect to any premises as a step which must be taken if the Executive is not to become entitled to cancel the fire certificate relating to the premises;

may within 21 days from the relevant date, appeal to a court in accordance with the following provisions of this Regulation; and on such appeal the court may make such order as it thinks fit.

(2) In this Regulation "the relevant date" means—

(*a*) in the case of a person who was given notice in accordance with these Regulations of the matter by which he is aggrieved, the date on which he was served with that notice (and for this purpose the date on which a certificate was issued or an application deemed to have been refused shall be treated as a date on which the applicant was served with a notice);

(*b*) in the case of any other person, the date on which the relevant notice, or amended notice, as the case may be, was posted under Regulation 5(6).

(3) In any proceedings for an offence of contravening Regulation 11 or a requirement of, or condition contained in, a fire certificate at any time during the period between the relevant date and the final determination of an appeal under this Regulation, it shall be a defence that the appeal was brought and had not been determined.

(4) In England and Wales, an appeal under this Regulation shall be brought by way of complaint to a magistrates' court for an order and the Magistrates' Courts Act 1952 shall apply to the proceedings; and in Scotland any such appeal shall be to the sheriff within whose jurisdiction the premises are situated.

(5) The Executive and any other person aggrieved by an order made by a magistrates' court on a complaint under this Regulation may appeal therefrom to the Crown Court.

Modification of local Acts

13. Local Acts passed before or in the same session as the Health and Safety at Work etc. Act 1974 shall have effect subject to the following modification, that is to say, that a person required by or under any such local Act to do anything in relation to any premises shall not be treated as having acted in contravention of that local Act by reason of his failure to do that thing in so far as the failure is attributable to the fact that remedying it would involve a contravention of these Regulations.

Power of county court and sheriff to modify agreements and apportion expenses

14.—(1) A person who, by reason of the terms of any agreement or lease relating to any premises to which these Regulations apply is prevented from carrying out or doing any structural or other alteration or other thing whose carrying out or doing is requisite in order to secure compliance with a provision of these Regulations or with a notice served under Regulation 4(4), 6(4), 7(4) or 8 or with a condition imposed under Regulation 5(2) may apply to the county court within whose jurisdiction the premises are situated and the court may make such an order setting aside or modifying any terms of the agreement or lease as it considers just and equitable in the circumstances of the case.

(2) Where the carrying out or doing on any such premises of any structural or other alterations or other thing whose carrying out or doing is requisite as mentioned in paragraph (1) of this Regulation involves a person having an interest in the premises in expense, or in increased expense, and he alleges that the whole or part of the expense or, as the case may be, the increase ought to be borne by some other person having an interest in the premises, the first-mentioned person may apply to the county court within whose jurisdiction the premises are situated and the court, having regard to the terms of any agreement or lease relating to the premises may by order give such directions with respect to the persons by whom the expense or increase is to be borne and in what proportions it is to be borne by them and, if need be, for modification of the terms of any such agreement or lease so far as concerns rent payable in respect of the premises as the court considers just and equitable in the circumstances of the case.

(3) In the application of this Regulation to Scotland, for references to a county court there shall be substituted references to the sheriff.

Exemptions

15.—(1) Where the Executive is satisfied in respect of any particular premises or part thereof or in respect of any description of premises that any requirement of Regulations 3 to 11 is inappropriate or is not reasonably practicable, it may by certificate in writing (which it may at its discretion revoke) exempt those premises or that part of them or that description of premises from such requirement to such extent and subject to such conditions as may be specified in the certificate.

(2) An inspector shall be exempt from the requirements of section 23(4) of the Health and Safety at Work etc. Act 1974 (which requires an inspector to consult the fire authority before serving certain improvement and prohibition notices) in relation to any notice in connection with any premises in respect of which a fire certificate is required under these Regulations.

Certain certificates under Fire Precautions Act 1971 *to be deemed to have been issued under these Regulations*

16. Where immediately before a fire certificate becomes required under these Regulations in respect of any premises a fire certificate issued under the Fire

Precautions Act 1971(**a**) ("Fire Precautions Act certificate") was in force in respect of those premises, the Fire Precautions Act certificate shall be deemed to be a fire certificate issued under these Regulations and accordingly any condition attached to the Fire Precautions Act certificate shall have effect as a condition imposed under Regulation 5(2) and the certificate may (in particular) be varied or cancelled in accordance with these Regulations.

Transitional provisions

17.—(1) In this Regulation—

"Factories Act certificate" means a certificate under section 40 of the Factories Act 1961(**b**);

"Offices Act certificate" means a certificate under section 29 of the Offices, Shops and Railway Premises Act 1963(**c**);

"existing certificate" means a Factories Act certificate or an Offices Act certificate;

and this Regulation applies where on the date of the coming into operation of these Regulations a fire certificate under these Regulations becomes required in respect of any premises and immediately prior to that date an existing certificate was in force in respect of those premises.

(2) The existing certificate shall continue in force (notwithstanding the repeal of the section under which it was issued) and—

(*a*) shall, as from the said date, be deemed to be a fire certificate issued under these Regulations and any condition attached to the existing certificate shall have effect as a condition imposed under Regulation 5 (2); and

(*b*) may (in particular) be varied or cancelled in accordance with these Regulations.

(3) The existing certificate, as it has effect by virtue of paragraph (2) of this Regulation, shall be treated as imposing in relation to the premises the like requirements as immediately prior to the said date were imposed in relation to it by the following provisions, that is to say—

(*a*) if the existing certificate is a Factories Act certificate, sections 41(1) and (3), 48 (except subsections (5), (8) and (9)), 49(1), 51(1) and 52(1) and (4) of the Factories Act 1961;

(*b*) if the existing certificate is an Offices Act certificate, sections 30(1) and (3), 33, 34(1) and (2), 36(1) and 38(1) of the Offices, Shops and Railway Premises Act 1963.

(4) Any application for a Factories Act certificate or an Offices Act certificate with respect to any premises which is pending at the date of the coming into operation of these Regulations shall, in the case of premises for which a fire certificate is required by these Regulations, be deemed to be an application for a fire certificate in respect of them duly made in accordance with these Regulations and accordingly the provisions of Regulation 4(2) to (4) shall apply.

(5) If any notice has been given to the appropriate authority of a proposal to undertake any of the changes of which notice is required to be given by section 41(3) of the Factories Act 1961 or section 30(3) of the Offices, Shops and Railway Premises Act 1963 such notice shall be deemed to be a notice duly given in accordance with Regulation 6(1) and accordingly the provisions of Regulation 6(2) to (4) shall apply.

(**a**) 1971 c. 40. (**b**) 1961 c. 34. (**c**) 1963 c. 41.

Regulation 3

SCHEDULE 1

PART I

PREMISES FOR WHICH A FIRE CERTIFICATE IS REQUIRED

1. Any premises at which are carried on any manufacturing processes in which the total quantity of any highly flammable liquid under pressure greater than atmospheric pressure and above its boiling point at atmospheric pressure may exceed 50 tonnes.

2. Any premises at which is carried on the manufacture of expanded cellular plastics and at which the quantities manufactured are normally of, or in excess of, 50 tonnes per week.

3. Any premises at which there is stored, or there are facilities provided for the storage of, liquefied petroleum gas in quantities of, or in excess of, 100 tonnes except where the liquefied petroleum gas is kept for use at the premises either as a fuel, or for the production of an atmosphere for the heat-treatment of metals.

4. Any premises at which there is stored, or there are facilities provided for the storage of, liquefied natural gas in quantities of,or in excess of,100 tonnes except where the liquefied natural gas is kept solely for use at the premises as a fuel.

5. Any premises at which there is stored, or there are facilities provided for the storage of, any liquefied flammable gas consisting predominantly of methyl acetylene in quantities of, or in excess of, 100 tonnes except where the liquefied flammable gas is kept solely for use at the premises as a fuel.

6. Any premises at which oxygen is manufactured and at which there are stored, or there are facilities provided for the storage of, quantities of liquid oxygen of, or in excess of, 135 tonnes.

7. Any premises at which there are stored, or there are facilities provided for the storage of, quantities of chlorine of, or in excess of, 50 tonnes except when the chlorine is kept solely for the purpose of water purification.

8. Any premises at which artificial fertilizers are manufactured and at which there are stored, or there are facilities provided for the storage of, quantities of ammonia of, or in excess of, 250 tonnes.

9. Any premises at which there are in process, manufacture, use or storage at any one time, or there are facilities provided for such processing, manufacture, use or storage of, quantities of any of the materials listed below in, or in excess of, the quantities specified—

Phosgene	5 tonnes
Ethylene oxide	20 tonnes
Carbon disulphide	50 tonnes
Acrylonitrile	50 tonnes
Hydrogen cyanide	50 tonnes
Ethylene	100 tonnes
Propylene	100 tonnes
Any highly flammable liquid not otherwise specified	4,000 tonnes

10. Explosives factories or magazines which are required to be licensed under the Explosives Act 1875(a).

11. Any building on the surface at any mine within the meaning of the Mines and Quarries Act 1954(b).

(a) 1875 c. 17. (b) 1954 c. 70.

12. Any premises in which there is comprised—

(*a*) any undertaking on a site for which a licence is required in accordance with section 1 of the Nuclear Installations Act 1965 (a) or for which a permit is required in accordance with section 2 of that Act; or

(*b*) any undertaking which would, except for the fact that it is carried on by the United Kingdom Atomic Energy Authority, or by, or on behalf of, the Crown, be required to have a licence or permit in accordance with the provisions mentioned in sub-paragraph (*a*) above.

13. Any premises containing any machine or apparatus in which charged particles can be accelerated by the equivalent of a voltage of not less than 50 megavolts except where the premises are used as a hospital.

14. Any premises at which there are in process, manufacture, use or storage at any one time, or there are facilities provided for such processing, manufacture, use or storage of, quantities of unsealed radioactive substances classified according to Schedule 3 of the Ionising Radiations (Unsealed Radioactive Substances) Regulations 1968 (b) in, or in excess of the quantities specified—

Class I radionuclides	—	10 curies
Class II and III radionuclides	—	100 curies
Class IV radionuclides	—	1,000 curies

15. Any building, or part of a building, which either—

(*a*) is constructed for temporary occupation for the purposes of building operations or works of engineering construction; or

(*b*) is in existence at the first commencement there of any further such operations or works

and which is used for any process or work ancillary to any such operations or works.

PART II

CONDITIONS TO BE SATISFIED IF A FIRE CERTIFICATE IS NOT TO BE REQUIRED FOR PREMISES SPECIFIED IN PARAGRAPH 15 ABOVE

16. Not more than 20 persons are employed at any one time in the building or part of the building.

17. Not more than 10 persons are employed at any one time elsewhere than on the ground floor of the building or part of the building.

18. No explosive or highly flammable material is stored or used in or under the building or part of the building.

19. The building or part of the building is provided with such means of escape in case of fire for the persons employed there as may reasonably be required in the circumstances of the case.

20. In the building or part of the building there is provided and maintained appropriate means for fighting fire which are so placed as to be readily available for use.

21. While any person is in the building or part of the building for the purpose of employment or meals, the doors of the building or part and of any room therein in which he is, and any doors which afford a means of exit for persons employed in the building or part are not locked or fastened in such a manner that they cannot easily and immediately be opened from the inside.

22. Any doors opening on to any staircase or corridor from any room in the building or part of the building in which more than 10 persons are employed, except in the case of sliding doors, are constructed to open outwards.

(a) 1965 c. 57. (b) S.I. 1968/780 (1968 II, p. 2153).

23. Every window, door or other exit affording a means of escape from the building or part of the building in case of fire or giving access thereto, other than the means of exit in ordinary use, is distinctively and conspicuously marked by a notice of adequate size.

24. The contents of any room in the building or part of the building in which persons are employed are so arranged or disposed that there is a free passage way for all persons employed in the room to a means of escape in case of fire.

PART III

DEFINITIONS

25. In this Schedule—

(*a*) "liquefied petroleum gas" means commercial butane, commercial propane or any mixture thereof;

"commercial butane" means a hydrocarbon mixture consisting predominantly of butane, butylene or any mixture thereof;

"commercial propane" means a hydrocarbon mixture consisting predominantly of propane, propylene or any mixture thereof;

"highly flammable liquid" means any liquid, liquid solution, emulsion or suspension (other than liquefied petroleum gas or liquefied natural gas) which when tested in the manner specified in Schedule 1 to the Highly Flammable Liquids and Liquefied Petroleum Gases Regulations 1972(a), gives off a flammable vapour at a temperature of less than 32 degrees Celsius and, when tested in the manner specified in Schedule 2 to those Regulations, supports combustion.

(*b*) Any reference to a provision made by an enactment or instrument shall be construed as a reference to that provision as amended by any enactment or instrument and as including a reference to any provision which re-enacts or replaces it.

SCHEDULE 2

Regulation 4

PARTICULARS TO BE INCLUDED IN AN APPLICATION FOR A FIRE CERTIFICATE

1. Address of the premises in respect of which the application is made.

2. Description of premises by reference to the list set out in Schedule 1 to these Regulations.

3. Nature of the processes carried on, or to be carried on, on the premises.

4. Nature and approximate quantities of any explosive or highly flammable substance kept, or to be kept, on the premises.

5. Maximum number of persons likely to be on the premises at any one time.

6. Maximum number of persons likely to be in any building of which the premises form part at any one time.

7. Name and address of the occupier of the premises.

8. Name and address of any other person who has control of the premises.

9. If the premises consist of part of a building, the name and postal address of the person or persons having control of the building or any part of it.

(a) S.I. 1972/917 (1972 II, p. 2870).

EXPLANATORY NOTE

(*This Note is not part of the Regulations.*)

These Regulations provide that a fire certificate issued by the Health and Safety Executive shall be required in respect of premises of the kind specified in Schedule 1 to the Regulations. They make provision as to applications for and the issue of such certificates, their contents and the conditions which they may impose. They specify the steps to be taken where material changes are proposed or have occurred and provide for the variation or cancellation of fire certificates. The Regulations also make provision for appeals in specified cases, including a refusal to issue or vary a fire certificate, and for the granting of exemptions.

Appendix XI

The Fire Precautions (Factories, Offices, Shops and Railway Premises) Order 1976 No. 2009

1. This Order may be cited as the Fire Precautions (Factories, Offices, Shops and Railway Premises) Order 1976 and shall come into operation on 1st January 1977.

2.—(1) In this Order—

"the Act" means the Fire Precautions Act 1971;

"covered market place" shall be construed generally and not as limited to a place where a market is held by virtue of a grant from the Crown or of prescription or under statutory authority;

"factory premises" means premises constituting, or forming part of, a factory within the meaning of the Factories Act 1961 **(a)**, and premises to which sections 123(1) and 124 of that Act (application to electrical stations and institutions, respectively) apply;

"office premises", "shop premises" and "railway premises" mean premises to which the Offices, Shops and Railway Premises Act 1963**(b)** applies and premises which are deemed to be such premises for the purposes of that Act, other than premises consisting of a covered market place wherein shop premises are aggregated.

(2) The Interpretation Act 1889**(c)** applies for the interpretation of this Order as it applies for the interpretation of an Act of Parliament.

(3) A reference in this Order to an enactment shall be construed as a reference to that enactment as amended or replaced by or under any other enactment.

3. The following uses of premises are hereby designated for the purposes of section 1 of the Act (which requires fire certificates for premises put to designated uses), that is to say use as factory premises, office premises, shop premises, or railway premises, being (in each case) premises in which persons are employed to work.

4.—(1) Notwithstanding the foregoing provisions of this Order, a fire certificate shall not by virtue of section 1 of the Act be required for factory premises, office premises, shop premises or railway premises being premises—

(a) 1961 c. 34. **(b)** 1963 c. 41. **(c)** 1889 c. 63.

(*a*) in which—

(i) not more than twenty persons are employed to work at any one time; or

(ii) not more than ten persons are so employed elsewhere than on the ground floor,

unless any one of the conditions specified in paragraph (2) below applies to the premises; or

(*b*) of a description specified in regulations which are from time to time made under the provisions of the Health and Safety at Work etc. Act 1974(**a**) and which provide for the issue of fire certificates by the Health and Safety Executive.

(2) The conditions mentioned in paragraph (1)(*a*) above are—

(*a*) that the premises are in a building which contains two or more of the premises described in Article 3 above and the aggregate of persons employed to work at any one time in all those premises exceeds twenty;

(*b*) that the premises are in a building which contains two or more of the premises so described and in all those premises the aggregate of persons employed to work at any one time elsewhere than on the ground floor of the building exceeds ten;

(*c*) that in the case of factory premises explosive or highly flammable materials (other than materials of such a kind and in such a quantity that the fire authority have determined that they do not constitute a serious additional risk to persons in the premises in case of fire) are stored or used in or under the premises.

EXPLANATORY NOTE

(This Note is not part of the Order.)

Section 1 of the Fire Precautions Act 1971 provides that a fire certificate is required in respect of any premises which are put to a designated use.

This Order designates generally for this purpose the use of factory, office, shop and railway premises in which persons are employed to work. By Article 4 a fire certificate is not required for premises in which the persons employed at any one time do not exceed the numbers specified in the Order or for certain other premises.

(**a**) 1974 c. 37.

The Fire Precautions (Non-Certificated Factory, Office, Shop and Railway Premises) Regulations 1976 No. 2010

1. These Regulations may be cited as the Fire Precautions (Non-Certificated Factory, Office, Shop and Railway Premises) Regulations 1976 and shall come into operation on 1st January 1977.

2.—(1) In these Regulations—

"the Act" means the Fire Precautions Act 1971;

"covered market place" shall be construed generally and not as limited to a place where a market is held by virtue of a grant from the Crown or of prescription or under statutory authority;

"factory premises" means premises constituting, or forming part of, a factory within the meaning of the Factories Act 1961**(a)** and premises to which sections 123(1) and 124 of that Act (application to electrical stations and institutions, respectively) apply;

"office premises", "shop premises" and "railway premises" mean premises to which the Offices, Shops and Railway Premises Act 1963**(b)** applies and premises which are deemed to be such premises for the purposes of that Act, other than premises consisting of a covered market place wherein shop premises are aggregated.

(2) A reference in these Regulations to an enactment shall be construed as a reference to that enactment as amended or replaced by or under any other enactment.

(3) The Interpretation Act 1889**(c)** shall apply to the interpretation of these Regulations as it applies to the interpretation of an Act of Parliament.

3.—(1) Subject to paragraph (2) below, these Regulations apply to factory premises, office premises, shop premises and railway premises, being (in each case) premises—

(*a*) in which persons are employed to work; and

(*b*) in respect of which a fire certificate is not required by virtue of an order under section 1 of the Act.

(a) 1961 c. 34. **(b)** 1963 c. 41. **(c)** 1889 c. 63.

(2) Nothing in these Regulations applies to any factory premises, office premises, shop premises or railway premises, being premises of a description specified in regulations which are from time to time made under the provisions of the Health and Safety at Work etc. Act 1974(**a**) and which provide for the issue of fire certificates by the Health and Safety Executive.

4.—(1) In factory premises to which these Regulations apply any doors opening on to any staircase or corridor from any room in which more than ten persons are employed to work, and all other doors affording a means of exit from the premises or giving access thereto, shall, except in the case of sliding doors, be constructed to open outwards.

(2) Every window, door or other exit affording means of escape from the factory premises in case of fire or giving access thereto, other than the means of exit in ordinary use, shall be distinctively and conspicuously marked by a notice in letters of adequate size.

(3) Every hoistway or liftway inside the factory premises shall be completely enclosed by construction having a fire resistance of not less than thirty minutes and all means of access to the hoistway or liftway shall be fitted with doors having a fire resistance of not less than thirty minutes; except that any such hoistway or liftway which is not provided with a vent at the top shall be enclosed at the top only by material easily broken by fire.

(4) A fire authority may by notice in writing grant, subject to any conditions specified in the notice, exemption from compliance with any of the requirements of paragraphs (1) and (3) above in any case where they are satisfied that compliance with those requirements is inappropriate or undesirable.

(5) If any person avails himself of an exemption granted under paragraph (4) above and fails to comply with any of the conditions specified in a notice granting such exemption, he shall be deemed to be responsible for a contravention of this Regulation.

5.—(1) While a person employed to work in premises to which these Regulations apply is in the premises for the purpose of doing his work or eating a meal, the doors of any doorways through which he might have to pass so as to get out of the premises shall not be so locked or fastened that they cannot be easily and immediately opened by him on his way out.

(2) The contents of any room in the premises, being a room wherein work is done by any of the persons employed to work in the premises, shall be so arranged or disposed as to afford to the persons who work in the room free passage-way to a means of escape in case of fire.

6.—(1) In all premises to which these Regulations apply there shall be provided and maintained appropriate means for fighting fire, which shall be so placed as to be readily available for use.

(2) In this regulation "maintained" means maintained in an efficient state, in efficient working order and in good repair.

(**a**) 1974 c. 37.

7.—(1) Subject to paragraphs (2) and (3) below, the requirements of Regulations 4(1), (2) and (3), 5 and 6 above are imposed on the occupier of the premises who shall be responsible for a contravention of any of those provisions.

(2) Where factory premises are held under a lease or an agreement for a lease or under a licence and consist of part of a building all parts of which are in the same ownership, the requirements of Regulation 4(1), (2) and (3) above are imposed on the owner of the relevant building and for a contravention, in relation to such premises, of any of those provisions the owner of the relevant building shall be responsible instead of the occupier of the premises.

(3) Where factory premises consist of part of a building in which different parts are owned by different persons, the requirements of Regulation 4(1), (2) and (3) above are imposed on the persons who between them own the relevant building and for a contravention, in relation to such premises, of any of those provisions each of the persons who between them own the relevant building shall be responsible instead of the occupier of the premises.

(4) If any provision of Regulation 4, 5 or 6 above is contravened, the person or each of the persons who, under Regulation 4(5) or this Regulation, is or are responsible for the contravention, shall be guilty of an offence under section 12 of the Act.

(5) Paragraph (4) above shall not apply to the Crown.

EXPLANATORY NOTE

(*This Note is not part of the Regulations.*)

These Regulations provide for certain fire precautions to be taken in the smaller factories, offices, shops and railway premises which do not require a fire certificate under the Fire Precautions Act 1971. The precautions to be taken relate to the means of escape from the premises in case of fire (Regulations 4 and 5) and the provision of fire-fighting equipment (Regulation 6). The Regulations impose requirements generally on occupiers of premises but, in the case of factory premises forming parts of buildings which are in single ownership or are plurally owned, the owners of the buildings are made responsible for the contravention of those requirements which apply only to factory premises (Regulation 7).

Appendix XIII

Bibliography

Report of the Committee on Safety and Health at Work 1970–1972. Cmnd. 5034, reprinted 1975.

1st, 2nd and 3rd Circulars of the Royal Commission on Civil Liability and Compensation for Personal Injury. April 1973 and February 1974.

Report of the Committee on Personal Injuries' Litigation. Cmnd. 3691, 1968.

Report of the Royal Commission on Civil Liability and Compensation for Personal Injury. Cmnd. 7054, 1978.

Accidents Compensation and the Law, P. S. Atiyah. Weidenfeld and Nicholson, 2nd Edn. 1975.

Employers Liability at Common Law, John Munkman. Butterworths, 8th Ed. 1975.

Settled out of Court, Laurence Ross. Aldine Publishing Co. Ltd. (Chicago), U.S.A., 2nd Edn. 1975.

Index